T0408336

APPLIED CHEMISTRY AND CHEMICAL ENGINEERING

Volume 1

Mathematical and Analytical Techniques

APPLIED CHEMISTRY AND CHEMICAL ENGINEERING

Volume 1

Mathematical and Analytical Techniques

Edited by

A. K. Haghi, PhD

Devrim Balköse, PhD

Omari V. Mukbaniani, DSc

Andrew G. Mercader, PhD

Apple Academic Press Inc.
3333 Mistwell Crescent
Oakville, ON L6L 0A2 Canada

Apple Academic Press Inc.
9 Spinnaker Way
Waretown, NJ 08758 USA

Printed in the United States of America on acid-free paper
International Standard Book Number-13: 978-1-77188-515-7 (Hardcover)
International Standard Book Number-13: 978-1-315-36562-6 (eBook)

Library and Archives Canada Cataloguing in Publication

Applied chemistry and chemical engineering / edited by A.K. Haghi, PhD, Devrim Balköse, PhD, Omari V. Mukbaniani, DSc, Andrew G. Mercader, PhD.
Includes bibliographical references and indexes.
Contents: Volume 1. Mathematical and analytical techniques --Volume 2. Principles, methodology, and evaluation methods --Volume 3. Interdisciplinary approaches to theory and modeling with applications --Volume 4. Experimental techniques and methodical developments --Volume 5. Research methodologies in modern chemistry and applied science.
Issued in print and electronic formats.
ISBN 978-1-77188-515-7 (v. 1 : hardcover).--ISBN 978-1-77188-558-4 (v. 2 : hardcover).--ISBN 978-1-77188-566-9 (v. 3 : hardcover).--ISBN 978-1-77188-587-4 (v. 4 : hardcover).--ISBN 978-1-77188-593-5 (v. 5 : hardcover).--ISBN 978-1-77188-594-2 (set : hardcover).
ISBN 978-1-315-36562-6 (v. 1 : PDF).--ISBN 978-1-315-20736-0 (v. 2 : PDF).-- ISBN 978-1-315-20734-6 (v. 3 : PDF).--ISBN 978-1-315-20763-6 (v. 4 : PDF).-- ISBN 978-1-315-19761-6 (v. 5 : PDF)
1. Chemistry, Technical. 2. Chemical engineering. I. Haghi, A. K., editor
TP145.A67 2017 660 C2017-906062-7 C2017-906063-5

Library of Congress Cataloging-in-Publication Data

Names: Haghi, A. K., editor.
Title: Applied chemistry and chemical engineering / editors, A.K. Haghi, PhD [and 3 others].
Description: Toronto ; New Jersey : Apple Academic Press, 2018- | Includes bibliographical references and index.
Identifiers: LCCN 2017041946 (print) | LCCN 2017042598 (ebook) | ISBN 9781315365626 (ebook) | ISBN 9781771885157 (hardcover : v. 1 : alk. paper)
Subjects: LCSH: Chemical engineering. | Chemistry, Technical.
Classification: LCC TP155 (ebook) | LCC TP155 .A67 2018 (print) | DDC 660--dc23
LC record available at https://lccn.loc.gov/2017041946

Apple Academic Press also publishes its books in a variety of electronic formats. Some content that appears in print may not be available in electronic format. For information about Apple Academic Press products, visit our website at **www.appleacademicpress.com** and the CRC Press website at **www.crcpress.com**

ABOUT THE EDITORS

A. K. Haghi, PhD

A. K. Haghi, PhD, holds a BSc in Urban and Environmental Engineering from the University of North Carolina (USA), an MSc in Mechanical Engineering from North Carolina A&T State University (USA), a DEA in applied mechanics, acoustics and materials from the Université de Technologie de Compiègne (France), and a PhD in engineering sciences from the Université de Franche-Comté (France). He is the author and editor of 165 books, as well as of 1000 published papers in various journals and conference proceedings. Dr. Haghi has received several grants, consulted for a number of major corporations, and is a frequent speaker to national and international audiences. Since 1983, he served as professor at several universities. He is currently Editor-in-Chief of the *International Journal of Chemoinformatics and Chemical Engineering* and the *Polymers Research Journal* and on the editorial boards of many international journals. He is also a member of the Canadian Research and Development Center of Sciences and Cultures (CRDCSC), Montreal, Quebec, Canada.

Devrim Balköse, PhD

Devrim Balköse, PhD, is currently a faculty member in the Chemical Engineering Department at the Izmir Institute of Technology, Izmir, Turkey. She graduated from the Middle East Technical University in Ankara, Turkey, with a degree in Chemical Engineering. She received her MS and PhD degrees from Ege University, Izmir, Turkey, in 1974 and 1977, respectively. She became Associate Professor in Macromolecular Chemistry in 1983 and Professor in process and reactor engineering in 1990. She worked as Research Assistant, Assistant Professor, Associate Professor, and Professor between 1970 and 2000 at Ege University. She was the Head of the Chemical Engineering Department at the Izmir Institute of Technology, Izmir, Turkey, between 2000 and 2009. Her research interests are in polymer reaction engineering, polymer foams and films, adsorbent development, and moisture sorption. Her research projects are on nanosized zinc borate production, ZnO polymer composites, zinc borate lubricants, antistatic additives, and metal soaps.

Omari V. Mukbaniani, DSc

Omari Vasilii Mukbaniani, DSc, is Professor and Head of the Macromolecular Chemistry Department of Iv. Javakhishvili Tbilisi State University, Tbilisi, Georgia. He is also the Director of the Institute of Macromolecular Chemistry and Polymeric Materials. He is a member of the Academy of Natural Sciences of the Georgian Republic. For several years he was a member of the advisory board of the *Journal Proceedings of Iv. Javakhishvili Tbilisi State University* (Chemical Series) and contributing editor of the journal *Polymer News* and the *Polymers Research Journal*. He is a member of editorial board of the *Journal of Chemistry and Chemical Technology*. His research interests include polymer chemistry, polymeric materials, and chemistry of organosilicon compounds. He is an author more than 420 publications, 13 books, four monographs, and 10 inventions. He created in the 2007s the "International Caucasian Symposium on Polymers & Advanced Materials," ICSP, which takes place every other two years in Georgia.

Andrew G. Mercader, PhD

Andrew G. Mercader, PhD, studied Physical Chemistry at the Faculty of Chemistry of La Plata National University (UNLP), Buenos Aires, Argentina, from 1995–2001. Afterwards he joined Shell Argentina to work as Luboil, Asphalts and Distillation Process Technologist, as well as Safeguarding and Project Technologist. His PhD work on the development and applications of QSAR/QSPR theory was performed at the Theoretical and Applied Research Institute located at La Plata National University (INIFTA). He received a post-doctoral scholarship to work on theoretical-experimental studies of biflavonoids at IBIMOL (ex PRALIB), Faculty of Pharmacy and Biochemistry, University of Buenos Aires (UBA). He is currently a member of the Scientific Researcher Career in the Argentina National Research Council, at INIFTA.

Applied Chemistry and Chemical Engineering, 5 Volumes

Applied Chemistry and Chemical Engineering,
Volume 1: Mathematical and Analytical Techniques
Editors: A. K. Haghi, PhD, Devrim Balköse, PhD, Omari V. Mukbaniani, DSc, and Andrew G. Mercader, PhD

Applied Chemistry and Chemical Engineering,
Volume 2: Principles, Methodology, and Evaluation Methods
Editors: A. K. Haghi, PhD, Lionello Pogliani, PhD, Devrim Balköse, PhD, Omari V. Mukbaniani, DSc, and Andrew G. Mercader, PhD

Applied Chemistry and Chemical Engineering,
Volume 3: Interdisciplinary Approaches to Theory and Modeling with Applications
Editors: A. K. Haghi, PhD, Lionello Pogliani, PhD, Francisco Torrens, PhD, Devrim Balköse, PhD, Omari V. Mukbaniani, DSc, and Andrew G. Mercader, PhD

Applied Chemistry and Chemical Engineering,
Volume 4: Experimental Techniques and Methodical Developments
Editors: A. K. Haghi, PhD, Lionello Pogliani, PhD, Eduardo A. Castro, PhD, Devrim Balköse, PhD, Omari V. Mukbaniani, PhD, and Chin Hua Chia, PhD

Applied Chemistry and Chemical Engineering,
Volume 5: Research Methodologies in Modern Chemistry and Applied Science
Editors: A. K. Haghi, PhD, Ana Cristina Faria Ribeiro, PhD, Lionello Pogliani, PhD, Devrim Balköse, PhD, Francisco Torrens, PhD, and Omari V. Mukbaniani, PhD

CONTENTS

LIST OF CONTRIBUTORS

Soltan Ali Ogli Aliyev
Department of Mathematics and Mechanics, National Academy of Science of Azerbaijan "AMEA," Baku, Azerbaijan. E-mail: Soltanaliyev@yahoo.com

D. S. Andreev
Graduate Student of Volgograd State Architecture Building University, Volgograd, Russia. E-mail: power_words@mail.ru

Hossein Hariri Asli
Civil Engineering Department, Faculty of Engineering, University of Guilan, Rasht, Iran. E-mail: hh_asli@yahoo.com

Kaveh Hariri Asli
Department of Mathematics and Mechanics, National Academy of Science of Azerbaijan "AMEA," Baku, Azerbaijan. E-mail: hariri_k@yahoo.com

V. A. Babkin
Doctor of Chemical Sciences, Professor, Academician of International Academy "Contenant", Academician of Russian Academy of Nature, Sebryakovsky Branch, Volgograd State University of Architecture and Engineering, Volgograd, Russia. E-mail: Babkin_v.a@mail.ru

Emili Besalú
Departament de Química & Institut de Química Computacional i Catàlisi, Universitat de Girona, Campus Montilivi, C/Maria Aurèlia Campmany, Girona, Spain

V. I. Binyukov
The Federal State Budget Institution of Science, N. M. Emanuel Institute of Biochemical Physics, Russian Academy of Sciences, 4 Kosygin Str., Moscow 119334, Russia

Mohammad Reza Farahani
Department of Applied Mathematics, Iran University of Science and Technology (IUST), Narmak, Tehran 16844, Iran. E-mail: MrFarahani88@gmail.com

Wei Gao
School of Information Science and Technology, Yunnan Normal University, Kunming 650500, China. E-mail: 1gaowei@ynnu.edu.cn

Alexei A. Gridnev
N. N. Semenov Institute of Chemical Physics, Russian Academy of Sciences, Moskva, Russia

Muhammad Kamran Jamil
Abdus Salam School of Mathematical Sciences, Government College University (GCU), Lahore, Pakistan E-mail: 3m.kamran.sms@gmail.com

Sam John
Research and Post Graduate Department of Chemistry, St Berchmans College, Kottayam, India

Peter Jurkovič
VIPO A.S., Partizánske, Gen. Svobodu 1069/4, 958 01 Partizánske, Slovakia

Svetlana N. Kholuiskaya
N. N. Semenov Institute of Chemical Physics, Russian Academy of Sciences, Moskva, Russia.
E-mail: soho@chph.ras.ru

G. Khovanets'
Department of Physical Chemistry of Fossil Fuels InPOCC, National Academy of Sciences of Ukraine, Naukova Str. 3a, 79060 Lviv, Ukraine

Beena Mathew
School of Chemical Sciences, Mahatma Gandhi University, Kottayam, India

L. I. Matienko
The Federal State Budget Institution of Science, N. M. Emanuel Institute of Biochemical Physics, Russian Academy of Sciences, 4 Kosygin Str., Moscow 119334, Russia.
E-mail: matienko@sky.chph.ras.ru

Ján Matyašovský
VIPO a.s., Partizánske, Gen. Svobodu 1069/4, 958 01 Partizánske, Slovakia

Yu. Medvedevskikh
Department of Physical Chemistry of Fossil Fuels InPOCC, National Academy of Sciences of Ukraine, Naukova Str. 3a, 79060 Lviv, Ukraine

E. M. Mil
The Federal State Budget Institution of Science, N. M. Emanuel Institute of Biochemical Physics, Russian Academy of Sciences, 4 Kosygin Str., Moscow 119334, Russia

Vadim V. Minin
N. S. Kurnakov Institute of General and Inorganic Chemistry, Russian Academy of Sciences, Moskva, Russia

L. A. Mosolova
The Federal State Budget Institution of Science, N. M. Emanuel Institute of Biochemical Physics, Russian Academy of Sciences, 4 Kosygin Str., Moscow 119334, Russia

V. V. Petrov
TiT Student of 11-d-15, Sebryakovsky Branch of Volgograd State University of Architecture and Civil Engineering, Volgograd, Russia. E-mail: motovlad2013@yandex.ru

Lionello Pogliani
Unidad de Investigación de Diseño de Fármacos y Conectividad Molecular, Departamento de Química Física, Facultad de Farmacia, Universitat de València, Burjassot, València, Spain, MOLware SL, Valencia, Spain. E-mail: liopo@uv.es

S. Rafiei
University of Guilan, Rasht, Iran

T. Sajini
Research and Post Graduate Department of Chemistry, St Berchmans College, Kottayam, India; School of Chemical Sciences, Mahatma Gandhi University, Kottayam, India

Ján Sedliačik
Technical University in Zvolen, Masaryka 24, 960 53 Zvolen, Slovakia. E-mail: sedliacik@tuzvo.sk

T. Sezonenko
Department of Physical Chemistry of Fossil Fuels InPOCC, National Academy of Sciences of Ukraine, Naukova Str. 3a, 79060 Lviv, Ukraine

Anamika Singh
Department of Botany, Maitreyi Collage, University of Delhi, New Delhi, India

Mária Šmidriaková
Technical University in Zvolen, Masaryka 24, 960 53 Zvolen, Slovakia

Ladislav Šoltés
Institute of Experimental Pharmacology and Toxicology, Slovak Academy of Sciences, 84104 Bratislava, Slovakia

O. V. Stoyanov
Doctor of Engineering Sciences, Professor of Department of "Technology of Plastic Masses" of Kazan State Technical University, Kazan, Russia. E-mail: stoyanov@mail.ru

Regina Ravilevna Usmanova
Ufa State Technical University of Aviation, Ufa 450000, Bashkortostan, Russia. E-mail: Usmanovarr@mail.ru

J. Vicente Julian-Ortiz
Unidad de Investigación de Diseño de Fármacos y Conectividad Molecular, Departamento de Química Física, Facultad de Farmacia, Universitat de València, Burjassot, València, Spain, MOLware SL, Valencia, Spain. E-mail: jejuor@uv.es

Gennady Efremovich Zaikov
Doctor of Chemical Sciences, Professor, Academician of International Academy of Science (Munich, Germany), Honored Scientist of Russian Federation, Institute of Biochemical Physics, Moscow, Russia; N. M. Emanuel Institute of Biochemical Physics, Russian Academy of Sciences, Moscow 119991, Russia. E-mail: chembio@sky.chph.ras.ru

V. Zakordonskiy
Ivan Franko National University of Lviv, Kyryla and Mefodiya Str. 6, 79005 Lviv, Ukraine. E-mail: zakordonskiy@franko.lviv.ua

M. Ziaei
University of Guilan, Rasht, Iran

LIST OF ABBREVIATIONS

ACF	activated carbon fiber
ACNFs	activated carbon nanofibers
AFM	atomic force microscopy
AMAAB	4-amino-4-methacrylatylazobenzene
ARD	acireductone dioxygenase
BPA	bisphenol A
CNFs	carbon nanofibers
DMF	dimethylformamide
ELL	economic level of leakage
EOR	extent of stabilization reaction
FDs	finite differences
FE	finite elements
GH	Gibbs–Helmholtz
GTP	group-transfer polymerization
HEMA	2-hydroxyethyl methacrylate
HEMA	2-hydroxyethylmethacrylate
HOIC	hybrid organic–inorganic composites
IARC	International Agency for Research on Cancer
ICI	Imperial Chemical Industries
MANFAB	(4-methacryloyloxy) nonafluoroazobenzene
MAPASA	4-[(4-methacryloyloxy)phenylazo]benzene sulfonic acid
MIPs	molecularly imprinted polymers
MPABA	4-[(4-methacryloyloxy)phenylazo]benzoic acid
OSG	overlying simple graph
PAN	polyacrylonitrile
PDE	partial differential equation
PhAAAn	p-phenylazoacrylanilide
PHAs	polyhydroxyalkanoates
PHB	poly-beta butyric acid
PHEMA	poly(2-hydroxyethyl) methacrylate
PHV	poly-beta-hydroxy valeric acid
PP	polypropylene
QSPR/QSAR	quantitative structure–property and structure–activity relationships

SCP	single cell protein
SEM	scanning electron microscopy
SRMs	stimuli-responsive materials
TB	Titius–Bode
TEOS	tetraethoxysilane
UF	urea–formaldehyde
WDNs	water distribution networks

PREFACE

This volume is the first of the 5-volume set on Applied Chemistry and Chemical Engineering. This volume brings together innovative research, new concepts, and novel developments on modern approaches to modeling and calculation in applied chemistry and chemical engineering as well as experimental designs.

The volume brings together innovative research, new concepts, and novel developments in the application of informatics tools for applied chemistry and computer science. It discusses the developments of advanced chemical products and respective tools to characterize and predict the chemical material properties and behavior. Providing numerous comparisons of different methods with one another and with different experiments, not only does this book summarize the classical theories, but it also exhibits their engineering applications in response to the current key issues. Recent trends in several areas of chemistry and chemical engineering science, which have important application to practice and industry, are also discussed.

The volume presents innovative research and demonstrates the progress and promise for developing chemical materials that seem capable of moving this field from laboratory-scale prototypes to actual industrial applications

Features

- Presents information on the important problems of chemical engineering modeling and nanotechnology. These investigations are accompanied by real-life applications in practice.
- Includes new theoretical ideas in calculating experiments and experimental practice.
- Looks at new trends in chemoinformatics.
- Introduces the types of challenges and real problems that are encountered in industry and graduate research.
- Presents computational chemistry examples and applications.
- Focuses on concepts above formal experimental techniques and theoretical methods.

Applied Chemistry and Chemical Engineering: Volume 1: Mathematical and Analytical Techniques provides valuable information for chemical engineers and industrial researchers as well as for graduate students. This book will be essential amongst chemists, engineers, and researchers in providing mutual communication between academics and industry professionals around the world.

Applied Chemistry and Chemical Engineering, 5-Volume Set includes the following volumes:

- Applied Chemistry and Chemical Engineering,
  Volume 1: Mathematical and Analytical Techniques
- Applied Chemistry and Chemical Engineering,
  Volume 2: Principles, Methodology, and Evaluation Methods
- Applied Chemistry and Chemical Engineering,
  Volume 3: Interdisciplinary Approaches to Theory and Modeling with Applications
- Applied Chemistry and Chemical Engineering,
  Volume 4: Experimental Techniques and Methodical Developments
- Applied Chemistry and Chemical Engineering,
  Volume 5: Research Methodologies in Modern Chemistry and Applied Science

PART I

Modern Approaches to Modelling and Calculation

CHAPTER 1

DIGRAPHS, GRAPHS, AND THERMODYNAMICS EQUATIONS

LIONELLO POGLIANI*

Facultad de Farmacia, Department de Química Física, Universitat de València, Av. V.A. Estellés s/n, 46100 Burjassot, València, Spain

**E-mail: lionello.pogliani@uv.es*

CONTENTS

ABSTRACT

The particular structure of many thermodynamic equations can be mimicked by the aid of graphs both directed and simple graphs. Starting points are two types of directed graphs (also digraphs), the energy-digraph, or *E*-digraph, and the entropy digraph, or *S*-digraph. The most important thermodynamic relationships can be modeled by the aid of these two tools plus a set of simple rules and a series of symmetry operations performed with simple graphs superposed on the previous digraphs. Actually, in this way, not only the most famous thermodynamic relations can be derived in a fully automatic way, but the "machinery" can also be used to solve some thermodynamic problems.

1.1 INTRODUCTION

The first attempt to derive in a direct way the thermodynamic equations was done in 1914 by the physicist Percy William Bridgam (1882–1961, Nobel Prize in 1946), who suggested an algebraic method to derive the more than 700 first derivatives encompassing 3 parameters chosen among a pool of 10 fundamental parameters and the more than 10^9 relations between the first derivatives. Bridgam's method further simplified by A. N. Shaw in 1935 is succinctly presented in Appendix 6 of Ref. [1]. His method is based on mathematical functions known as Jacobians and on a short-hand way to encode differentials. Nevertheless, the method was hardly a success. In fact, Brian Smith in his preface to *Basic Chemical Thermodynamics*[2] could write the following words about the feelings students developed when they had to go through the study of thermodynamics: "The first time I heard about chemical thermodynamics was when a second-year undergraduate brought me the news early in my freshman year. He told me a spine-chilling story of endless lectures with almost three-hundred numbered equations, all of which, it appeared, had to be committed to memory and reproduced in exactly the same form in subsequent examinations. Not only did these equations contain all the normal algebraic symbols but in addition they were liberally sprinkled with stars, daggers, and circles so as to stretch even the most powerful of minds."

1.2 GRAPH-BASED APPROACH

A diagrammatic scheme was not long ago proposed to derive many thermodynamic relationships,[3,4] and it was based on an approach used to derive the

Maxwell relations, which is described in Callen's book on thermodynamics.[5] It seems that it was first proposed in 1929 by Max Born (1882–1970, Nobel Prize in 1954). This diagrammatic approach underwent further improvements by the aid of graph and vector concepts.[6–10] In the following sections, the graph-based approach for the thermodynamic relations will be discussed in detail.

1.2.1 *THE DIGRAPH*

A directed graph, or digraph, consists of a set V of *vertices* (or *nodes*) together with a set E of ordered pairs of elements of V-called *edges* (or *arcs*).

In a directed graph, a *vertex* is represented by a point, and each ordered pair is represented using an edge with its direction indicated by an arrow (Fig. 1.1). In the simple directed acyclic graph $a \rightarrow b$, vertex a is called the initial vertex (or tail) of the edge (a, b), and vertex b is called the terminal vertex (or head) of this edge. Vertex a is said to be adjacent to b and b is said to be adjacent from a.[11] A vertex can also be an isolated unconnected vertex, that is, a zero vertex.

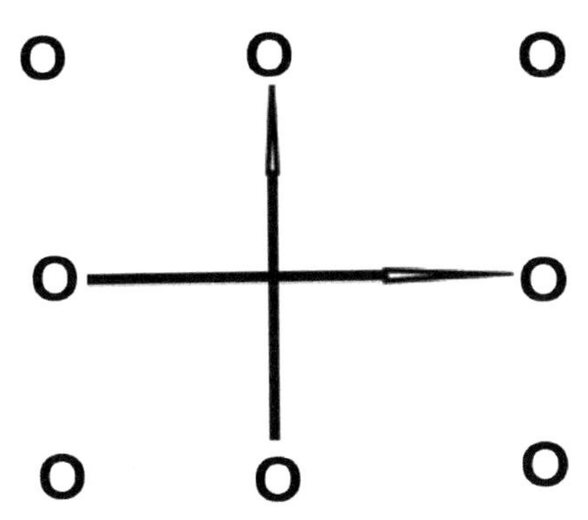

FIGURE 1.1 A digraph with four zero vertices and two head and two tail vertices.

1.2.2 *THE ENERGY AND THE ENTROPY DIGRAPHS*

To get into thermodynamics, we have to label the vertices of digraph of Figure 1.1 with the following set of eight fundamental thermodynamic properties: $\{A, G, H, U, P, S, T, V\}$. These eight properties can be arranged into two subsets, an energy-dimensioned subset of the four *zero*-degree vertices labeled with energy functions $\{A, G, H, U\}$, and a subset of *head–tail* vertices labeled with the natural variables {*P*, *S*, *V*, *T*}. Of these, two labels *S* and *P* are the *tail* vertices and *V* and *T* are the *head* vertices. The resulting digraph is called the energy digraph or *E*-digraph (Fig. 1.2, left). The eight fundamental thermodynamic properties are the Helmholtz energy, *A*; the Gibbs energy, *G*; the enthalpy, *H*; and the internal energy, *U*. The natural variables of the digraph subset are the pressure, *P*, the entropy, *S*, the absolute temperature, *T*, and the volume, *V*.

The *E*-digraph is the digraph of the relation *R* = {*A*, *G*, *H*, *U*, (*P*, *V*), (*S*, *T*)}. Multiplication of *P* (*tail*) with *V* (*head*), and *S* (*tail*) with *T* (*head*) allow to obtain two energy-dimensioned quantities: PV and ST. Notice that the thermodynamic labels of the zero-degree vertices are ordered clockwise (clockwise rotations are here considered positive), while the labels of the natural variables have a slanted *Z* alphabetical order.

The second set of fundamental thermodynamic properties that are going to label the vertices of Fig. 1.1, where the vertical arrow has been inverted, are the following: {M_1, M_2, M_3, *S*, *P*/*T*, *U*, 1/*T*, *V*}. The resulting digraph can be named an entropy digraph or *S*-digraph (Fig. 1.2, right). M_1, M_2, and M_3 denote the Massieu entropic functions, which are useful in the theory of irreversible thermodynamics and in statistical mechanics, while the other quantities have been defined in the previous paragraph. M_1, M_2, and M_3 are due to the French mineralogist François-Jacques Dominique Massieu.[5]

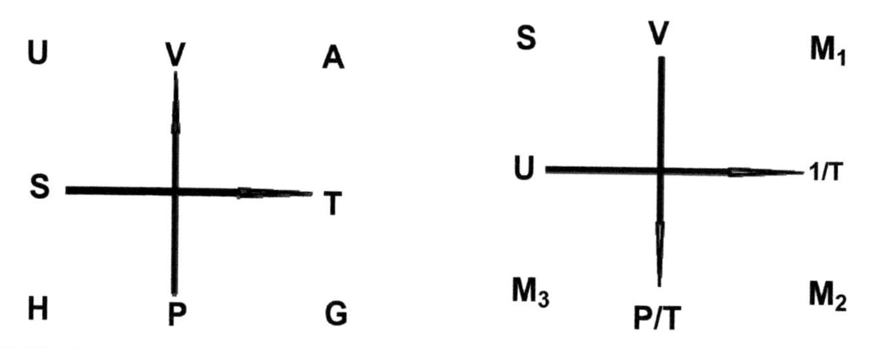

FIGURE 1.2 The *E*-digraph (left) and the *S*-digraph (right).

Even here, these eight properties can be arranged into two subsets, a zero-degree vertices labeled with entropy-dimensioned entropic functions {M_1, M_2, M_3, *S*}, and a subset of *head–tail* vertices labeled with the natural variables, {*P*/*T*, *U*, *V*, 1/*T*}. Of these two (*U*, *V*) are *tail* vertices, and two (*P*/*T*, 1/*T*) are *head* vertices.

The S-digraph is the digraph of the relation: $R = \{M_1, M_2, M_3, S, (V, P/T), (U, 1/T)\}$. Multiplying the *tail* V, with the *head* P/T, and the *tail* U with the *head* $1/T$, two entropy-dimensioned quantities are obtained: PV/T and U/T. Even in this digraph, the thermodynamic labels of the zero-degree vertices are ordered clockwise, while the labels of the natural variables have, practically, a slanted Z alphabetical order (P/T, $1/T$, U, V).

1.2.3 PROPERTIES OF THE E- AND S-DIGRAPHS

These two digraphs share the following three properties that allow to build the thermodynamic many equations of the standard exposition of classical thermodynamics of simple systems.

Property 1 (functional property): The corner parameters are functions of their nearby natural variables,

$$A = A(V, T),\ G = G(T, P),\ H = H(P, S),\ U = U(S, V) \quad (1.1)$$

$$M_1 = M_1(V, 1/T),\ M_2 = M_2(1/T, P/T),\ M_3 = M_3(P/T, U),\ S = S(U, V) \quad (1.2)$$

These functional relations allow to derive the total differentials of the corner properties.

Property 2 (orthogonal property): Variables belonging to the same arrow can be multiplied with each other to obtain either an energy-dimensioned term (PV and ST) or an entropy-dimensioned term (U/T and PV/T). Variables belonging to orthogonal arrows cannot be multiplied with each other. Zero-degree quantities cannot be multiplied with each other.

Property 3 (directional property): Flow toward an arrowhead (from a *tail* to a *head*) is positive, while flow toward an arrow tail (from a *head* to a *tail*) is negative.

1.3 OVERLYING SIMPLE GRAPHS

Now, all we have to do is to superimpose on these two digraphs a series of simple graphs (overlying simple graph [OSG]), which mimic the shape of a capital letter: F, M, N, and P, where at the vertices of these letter-shaped graphs are placed the thermodynamic properties. Each superposition gives rise to a simple graph relation, R(I-OSG:E or S), among the encompassed vertices of the I-OSG (I stands for any of the letter-shaped simple graphs) that gives rise to a thermodynamic equation.

1.3.1 THE N EQUATIONS CONCERNING THE ZERO-DEGREE VERTICES

The E-digraph and the N-shaped OSG, that is, N-OSG (Fig. 1.3), allow to obtain the eight N relationships. The well-known relation between energy functions H and U, $H = U + PV$, which obeys properties 2 and 3, could succinctly be rewritten in the following way,

$$R(N\text{-OSG}:E) = \{H{:}U, (P, V)\} \rightarrow H = U + PV \quad (1.3)$$

Relations of the type $R(N\text{-OSG}:E) = \{V{:}P, (U, H)\}$ are dimensionally wrong, and relations that start at the diagonal vertices are not allowed. Furthermore, they would give rise to meaningless relations.

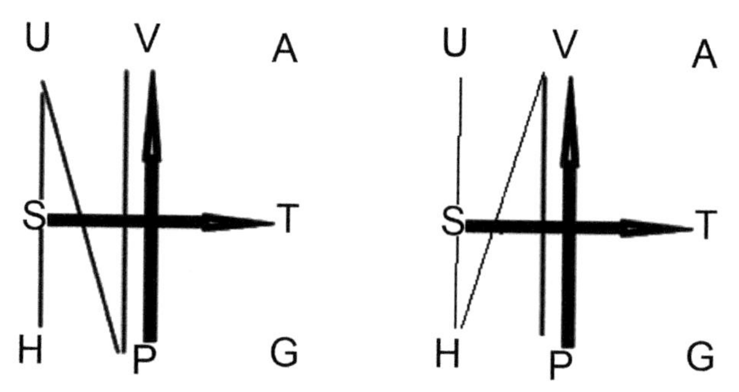

FIGURE 1.3 The N-OSG and E-digraph, *left*: for eq 1.3; *right*: for eq 1.4.

Symmetry operations like C_4 rotations (90° clockwise), C_2 rotations (180° clockwise), σ_{PV} reflections through the PV axis and $_{ST}$ reflections through ST axis of the $R(N\text{-OSG}:E)$ of eq 1.3 allow to obtain all other relations among the potentials (two relations for each vertex). Composite symmetry operations, starting from right operation, are also allowed like, $\sigma_{ST}\cdot\sigma_{PV}, \sigma_{ST}\cdot C_4$, and others.

$$\sigma_{ST}[R(N\text{-OSG}:E)] = \{U{:}H, (V,P)\} \rightarrow U = H - PV \quad (1.4)$$

$$\sigma_{PV}[R(N\text{-OSG}:E)] = \{G{:}A, (P,V)\} \rightarrow G = A + PV \quad (1.5)$$

$$-C_2[R(N\text{-OSG}:E)] = \{A{:}G, (V, P)\} \rightarrow A = G - PV \quad (1.6)$$

$$C_4[R(N\text{-OSG}:E)] = \{U{:}A, (S, T) \rightarrow U = A + ST \quad (1.7)$$

$$\sigma_{PV}\cdot C_4[R(N\text{-OSG}:E)] = \{A{:}U, (T, S)\} \rightarrow A = U - TS \quad (1.8)$$

$$-C_4[R(N\text{-OSG}:E)] = \{G{:}H, (T, S)\} \rightarrow G = H - ST \quad (1.9)$$

$$-C_4 \cdot \sigma_{ST}[R(N\text{-OSG}:E)] = \{H{:}G, (S, T)\} \rightarrow H = G + ST \quad (1.10)$$

The *N*-OSG on the *S*-digraph (Fig. 1.4), and the given properties 2 and 3, let us derive the eight *N* entropic relations, among which the most important are the following three relationship for M_1, M_2, and M_3.

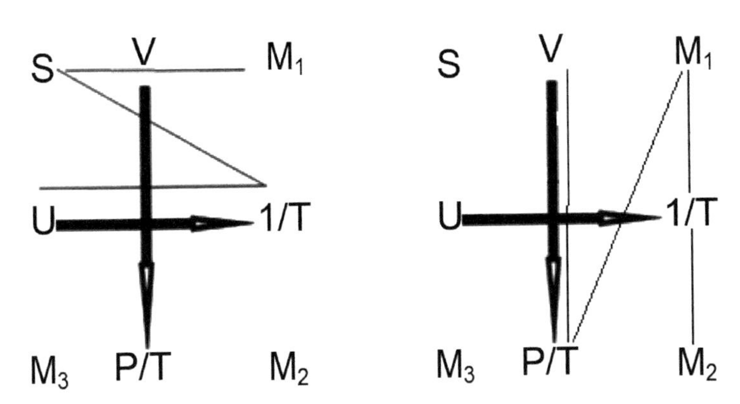

FIGURE 1.4 The *N*-OSG and the *S*-digraph, *left*: eq 1.11; *right*: eq 1.12.

$$R(N\text{-OSG}:S) = \{M_1{:}S, (1/T, U)\} \rightarrow M_1 = S - U/T$$
$$= (TS - U)/T \; M_1 = -A/T \quad (1.11)$$

The very last result was obtained by the aid of eq 1.8. A C_4 rotation of $R(N\text{-OSG}:S)$ allows to derive the thermodynamic relation for M_2 [after insertion of $M_1 = -A/T$, and G from eq 1.5],

$$C_4[R(N\text{-OSG}:S)] = \{M_2{:}M_1, (P/T, V)\} \rightarrow M_2 = M_1 - PV/T =$$
$$-(A + PV)/T \; M_2 = -G/T \quad (1.12)$$

This is the well-known Planck function[5]: $Y = M_2 = -G/T$. A C_2 operation on R_{SN} let us uncover the thermodynamic meaning of M_3, which, after insertion for $M_2 = -G/T$ from eq 1.12, cannot be further simplified,

$$C_2[R(N\text{-OSG}:S)] = \{M_3{:}M_2, (U, 1/T)\} \; M_3 = M_2 + U/T$$
$$M_3 = (U - G)/T \quad (1.13)$$

1.3.2 THE P EQUATIONS CONCERNING THE ZERO-DEGREE VERTICES

Be the *P*-OSG and the *E*-digraph of Figure 1.5, by the aid of the superposition together with properties 2 and 3, it is possible to encode the central relationship of thermodynamics. The encoding relation is,

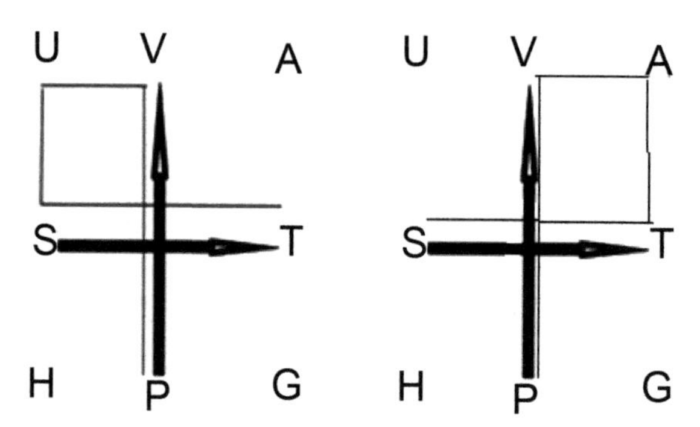

FIGURE 1.5 The *P*-OSG and the *E*-digraph, *left*: eq 1.14; *right*: eq 1.15.

$$R(P\text{-OSG}:E) = \{U:(S, T), (V, P)\} \rightarrow \mathrm{d}U = \mathrm{d}S \cdot T - \mathrm{d}V \cdot P \quad (1.14)$$

that is, rearranging, $\mathrm{d}U = T\mathrm{d}S - P\mathrm{d}V$.

With a σ_{PV} on $R(P\text{-OSG}:E)$ and then a σ_{ST} on $\sigma_{PV}[R(P\text{-OSG}:E)]$, the following thermodynamic relationships can be derived,

$$\sigma_{PV}[R(P\text{-OSG}:E)] = \{A:(T, S), (V, P)\} \rightarrow \mathrm{d}A = -\mathrm{d}T \cdot S - \mathrm{d}V \cdot P$$
$$\mathrm{d}A = -S\mathrm{d}T - P\mathrm{d}V \quad (1.15)$$

$$-C_2[R(P\text{-OSG}:E)] = \{G:(T, S), (P, V)\} \rightarrow \mathrm{d}G = -\mathrm{d}T \cdot S + \mathrm{d}P \cdot V$$
$$\mathrm{d}G = -S\mathrm{d}T + V\mathrm{d}P \quad (1.16)$$

The *P*-OSG and the *S*-digraph of Figure 1.6 give rise to the following thermodynamic relationship (notice that P/T is the arrowhead vertex),

$$R(P\text{-OSG}:S) = \{S:(U, 1/T), (V, P/T)\} \rightarrow \mathrm{d}S = \mathrm{d}U \cdot (1/T)$$
$$+ \mathrm{d}V \cdot (P/T)\ \mathrm{d}S = \mathrm{d}U/T + (P/T)\mathrm{d}V \quad (1.17)$$

Successive *P*-OSG operations give rise to other relationships, among which (upon solving $\mathrm{d}(P/T)$ and rearranging) eq 1.19, a no-easy relation to arrive at with purely algebraic methods,

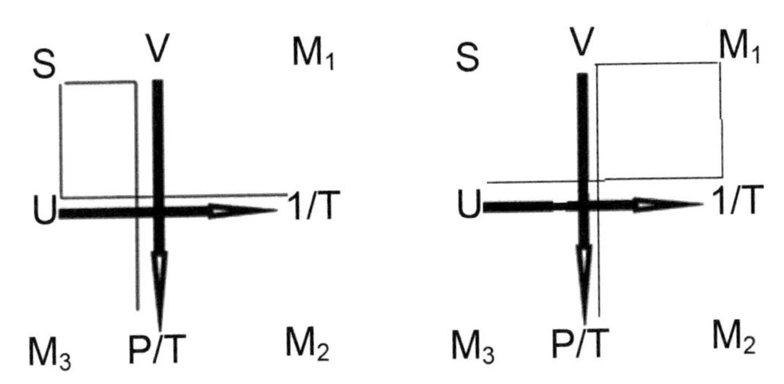

FIGURE 1.6 The *P*-OSG and the *S*-digraph, *left*: eq 1.17; *right*: eq 1.18.

$$\sigma_{PV/T}[R(P\text{-OSG}{:}S)] = \{M_1{:}(1/T, U), (V, P/T)\}$$
$$dM_1 = -d(1/T)\cdot U + dV\cdot(P/T) = -(U/T^2)dT + (P/T)dV \tag{1.18}$$

$$\sigma_{U/T}[R(P\text{-OSG}{:}S)] = \{M_3{:}(U, 1/T), (P/T, V)\} \rightarrow dM_3$$
$$= dU\cdot(1/T) - d(P/T)\cdot V = (1/T)dU - Vd(P/T) = (1/T)$$
$$dU + (VP/T^2)dT - (V/T)dP \tag{1.19}$$

1.3.3 THE F EQUATIONS CONCERNING THE ZERO AND TAIL–HEAD VERTICES

The *F*-OSG and the *E*-digraph of Figure 1.7 together with a result obtained from eq 1.15, that is, $(\partial A/\partial V)_T = -P$, allow to derive the following encoding relation, where the second and fourth property of *R*(*F*-OSG:*E*) determine the sign of the relation (from *V* to *P*, i.e., here the only allowed flow),

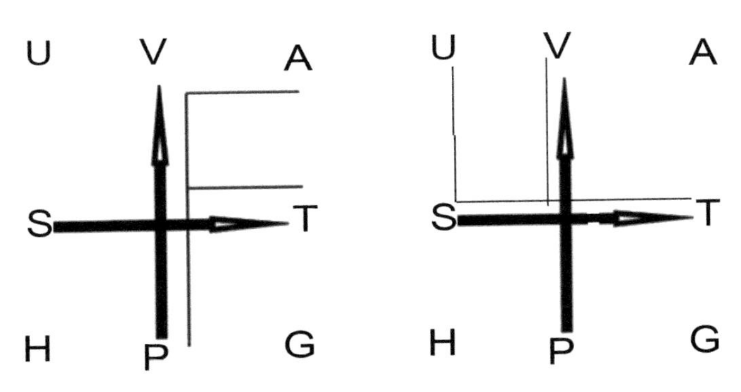

FIGURE 1.7 The *F*-OSG on the E-digraph, *left*: eq 1.20; *right*: eq 1.21.

$$R(F\text{-OSG}:E) = \{A, V, T{:}P\} \rightarrow (\partial A/\partial V)_T = -P \quad (1.20)$$

$A - C_4$ operation on $R(F\text{-OSG}:E)$ and a σ_{ST} operation on $-C_4R(F\text{-OSG}:E)$ give us two new relations,

$$-C_4R(F\text{-OSG}:E) = \{U, S, V{:}T)\} \rightarrow (\partial U/\partial S)_V = T \quad (1.21)$$

$$\sigma_{ST}[-C_4R(F\text{-OSG}:E)] = \{H, S, P{:}T)\} \rightarrow (\partial H/\partial S)_P = T \quad (1.22)$$

A σ_{PV} operation on $R(F\text{-OSG}:E)$ let us derive the following equation:

$$\sigma_{PV}[R(F\text{-OSG}:E)] = \{U, V, S{:}P)\} \rightarrow (\partial U/\partial V)_S = -P \quad (1.23)$$

When the F-OSG is applied to the S-digraph in Figure 1.8, the following thermodynamic relationship can be obtained (compared with eq 1.21),

$$R(F\text{-OSG}:S) = \{S, U, V{:}1/T)\} \rightarrow (\partial S/\partial U)V = 1/T \quad (1.24)$$

Performing some reflection operations on $R(F\text{-OSG}:S)$, we obtain two other not at all evident relations,

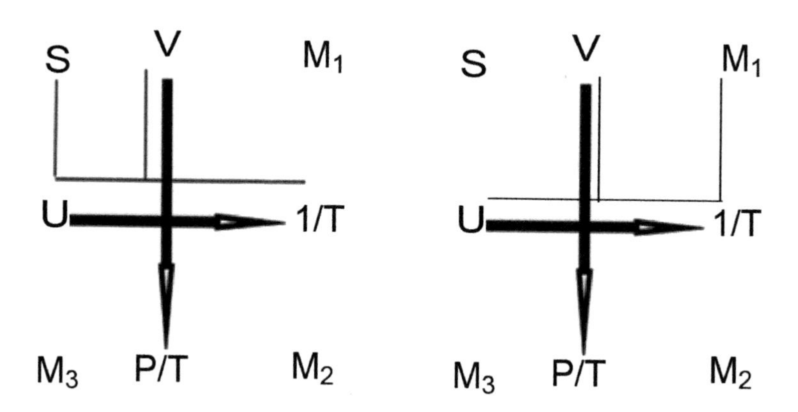

FIGURE 1.8 The F-OSG on the S-digraph, *left*: eq 1.24; *right*: eq 1.25.

$$\sigma_{VP/T}[R(F\text{-OSG}:S)] = \{M_1, 1/T, V{:}U)\} \rightarrow (\partial M_1/\partial(1/T))_V = -U \quad (1.25)$$

$$\sigma_{U/T}[R(F\text{-OSG}:S)] = \{M_3, U, P/T{:}1/T)\} \rightarrow (\partial M_3/\partial U)_{P/T} = 1/T \quad (1.26)$$

1.3.4 THE M MAXWELL RELATIONS CONCERNING THE HEAD–TAIL VERTICES

The history of the diagrammatic method for the thermodynamic equations started with the Maxwell relations as Callen suggested.[5] These relations concern the E-digraph only. The M-OSG and digraph of Figure 1.9 and the fact that these relations concern the partial derivatives of the *head–tail* properties, where the third property is held constant, allows to write the R(M-OSG:E) relation and its corresponding Maxwell equation,

$$R(M\text{-OSG}:E) = \{(P, T, V):(S, V, T))\} \rightarrow (\partial P/\partial T)_V = (\partial S/\partial V)_T \quad (1.27)$$

The sign is under the control of the only allowed flow, that is, from the first to the third property in each parenthesis. The other three relations can be derived with C_4 rotations of R(M-OSG:E),

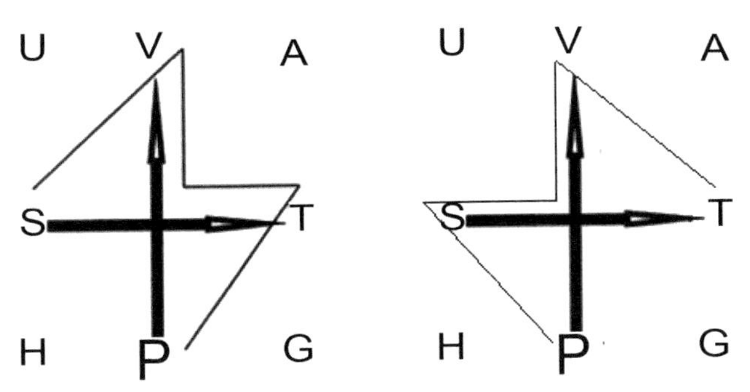

FIGURE 1.9 The M-OSG on the E-digraph, *left*: eq 1.27; *right*: eq 1.28.

$$-C_4[R(M\text{-OSG}:E)] = \{(T, V, S):(P, S, V)\} \rightarrow -(\partial T/\partial V)_S = (\partial P/\partial S)_V \quad (1.28)$$

$$-C_2[R(M\text{-OSG}:E)] = \{(V, S, P):(T, P, S)\} \rightarrow -(\partial V/\partial S)P = -(\partial T/\partial P)_S \quad (1.29)$$

$$C_4[R(M\text{-OSG}:E)] = \{(S, \mathrm{P}, T):(V, T, P)\} \rightarrow (\partial S/\partial P)_T = -(\partial V/\partial T)_{\mathrm{P}} \quad (1.30)$$

1.4 PROBLEMS

Problem 1. Be the Gibbs–Helmholtz (GH) equation at P = cost, $[\partial(G/T)/\partial(1/T)]_P = H(GH)$. Find the answer for the following question: $[\partial(A/T)/\partial(1/T)]_V = ?$

Answer: Draw the simple *GH*-OSG connecting all parameters of the GH equation on the *E*-digraph (Fig. 1.10, left), perform then a σ_{ST} operation (Fig. 1.10, right) and the result is $[\partial(A/T)/\partial(1/T)]V = U$

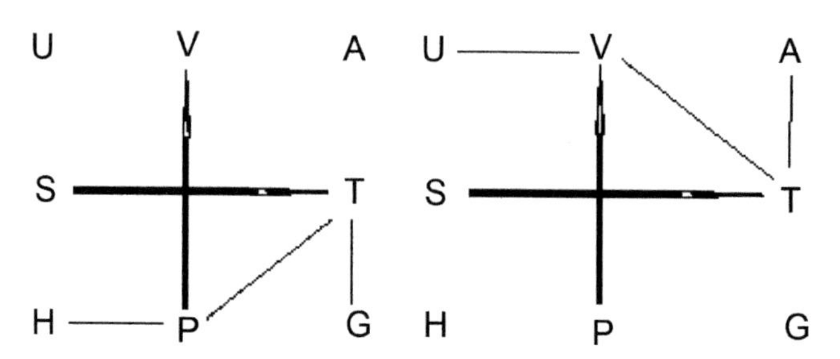

FIGURE 1.10 The *GH*-OSG and the *E*-digraph, *left*: the problem; *right*: the solution.

Problem 2. Be the expression for the *internal pressure* at T = cost (isothermal), $\pi_T = (\partial U/\partial V)_T = T(\partial S/\partial V)_T - P$. Find the corresponding expression for $(\partial H/\partial P)_T$, known in thermodynamics as the *isothermal Joule–Thomson coefficient*, μ_T.

Answer: Draw the simple graph connecting the T, S, V, T, and P parameters of π_T on the *E*-digraph of Figure 1.11, left, perform a σ_{ST} operation of the π_T-OSG and obtain (Fig. 1.11, right): $\mu_T = (\partial H/\partial P)_T = T(\partial S/\partial P)_T + V$ (the only change: P replaces V and vice versa).

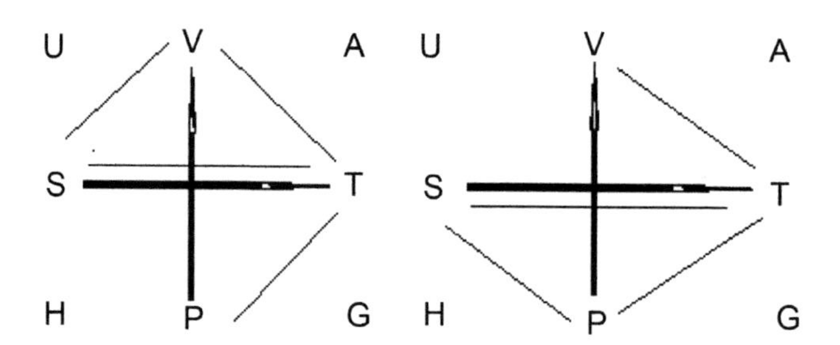

FIGURE 1.11 The π_T-OSG and the *E*-digraph, *left*: the problem; *right*: the solution.

1.5 CONCLUSION

A metalanguage, in logic and in linguistic, is a language used to make statements about statements in another language, which is called the object language. More broadly, it can refer to any terminology used to discuss the language itself, a written grammar for instance. Thermodynamics with all its mathematical equations does awake the feeling that there ought to exist a formalism that could encompass and order the different types of relations. What we have outlined in the previous sections can then be considered as a sort of metalanguage for thermodynamics. It is based on the use of directed graphs and simple graphs, and it allows us to derive in a completely "geometric" way many equations of thermodynamics.

The reader has surely noticed that the starting move was normally done by the aid of a well-known thermodynamic relationship. Actually, the starting relation can easily be guessed by the aid of the digraph together with the OSG and properties 1–3. The usefulness of the method is finally underlined by its ability to solve problems. It is worth mentioning that graph methods have also been applied to encode of phase diagrams.[12]

Clearly, being able to derive many thermodynamic equations doesn't mean to understand thermodynamics. This is a quite serious problem as certified by the "battle" that continues to rage about the real meaning of entropy.[13–15] On the subject, the words by A. Sommerfed[16] are illuminating: "Thermodynamics is a funny subject. The first time you go through it, you don't understand it at all. The second time you go through it, you think you understand it, except for one or two small points. The third time you go through it, you know you don't understand it, but by that time you are so used to it, it doesn't bother you anymore."

KEYWORDS

- **digraphs**
- **energy- and entropy digraphs**
- **overlying simple graphs**
- **thermodynamic equations**
- **machinery**

REFERENCES

1. Newton, G. N.; Randall, M. (Revised by Pitzer, K. S.; Brewer, L.), *Thermodynamics*, McGraw-Hill: New York, 1961.
2. Brian, S. E. *Basic Chemical Thermodynamics*, Oxford University Press: Oxford, 2004.

3. Phillips, J. M. Mnemonic Diagrams for Thermodynamic Systems. *J. Chem. Ed.* **1987,** *64*, 674–675.
4. Rodriguez, J.; Brainard, A. An Improved Mnemonic Diagram for Thermodynamic Relationships. *J. Chem. Ed.* **1989,** *66*, 495–496.
5. Callen, H. B. *Thermodynamics and Introduction to Thermostatistics*. Wiley: New York, 1985.
6. Pogliani, L. Pattern Recognition and Alternative Physical Chemistry Methodologies. *J. Chem. Inf. Comput. Sci.* **1998,** *38*, 130–143.
7. Pogliani, L. *Magic Squares and the Mathematics of Thermodynamics. MATCH Commun. Math. Comput. Chem.* **2003,** *47*, 153–166.
8. Pogliani, L. The Diagrammatic method, and the Planck and Massieu Functions. *J. Chem. Ed.* **2001,** *78*, 680–681.
9. Pogliani, L. A Vector Representation for Thermodynamic Relationships. *J. Chem. Educ.* **2006,** *83*, 155–158.
10. Pogliani, L. Graphs and Thermodynamics. *J. Math. Chem.* **2009,** *46*, 15–23.
11. Rosen, K. H. *Discrete Mathematics and its Applications*. McGraw-Hill: New York, 1995.
12. Pogliani, L. Ordered Sequences of Thermodynamic Objects. In *Topics in Chemical Graph Theory*; Gutman, I. Ed.; *Mathematical Chemistry Monographs*, MCM: Kragujevac, 2014; vol 16a, pp 229–240.
13. Ben-Naim, A. *Entropy Demystified: The Second Law Reduced to Plain Common Sense*. World Scientific: London, 2007.
14. Ben-Naim, A. *A Farewell to Entropy: Statistical Thermodynamics Based on Information*, World Scientific: London, 2008.
15. Albert, D. Z. *Time and Chance*, second ed. Harvard University Press: Cambridge, 2003.
16. http://www.eoht.info/page/Thermodynamics+quotes. (Arnold Sommerfeld, 1868–1951, one of the founders of Quantum Mechanics.)

CHAPTER 2

USEFULNESS AND LIMITS OF PREDICTIVE RELATIONSHIPS

EMILI BESALÚ[1], LIONELLO POGLIANI[2,3*], and J. VICENTE JULIAN-ORTIZ[2,3]

[1]*Departament de Química & Institut de Química Computacional i Catàlisi, Universitat de Girona, Campus Montilivi, C/Maria Aurèlia Campmany, Girona, Spain*

[2]*Unidad de Investigación de Diseño de Fàrmacos y Conectividad Molecular, Departamento de Química Física, Facultad de Farmacia, Universitat de València, Burjassot, València, Spain*

[3]*MOLware SL, Valencia, Spain*

[*]*Corresponding author. E-mail: liopo@uv.es*

CONTENTS

ABSTRACT

Three rules, the Titius–Bode, the Dermott rules, and a classical linear molecular quantitative structure–property relationship, are revisited discussing some of their main characteristics and revealing up to which level the models have real predictive power or simply a descriptive one. A careful choice of the experimental values to be included in the model seems to be essential to the usefulness of a relationship. Furthermore, a predictive relationship is not always free from flaws.

2.1 INTRODUCTION

Some years ago, a series of studies started to reconsider some aspects on which linear quantitative structure–property and structure–activity relationships (QSPR/QSAR), and their graphical displays were based and how this could affect their predictive power.[1–3] More recently taking as starting point the *Titius–Bode* (TB) rule, the limits of validity of quantitative structure relationships[4–6] were discussed. In the present chapter, we would like to emphasize some characteristics of predictive relationships, which have become a subject of paramount importance in QSAR/QSPR. For this purpose, we will center our attention on the utility and validity of two cosmological and a physicochemical relationship: the Titius–Bode, and the Dermott rules,[7] and a classic QSPR.[8] The Titius–Bode rule applies to our planetary system, the Dermott rule applies to the moons of Jupiter, Saturn, and Uranus, while a classic QSPR, based on the Randić *branching index*, and actually known as $^1\chi$ index,[9] describes the boiling points of some alkanes.

Let us start presenting the Titius–Bode (eq 2.1) and the Dermott rules (eq 2.2),

$$d_{TB}(n) = 0.4 + 0.3 \cdot (2^n), \text{ with } n = -\infty, 0, 1, 2, 3, \ldots \quad (2.1)$$

$$T(n) = T(0)C^n, \text{ with } n = 1, 2, 3, 4, \ldots \quad (2.2)$$

Here, $d_{TB}(n)$ is the distance at which the planets of the solar system are located from the Sun, that is, the semimajor axis of each planet outward from the Sun in units such that the Earth's semimajor axis is equal to one. $T(n)$ is the orbital period of the nth satellite in days, $T(0)$ is a fraction of a day and C is a constant of the satellite system in question. The specific values for these constants are (d = days):

$$\textit{Jovian system}: T(0) = 0.444d, C = 2.03$$

$$\textit{Saturnian system}: T(0) = 0.462d, C = 1.59$$

$$\textit{Uranian system}: T(0) = 0.488d, C = 2.24$$

Notice the similarities of the $T(0)$ and C values with the corresponding constants of the Titius–Bode rule (0.3 and 2). Dermott rule is an empirical formula for the orbital period of the Jovian, Saturnian, and Uranian satellites orbiting the planets in the solar system. It was identified by the celestial mechanics researcher Stanley Dermott in the 1960s. We could reshape the Titius–Bode rule to be formally similar to the Dermott's rule but with a consistent loss of precision. Do not forget that the two rules differ in dimensions, the first one has AU dimensions (1 AU = 149,597,870.700 km or approximately the mean Earth–Sun distance), while the second has days (d) as dimension.

It has been said that such power rules may be a consequence of collapsing-cloud models of planetary and satellite systems possessing various symmetries, and that they may also reflect the effect of resonance-driven commensurabilities in the various systems. As pointed us by Georgi Gladyshev, a Russian physical chemistry professor, Liesegang's theory of periodic condensation[10] has also been used to explain the empirical Titius–Bode rule of planetary distances, according to which the distance of the nth planet from the Sun satisfies the relationship. Nevertheless, the cosmological aspects of the two predictive rules will not be our concern here.

Finally, what could also be considered as a rule is the ${}^1\chi$ index relationship for the description of the boiling points of alkanes:

$$T_b = a \cdot {}^1\chi + b, \text{ and } {}^1\chi = \Sigma(\delta_i\delta_j)^{-0.5} \tag{2.3}$$

The sum runs over the connections of the hydrogen-deleted graph that encodes the molecule[8,9] and a and b depend on the number and type of the chosen alkanes. Parameters δ_i and δ_i stand for the number of connections of two adjacent i–j atoms in the hydrogen-deleted graph.

2.2 RESULTS

Table 2.1 shows the results obtained with the Titius–Bode rule, while in Table 2.2 are the results obtained with the Dermott rule for the three different moon systems. A statistical linear regression of the TB rule is shown in eq

2.4, obtained with least-square analysis, plotting d versus k from Mercury till Uranus. It agrees pretty well with eq 2.1. It describes correctly the $N = 8$ planet orbits (d_{TB}) at semimajor axes as a function of the planetary sequence. The accuracy of the description is shown in Table 2.1 throughout the percent residual error column [100(Exp. − Calc.)/Exp].

$$d_{TB} = 0.4(\pm 0.05) + 0.3(\pm 0.002)k, \text{ where } k = 2^n, \text{ with } n = -\infty, 0, 1, 2, 3, \ldots \quad (2.4)$$

$$N = 8,\ Q^2 = 0.999,\ r^2 = 0.9998,\ s = 0.1,\ F = 24{,}018$$

TABLE 2.1 The Observed d/AU Values of the Semimajor Axis for the Planets of Our System, the Calculated d_{TB}/AU Values with the TB Rule, and the Corresponding Percent Residual Error.

Planet	2^n	d/AU	d_{TB}/AU	% Res. error
Mercury	0	0.39	0.4	−2.6
Venus	1	0.72	0.7	2.8
Earth	2	1.00	1.0	0.00
Mars	4	1.52	1.6	−5.3
Ceres[a]	8	2.77	2.8	−1.1
Jupiter	16	5.20	5.2	0.0
Saturn	32	9.54	10.0	−4.8
Uranus	64	19.2	19.6	−2.1
Neptune	128	30.06	38.8	−29
Pluto[a]	256	39.44	77.2	−96
Haumea[a]	512	43.13	154	−257
Makemake[a]	1024	45.79	307.6	−572
Eris[a]	2048	68.01	614.8	−804

[a]Ceres, Pluto, Haumea, Makemake, and Eris are dwarf planets. In bold the original planets used in the model, in italics the predicted planets.

The value N is the number of planets used in the model, r^2 is the square correlation coefficient, s is the standard deviation of estimates, F is the Fischer–Snedecor value, and Q^2 is the prediction coefficient for the leave-one-out method[11] (a kind of internal predictive parameter).

Table 2.2 shows the calculated values of some of the Jovian, Saturnian, and Uranian moons. The Dermott rule of eq 2.2 with the given fixed values for $T(0)$ and C seems to describe fairly well the four *Medicean* moons (these moons were discovered by Galileo), from *Io* to *Callisto*, and *Himalia*, while the description of *Amalthea* is quite poor. Four Saturnian moons are described fairly well, while the other three a bit lesser. Only two Uranian moons are rather finely described, while for the other two results are deceiving (for *Titania* see discussion). Least squares analysis gives rise to slightly different equations for the rule for each system of moons, which show very good statistics. There is only a deceiving parameter, and it concerns the *Uranian* moons with, $Q^2 = 0.652$.

TABLE 2.2 The Experimental Orbital Periods of Some of the Jovian, Saturnian, and Uranian Moons in Days (*d*) and the Corresponding Calculated Period with the Dermott Rule.

Jovian moons	*n*	Orbital period (day)	Dermott rule	% Res. error
Amalthea	1	0.4982	0.9013	−81
Io	2	1.7691	1.8297	−3.4
Europa	3	3.5512	3.7142	−4.6
Ganymede	4	7.1546	7.5399	−5.4
Callisto	5	16.689	15.306	8.3
Himalia	9	249.72	259.92	−4.1
Saturnian moons				
Mimas	1	0.9	0.735	18
Enceladus	2	1.37	1.168	15
Tethys	3	1.888	1.857	2.1
Dione	4	2.737	2.953	−7.9
Rhea	5	4.518	4.695	−3.9
Titan	8	15.95	18.87	−18
Iapetus	11	79.33	75.86	4.4
Uranian moons				
Miranda	1	1.414	1.093	23
Ariel	2	2.52	2.449	2.8
Umbriel	3	4.144	5.485	−32
Titania	*4*	*8.706*	*−12.286*	*−41*
Oberon	4	13.46	12.29	8.7

In Table 2.3 are the boiling points in K, of some alkanes, together with their $^1\chi$ index, and their number of carbon atoms [No. Cs]. Values are taken from Randić's seminal paper (methane excluded)[8] and from Ref. [12]. The following relationships describe the boiling points as a function of the $^1\chi$ index and of the number of carbon atoms [No. Cs], with ((2.5) and (2.6)) and without ((2.7) and (2.8)) methane.

TABLE 2.3 No. of Carbons, ${}^1\chi$ Index, and Experimental Boiling Points, T_b (K) for Some Alkanes.

Compounds	No. Cs	${}^1\chi$	T_b (K)
Methane	1	0	110.95
Ethane	2	1.0000	184.55
Propane	3	1.4142	230.95
n-Butane	4	1.9142	273.05
2-Methylpropane	4	1.7321	261.95
n-Pentane	5	2.4142	309.25
2-Methylbutane	5	2.2701	300.15
2,2-Dimethylpropane	5	2.0000	282.65
n-Hexane	6	2.9142	341.95
2-Methylpentane	6	2.7702	334.05
3-Methylpentane	6	2.8082	336.45
2,2-Dimetilbutane	6	2.5607	322.95
2,3-Dimethylbutane	6	2.6425	331.25
n-Heptane	7	3.4142	371.65
3-Ethylpentane	7	3.3461	366.65
2,2,3-Trimethylbutane	7	2.9432	354.1
2,2-Dimethylpentane	7	3.0607	352.35
2,3-Dimethylpentane	7	3.1807	362.95
2,4-Dimethylpentane	7	3.1259	353.75
2-Methylhexane	7	3.2700	363.25
3,3-Dimethylpentane	7	3.1213	359.2
3-Methylhexane	7	3.3081	364.95
n-Octane	8	3.9142	398.75
2,2,3,3-Tetramethylbutane	8	3.2500	379.65
2,3,3-Trimethylpentane	8	3.5040	387.85
2,2,3-Trimethylpentane	8	3.4814	383.1
2,3,4-Trimethylpentane	8	3.5534	386.85
2,2,4-Trimethylpentane	8	3.4165	372.45
2,2-Dimethylhexane	8	3.5607	380.0
3,3-Dimethylhexane	8	3.6213	385.1
2,5-Dimethylhexane	8	3.6259	382.1
2,4-Dimethylhexane	8	3.6639	382.0
2,3-Dimethylhexane	8	3.6807	389.0
3-Methy-3-ethylpentane	8	3.6819	391.5

TABLE 2.3 *(Continued)*

Compounds	No. Cs	$^1\chi$	T_b (K)
2-Methyl-3-ethylpentane	8	3.7188	388.8
3,4-Dimethylhexane	8	3.7188	391.1
2-Methylheptane	8	3.7701	390.7
3-Methyheptane	8	3.8081	392.0
4-Methyheptane	8	3.8081	390.9
3-Ethylhexane	8	3.8510	391.7
n-Nonane	9	4.4142	423.85
2-Methyloctane	9	4.2701	416.15
n-Decane	10	4.9142	447.35
2-Methylnonane	10	4.7701	440.05

$$T_b = 142.1(\pm 5.6) + 66.85(\pm 1.7)^1\chi\text{: } N = 44,\ Q^2 = 0.965,\ r^2 = 0.973,\ s = 11,\ F = 1535 \quad (2.5)$$

$$T_b = 124.9(\pm 6.5) + 33.10(\pm 0.9)[\text{No. Cs}]\text{: } N = 44,\ Q^2 = 0.958,\ r^2 = 0.969,\ s = 11,\ F = 1317 \quad (2.6)$$

$$T_b = 154.5(\pm 5.7) + 63.26(\pm 1.7)^1\chi\text{: } N = 43,\ Q^2 = 0.963,\ r^2 = 0.971,\ s = 9.0,\ F = 1384 \quad (2.7)$$

$$T_b = 142.7(\pm 5.2) + 30.73(\pm 0.7)[\text{No. Cs}]\text{: } N = 43,\ Q^2 = 0.974,\ r^2 = 0.978,\ s = 7.8,\ F = 1860 \quad (2.8)$$

Figures 2.1 and 2.2 show the residual plots (residuals as ordinate vs. calculate T_b) in the four cases (Fig. 2.1, left and right is related to eqs 2.5 and 2.6, etc.). It should be remembered that these kind of plots are crucial in interpreting QSPR/QSAR results,[1-3,13] even if they are rarely considered.

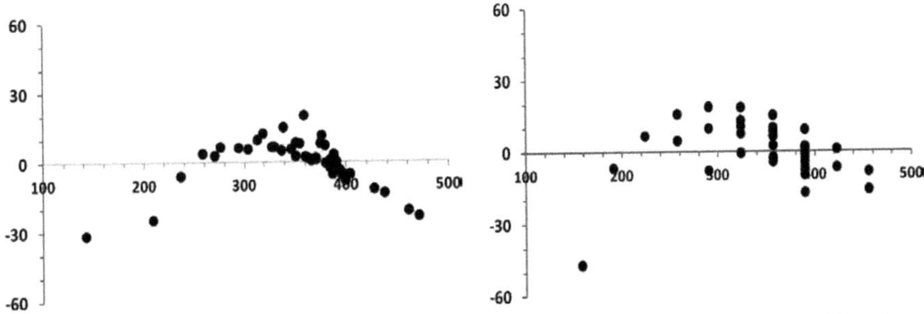

FIGURE 2.1 *Left*: Residual plot resulting from eq 2.5; *right*: residual plot resulting from eq 2.6.

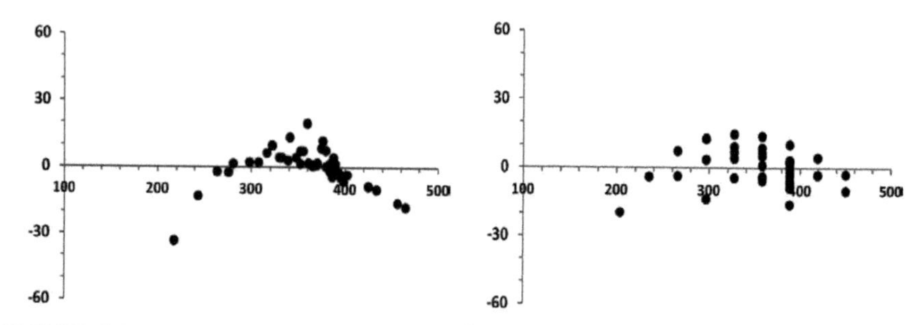

FIGURE 2.2 *Left*: Residual plot resulting from eq 2.7; *right*: residual plot resulting from eq 2.8.

2.3 DISCUSSION

Practically, the predictive quality of a relationship is its usefulness. If in the Titius–Bode rule (or in the Dermott rule) the sequence n is being read as the descriptor of the structure of our planetary system (or of the moon systems under study) and the distance (or orbital period) as their property, then some unknown properties should be guessed. Being this the case, we are facing a normal QSPR/QSAR problem and this is our concern here.

The dwarf planet Ceres (in italics) has been discovered thanks to the Titius–Bode rule of eq 2.4 in 1801 by the Italian astronomer Giuseppe Piazzi. It is the biggest body in the asteroid belt between Mars and Jupiter, making a third of the mass of the belt. Actually, the then newly proposed Titius–Bode rule was first brilliantly confirmed by the discovery of Uranus in 1781 by William Herschel. The high quality of the prediction is confirmed by the very good Q^2 vale for $N = 8$ (eq 2.4). Ceres and Uranus are the only two predictions of the rule as it fails to predict Neptune's orbit (29% error), and even more Pluto's orbit (96% error). Failure becomes enormous with the recently discovered dwarf planets: Haumea, discovered in 2004, with semimajor axis at 43.13 AU, is predicted at 154 AU with a 257% error. Makemake, discovered in 2005, with semimajor axis at 45.79 AU is predicted at 307.6 AU with a 572% error. Finally, Eris, discovered also in 2005, with semimajor axis at 68.01 AU is predicted at 614.8 AU with an 804% error. In fact, eq 2.9 shows how things get consistently worse (especially at the predictive leave-one-out Q^2 level) when we try to fit the 13 planets,

$$d_{TB} = 10.3(\pm 3.7) + 0.03(\pm 0.006)k, \text{ where } k = 2^{m},$$
$$\text{with } m = -\infty, 0, 1, 2, 3, \ldots \qquad (2.9)$$
$$N = 13 \text{ (all bodies)}, Q^2 = 0.46, r^2 = 0.749, s = 12, F = 33$$

The first-rate statistics of the Titius–Bode relationship shown in eq 2.4 reveals a bizarre and surprising flaw: $s = 0.1$ AU means a deviation of 15 million km nearly. Laskar's work[14,15] showed that Earth's orbit, as well as the orbits of the inner planets, is chaotic and that an error as small as 15 m in measuring the position of the Earth today would render it unpredictable over 100 million years' time. This poor s value tells us that the predictive character of eq 2.4 is imperfect, as its present value (0.1 AU) would render the unpredictability of the position of the planets quite drastic. Another strange aspect of this rule is the strong dependence of the guessed property on the chosen zero point. Had the Sun been included in the count, that is, $d = 0$ at $k = 0$ (for $n = -\infty$), the relationship would collapse. Furthermore, Mercury is not predicted as it is the bias of the rule, that is, Mercury is fixed at $k = 0$, which is a consequence of the fact that the Earth was fixed at $d = 1$. Actually, the value $n = -\infty$ for Mercury and its sudden jump to $n = 1$ for Venus is hardly understandable.

The TB relationship was built by the aid of the six points, Mercury, Venus, Earth, Mars, Jupiter, and Saturn, and "filled the blanks" of Ceres and Uranus with an astounding predictive character (overlooking the abnormal s value). Notice that if, and only if, Neptune is "overlooked," the TB distance of 38.8 is quite close to Pluto's real distance with an error of 1.6% only. This last "pseudo-guess" shows how deviant could be the practice to "silence" outliers to keep the predictive utility of a relationship working: overlooking points can sometimes, and even consistently, either improve or worsen the predictive quality of a relationship.

Regarding the Dermott rule, sometimes called the Dermott law, it should be said that it does not concern all moons of the three systems, and that the left-out moons are really a lot, that is, the careful choice of points with which to build a descriptive relationship is here extreme. In fact, there are more than 60 Jovian moons (67 confirmed moons), and only the 4 Galilean moons are massive enough (similar to our Moon) for their surfaces to have collapsed into a spheroid. The Saturnian moons are more than 50, but Titan alone makes around 96% of their mass while the 6 others ellipsoidal moons constitute a bit less than 4% of their mass. Of the known 27 Uranian moons, we reported only the more massive ones whose surface collapsed into a spheroid. If, in this last systems of moons *Titania* is included in the pack (in italics in Table 2.2) the only possible value for n would be $n = 4$, but with a deceiving result, as Q^2 (calculated by least-square analysis) becomes equal to 0.574. Furthermore, this inclusion would oblige to use the value $n = 5$ for Oberon with a consistent worsening of its prediction. Last but not least, the Dermott rule, unlike the Titius–Bode rule, was no help to find any new moon. Due to its poor number of points and the poor description of some moons, it should be said that the Dermott rule has practically only a descriptive character.

Let us go over to a classical chemical problem, the description of the boiling points of alkanes of Table 2.3 that is shown throughout eqs 2.5–2.8, and Figures 2.1 and 2.2. Concerning eqs 2.5 and 2.6, we notice that ${}^1\chi$ is a slightly better descriptor than the number of carbons [No. Cs]. Nevertheless, a careful look at this table and at the algorithm for ${}^1\chi$ (eq 2.3) let us notice a flaw, that is, following this algorithm methane, CH_4, should have either ${}^1\chi = \infty$ or be undefined as it has no carbon–carbon connections. Its zero value has here been fixed practically "ad hoc" to render things easier. Exclusion of methane from the calculations brings us to the predictive eqs 2.7 and 2.8, where [No. Cs], with a quite good predictive Q^2 value (s is also good), shows up as the best descriptor for the boiling points of this set of alkanes. Figures 2.1 and 2.2 reveal some other flaws of the model having in mind that residual plots should have their points randomly and equally distributed around the zero line. Actually, all four figures show a clear pattern with a maximum on the positive side of the residuals and an asymmetric distribution of points around the zero line. This last feature is less drastic with the [No. Cs] descriptor, when no methane is considered.

The advantage of ${}^1\chi$ seems to show up with the description of a set of alkanes with the same number of carbons. Let us take the 18 C8 alkanes of Table 2.3. Evidently, [No. Cs] is here no more a valid descriptor, while ${}^1\chi$ has good chances to be one. A least-square analysis of this set of alkanes delivers the following equation and quite deceiving predictive statistics,

$$T_b = 277.1(\pm 19.6) + 30.10(\pm 5.4){}^1\chi: N = 18, Q^2 = 0.520,$$
$$r^2 = 0.663, s = 3.7, F = 31 \quad (2.10)$$

There is another problem with graphs and eq 2.3 and it is the graph meaning of parameter b. It could be read as the least-square parameter of a unitary ${}^1\chi$ index (${}^1\chi = 1$) due to some unit undefined graph encoding an unknown compound whose temperature equals b. Actually, eq 2.3 could elegantly be reshaped into the following power series, where only the first two terms, $i = 0$ and 1, are taken into consideration (remember that ${}^1\chi^0 = 1$),

$$T_b = \Sigma_{i=0-n}\, a_i({}^1\chi)^i \quad (2.11)$$

Now, let us try to model the boiling points, without CH_4 (for the mentioned difficulties to define a ${}^1\chi$ for this compound), with a multilinear form of eq 2.11, that is, with $i = 0, 1, 2$. The results are excellent inclusive the predictive leave-one-out Q^2 and the s value.

$$T_b = 89.66\ (\pm 6.4) + 110.5\,{}^1\chi\ (\pm 4.3) - 7.927({}^1\chi)^2(\pm 0.7) \quad (2.12)$$
$$N = 44,\ Q^2 = 0.991,\ r^2 = 0.993,\ s = 4.5,\ F = 2867$$

The description due to a quadratic form for the number of carbons is the following:

$$T_\mathrm{b} = 110.5(\pm 10.4) + 42.14(\pm 3.4)[\mathrm{No.\ Cs}] - 0.918(\pm 0.3)[(\mathrm{No.\ Cs})]^2 \quad (2.13)$$
$$N = 44,\ Q^2 = 0.980,\ r^2 = 0.983,\ s = 6.9,\ F = 1180$$

The improvement of the modeling brought about by ${}^1\chi$ relatively to [No. Cs] is evident (the worsening of F relatively to eqs 2.5–2.9 is due to the fact that the independent variables are now two) and becomes even more evident with the residual plot of Figure 2.3, where at the left, we have the modeling due to eq 2.12. No mixed ${}^1\chi$ and [No. Cs] descriptors do a better job in describing the boiling points of the present alkanes.

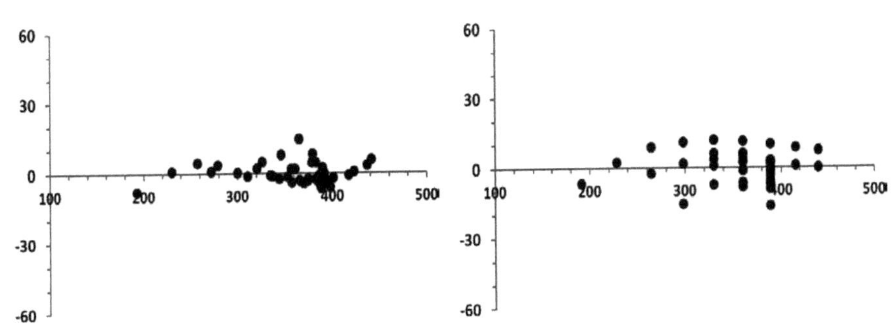

FIGURE 2.3 *Left*: Residual plot resulting from eq 2.12; *right*: residual plot resulting from eq 2.13.

2.4 CONCLUSION

Let us now imagine that the values of C and n for the Dermott and TB (here 2 and n) rules were not known, but that we have a lot of different values for them derived with different algorithms, from which we had to choose two. Combinatorial computation shows us that with 1000 values, the couples to choose from are 499,500, and with 10,000 values, the couples to choose from are 49,995,000. With such a huge number of couples, it would be not too difficult to find a pair that matches the known values for these two rules. A similar reasoning could be applied to different sets of values, that is, of descriptors for the boiling points, that is, it would be not that difficult to find two descriptors that would give rise to a good modeling bilinear relationship for alkanes.

The real problem to solve with the inferential and usefulness character of relationships is to pick up the optimal descriptor for the optimal set of compounds and to find it with rather easy methods. The description of the set of boiling points of alkanes confirms this point: [No. Cs] is a better descriptor than ${}^1\chi$ in a simple linear relationship. The fine modeling achieved with the second-degree relationship (eq 2.12) by the ${}^1\chi$ index underlines the fact that multilinear relationships can improve the model a lot using well known and easy to derive descriptors. Actually, the simple linear relationship should be considered an approximation of a power series equation that furthermore accomplishes also the task to give a deeper meaning to the relation.

The mathematician Charles S. Peirce discussed Bode's law as an example of fallacious reasoning in Lecture Five (pages 194–196) of his 1898 lectures: Reasoning and the Logic of Things (The 1898 Lectures in Cambridge, MA).[16] Surely, his judgment would have been even more drastic with the three Dermott rules. Would Peirce have been that drastic with quantitative structure property–relationships (and QSAR), especially with eq 2.12 that encompasses more than 40 points? Who knows? We have the impression that he was downplaying an important point: how helpful could predictive rules be.

KEYWORDS

- **Titius–Bode rule**
- **Dermott rule**
- **QSPR**
- **usefulness of relationships**
- **structure–property**

REFERENCES

1. Besalú, E.; de Julián-Ortiz, J. V.; Pogliani, L. An Overlooked Property of Plot Methods. *J. Math. Chem.* **2006,** *39*, 475–484.
2. Besalú, E.; de Julián-Ortiz, J. V.; Pogliani, L. Ordinary and Orthogonal Regressions in QSAR/QSPR and Chemistry-Related Studies. *MATCH Commun. Math. Comput. Chem.* **2010,** *63*, 573–583.
3. de Julián-Ortiz, J. V.; Besalú, E.; Pogliani L. Two-Variable Linear Regression: Modeling with Orthogonal Least-Squares Analysis. *J. Chem. Educ.* **2010,** *87*, 994–995.
4. Pogliani, L.; de Julián-Ortiz, J. V.; Besalú, E. The Titius–Bode Relationship and the Liability of QSAR/QSPR Models. *MATCH Commun. Math. Comput. Chem.* **2014,** *71*, 143–148.
5. Pogliani, L.; de Julián-Ortiz, J. V.; Besalú, E. How useful is a relationship? *Int. J. Chem. Mod.* **2013,** *5*, 295–302.
6. https://en.wikipedia.org/wiki/Titius%E2%80%93Bode_law (accessed on May 5, 2016).
7. http://en.wikipedia.org/wiki/Dermott's_law; https://it.wikipedia.org/wiki/Legge_di_Dermott (accessed on May 5, 2016).
8. Randić, M. On Characterization of Molecular Branching. *J. Am. Chem. Soc.* **1975,** *97*, 6609–6615.
9. Kier, L. B.; Hall, L. H. *Molecular Connectivity in Structure–Activity Analysis*, RSP-Wiley: Chichetser (UK), 1986.
10. Reeves, H. The Origins of the Solar System. In *The Origin of the Solar System*; Dermott, S.F., Ed.; Wiley: New York, 1978.
11. Carbó-Dorca, R.; Robert, D.; Amat, L.; Girones X.; Besalú, E. *Molecular Quantum Similarity in QSAR and Drug Design*. Springer: Berlin, 2000.
12. http://www.chemspider.com/; http://webbook.nist.gov/ (accessed on May 5, 2016).
13. Besalú, E.; Julian-Ortiz, J. V.; Pogliani, L. Trends and Plot Methods in MLR Model Studies. *J. Chem. Inf. Mod.* **2007,** *47*, 751–760.
14. Laskar, J. Large-Scale Chaos in the Solar System. *Astron. Astrophys.* **1994,** *287*, L9–L12.
15. Laskar, J.; Robutel, P.; Joutel, F.; Gastineau, M.; Correia, A. C. M.; Levrard, B. A Long-Term Numerical Solution for the Insolation Quantities of the Earth. *Astron. Astrophys.* **2004,** *428*, 261–285.
16. https://en.wikipedia.org/wiki/Talk:Titius%E2%80%93Bode_law#Charles_Sanders_Peirce (accessed on May 5, 2016).

CHAPTER 3

COMPUTATIONAL MODEL FOR BYPRODUCT OF WASTEWATER TREATMENT

SEYEDE MARYAM VAHEDI[1*], HOSSEIN HARIRI ASLI[2], and KAVEH HARIRI ASLI[3]

[1]*Tehran University of Medical Sciences, Tehran, Iran*

[2]*Civil Engineering Department, Faculty of Engineering, University of Guilan, Rasht, Iran*

[3]*Department of Mathematics and Mechanics, National Academy of Science of Azerbaijan "AMEA," Baku, Azerbaijan*

[*]*Corresponding author. E-mail: Dr_R_Vahedi@yahoo.com*

CONTENTS

ABSTRACT

Polyhydroxyalkanoates or PHAs are linear polyesters produced in nature by bacterial fermentation of sugar or lipids. The polyester is extracted and purified from the bacteria by optimizing the conditions of microbial fermentation of sugar or glucose. The simplest and most commonly occurring form of PHA is the fermentative production of poly-beta-hydroxybutyrate (poly-3-hydroxybutyrate, P3HB), which consists of 1000–30,000 hydroxy fatty acid monomers. The biosynthesis of PHA is usually caused by certain deficiency conditions (e.g., lack of macro-elements such as phosphorus, nitrogen, trace elements, or lack of oxygen) and the excess supply of carbon sources. In this work, pure oxygen was introduced in wastewater by a 40-l pure O_2 container. Microorganisms receiving pure O_2 grow quickly. By submerged porous diffusers and air nozzles, pure O_2 was inserted to an aeration tank in transient flow condition. An activated sludge system (anaerobic and aerobic) continuously acts as a source of primary sludge. Anaerobic and aerobic activated sludge reactors were used to supply the required raw sludge. The important results in this work were to decrease BOD5 amount equal to 80–90% and the industrial production of PHA to produce plastics, which are biodegradable and are used in the production of bioplastics.

3.1 INTRODUCTION

Biodegradable plastics are prepared under the title polyhydroxyalkanoates (PHAs). Its copolymers are over 40 compounds which have degradation to full capacity (100%). The low cost and relative ease of processing compared to other polymers attracts more attention. The molecular weight of polyhydroxylalkanoates or polyester hydroxyl is in the range of 105 and 106 depending on the type of microorganisms and growth conditions. This combination of physical properties mechanical and biocompatible is also applicable.

As mentioned earlier, the most important property of PHA is complete biodegradability. This compound is biodegraded by microorganisms, aerobic, anaerobic, and amphibians into CO_2 and water vapor that can be seen. A group of bacteria, as well as a bunch of these polymers or copolymers have ability to build them as a source of carbon storage in the cell, respectively. PHA granules in the cells of microorganisms can be stored easily by staining with Sudan Black and Blue Nile (blue).

The *Escherichia coli* type of plastic has properties that are comparable to synthetic plastics. Some of these can be 100% water-resistant, have low production costs compared to other biodegradable plastics, and flexibility. Generally, factors affecting the price of PHA production could be in the form of four parts including: productivity, production efficiency, cost of carbon source, and finally the costs of extraction and purification.

The PHA production in many studies was carried out. The necessary conditions for optimal production of the polymer in that studies were determined. But in the case of PHA production by intercropping (activated sludge), due to the newness of the issue, it is still not completed and the need for action is more comprehensive.

Satoh et al.[108] in 1998 described polymer production using sodium acetate as activated sludge as the feed (carbon source). In this study, the use of micro-aerohpilic–aerobic system (limited by the oxygenation in the anaerobic) increased production of PHAs by the activated sludge experience. It should be noted that there are wastewater industries that have a high percentage of volatile fatty acids (VFAs). It was also mentioned that the wastewater from factories contains cellulose acetate and acetic acid, with a mean 1–1200 ml, perfect combination to produce PHAs. The fermentation of waste such as municipal wastewater can use a significant amount of VFAs produced as an inexpensive substrate for the production of PHA.

As discussed above and taking into account the problems relating to the disposal of plastics used in packaging and containers and due to costs related to the disposal of sludge from wastewater treatment plants, the production of PHAs using activated sludge and effective factors have been selected as the subject of this study. First time in 1925 (92 years ago), a microbiologist at the Pasteur Institute in Paris called Lemon discovered and described poly-beta-butyric acid (PHB). He observed that in granular bodies in the cytoplasm of *Bacillus megaterium*, the ether was not resolved. He realized that the compounds are of general formula $(C_4H_6O_2)_n$ and are polyester. From the difference between the melting point of this compound, Lemon discovered the two parts of the polymer. The polymer was isolated by chloroform.[41]

Twenty-seven years later, in 1952, Capps[12] observed that both the polyester components separated by Lemon hydrolysis are the products of high molecular weight linear polyester, and its melting point is 180°C.

In 1958, Williamson and Wilkinson[12] data on the molecular weight and physical properties of PHB were reported. In the same year, Mkrayy and Wilkinson[41] observed that intracellular PHB accumulation of nitrogen limitation in the medium increases.

In the late 1950s and early 1960s, the company called W. R. Grace and other companies in the United States, Verberie and Baptist[12] produced some PHB with their business objectives. They obtained the patent to increase the production and isolation of PHB. They had a low efficiency of fermentation and extraction method using a solvent for PHB from the cells, thus were expensive. In addition, in the polymer produced, a bacterial infection was found in the separation which made the melting process difficult. The company faced the above-mentioned problems and abandoned the project as the industrial production of this polymer was delayed for a decade. It should be noted that this project is a prelude to further research on biodegradable plastic PHB and compatible with vital systems.

Imperial Chemical Industries (ICI) was a British chemical company and was, for much of its history, the largest manufacturer in Britain. In 1968, ICI in England began to spread technology of single cell protein (SCP). The SCP food products as protein were produced successfully. The price was much higher than the initial substrate for bacterial fermentation of proteins that can be presented in the business. Thus, the project was abandoned and further study of PHB was necessary. The technical information about large-scale fermentation and skills in the field of polymer increased significantly. In this opportunity, ICI equipment business provides a way to produce the PHB. ICI found bacteria that were able to accumulate PHB were stunned by more than 70% by weight For pure PHB, it is a special advantage compared to polypropylene (PP). But PHB copolymers had more interesting properties and therefore were more suitable for industrial applications.[41]

So Holmes et al. in 1981 presented controlled fermentation process in which bacteria produce copolymers P (3HA) of different carbon sources. The copolymer under the name "Biopol" of *Alcaligenes eutrophus* was grown for propionic acid and glucose production. Changes in glucose caused a change in the composition of propionic acid and copolymer, which contributes to the mechanical properties of thermoplastic with different melting points.[41]

Finally, in 1990, Wella Company in Germany produced the first commercial biodegradable product Biopol as packing shampoo bottles.[41]

Another study in 1997 in bioplastic production of sludge from municipal wastewater treatment has been studied.[72] The study was conducted in laboratory scale (PHAs) of sludge from municipal wastewater treatment, recovered in two stages. In the first stage for sludge, thermophilic anaerobic digestion conditions and then (PHAs) of organic compounds in fluid supernatant were produced by microorganism *A. eutrophus*. In this study, sewage sludge (5.3%), dry solids (70–50%), volatile solids, temperature of 65–500°C, and anaerobically fermented fatty acids and other compounds were used. The reductions of volatile solids and dissolved organic compounds in temperature, pH, and retention times were also noted from the study. In this research, result is that at 65°C and 5/5 pH, more dissolved organic matter is formed.

In this study, PHAs are produced in about 34% cell mass compared with the VFAs are as pure as the carbon source used to show an increase.

In this process, about 78% of the organic carbon in the supernatant fluid is used. According to investigations made into four main acids present in the supernatant containing acetic acid, propionic acid, butyric acid, and valeric acid were 6/87%, 6/62%, 8/56%, and 32%, respectively. PHAs produced in this method (74% weight) of C4 monomers and 9°C lower melting point 167°C of poly-3-hydroxybutyrate (P3HB) were reported. Bioplastics as a by-product of PHA produced in this work in the soil in 5 weeks to the amount (27–22%) in the sludge decomposes when the Shnavrgrdydh within 6 weeks in anaerobic conditions–aerobic decomposition rate was about 70%.

In 1997, PHA production of date syrup (Palm oil) have been studied by using the microorganism *Rhodobacter sphaeroides* (IFO 12203) on 15 g of organic acid which was about 2 g (i.e., more than 60% of the dry weight of the sludge PHA produced).[46]

Chuang and colleagues[25] in 1997 described the effects of HRT as well as dissolved oxygen and nutrient-removal process of combined AS-biofilm studies and have concluded that the anaerobic–anoxic and aerobic phosphorus removal are in accordance with the PHA production. The researchers also experimented in 1998 to determine the removal of phosphorus and PHA production in a system (anaerobic–anoxic–aerobic) in pilot scale and achieved similar results. In 1998, Lamu and colleagues[74] studied the effect of the carbon source in the production of PAHs by biological phosphorus removal with the main purpose of optimizing the removal of phosphorus and polymer production. In this study, acetate, propionate, and butyrate were used as the carbon source and as a result, the production of the polymer composition varied depending on the type of substrate used. For example, acetate copolymer has a tendency to produce a combination of hydroxyburyrate (HB) and hydroxyvalerate (HV) and usually HB is dominant. Propionate tends to produce more HV and value HB in it, and finally, when butyrate is used, the production of HB and HV is in a ratio of 5/1. In this study, the production efficiency of the carbon polymer used in cases where the carbon sources acetate, propionate, and butyrate were 97/0, 6.10, and 21/0, respectively.

Production of biodegradable plastics from chemical wastes by Chu and Yu[21] were also studied. In this study, *Alcaligenes* sp. is used as the microorganism capable of storing PHAs in cells. In this work, the production of polymers increased with the increase of C:N ratio. In this research, bioplastic production of excess sludge production has also reduced the need for consolidation. Polymer stored by microorganisms sp. beta-hydroxy-butyric acid (poly-beta-hydroxy butyric acid) and beta-hydroxy-butyric acid copolymers (3-HB) and beta-hydroxy valeric acid (3-HV), respectively. Wastewater of xenobiotic-branched carboxylic acids and ketones was used in this study. Production of new PHAs from food industrial wastewater by Yu et al.[24] has also been studied. In this study, wastewater from the brewing industry and processing factory and soybean were used, and it was observed that the activated sludge process also allows for the PHA productions are there in this study which also found that in the lesions of proper food (wastewater food technology) as the carbon source, it is possible to prepare polymer physical properties needed to be built up.

A study prepared in 1999 introduced the process of PHA production at wastewater treatment to reduce the amount of organic waste.[71]

Production of PHAs using activated sludge by Takabatk et al.[124] in 2000 was studied. In this study, control of polymer compounds and biomass enrichment of polymer are recommended. The result is that wastewater-produced product quality can be controlled.

In 2001, the kinetics of taking a combination of VFAs by bacteria *Ralstonia eutropha* PHA production process has been studied.[58] In this study, the rate of PHA production of acetate, propionate, and butyrate individually, or in combination, has been studied, and in the best conditions (a combination of acetic acid and butyric acid), PHA production rate is about 60 mg/g of biomass production hours. In this study, when combined acetic acid and propionic acid is used, the rate of PHA production reduced to 35 mg/g of biomass produced per hour.

The acetate concentration, pH, and the cell retention time of sludge in an activated sludge PHA production are examined by Adeline et al.[1] in 2003. In this study, pH between 7 and 8 on the implementation of the most appropriate sludge for maximum polymer production rate was determined. In this experiment, the ability to reduce sludge in wastewater treatment plant is increasing SRT. According to the results of this study, the addition of acetate to sewage sludge into the implementation phase increased the production of the polymer by 10%.

3.1.1 POLYMER STORAGE OF MICROORGANISMS

According to the study,[29] four groups of polymers, including lipids, carbohydrates, polyphosphates, and nitrogen compounds (as stored nitrogen), can be stored in microorganisms.

Some microorganisms are able to store more than one type of polymer that is within the cells. These type of microorganisms are capable of storing Gylkvzhn, PHA, and polyphosphate.

Dawes and Senior[28] stated that environmental conditions and mechanisms used on polymeric composition effect will be saved. In addition, the carbon source used in the polymer will be effectively stored in the cell. For example, when the source of carbon from glucose to acetate has changed, *E. coli* reduced carbohydrates and increased the amount of lipid stored.

According Sasykala and Ramana,[106] the acetyl-CoA-metabolized carbon source and pyruvate intermediate product does not usually lead carbon into PHA production. But the carbon source by following the metabolism of other compounds will increase the amount of glycogen in cells. Dawes[29] stated that the important role is the polymer storage by microorganisms. The carbon and energy needs for bacteria and microorganisms that allows using it as a source of carbon and energy. The polymers have the ability to save microorganisms in the face of limited nutrients, such as protein and RNA, tend to use cell content of the above-mentioned.

3.1.1.1 LIPIDS (POLYHYDROXYLALKANOATES)

In this section, conjunction with the PHA production by microorganisms and the factors affecting the production and composition of the material are presented. Also, in relation to biological synthesis pathways (biosynthesis pathways) and physical properties, biodegradable plastics recycling and potential applications of PHAs will be discussed.

Researchers[4,15,35,69,71,106,120] have done extensive research on PHA production by microorganisms. These compounds, which can accumulate in the cells of microorganisms, are affordable by a wide variety of bacteria. Most important of them is *R. eutropha*, previously known as *Alcaligenes eutropha*. Due to the ability of the bacteria to store PHA, these microorganisms have been the subject of many published research. It should be noted that this type of bacteria is capable of storing up to about 80% dry weight PHA cell.[71] Lists of a number of microorganisms capable of PHA accumulation within the cell are shown in Table 3.1.

According to the study of Lafferty et al.,[69] PHB has similar properties like tensile strength and flexibility of polystyrene and polyethylene. Production of PHA polymers in both aerobic and anaerobic conditions is completely (100%) biodegradable,[89] which is one of the advantages of this class of polymers in comparison with biodegradable plastics that are produced artificially..

TABLE 3.1 Microorganisms with Potential of Polyhydroxyalkanoates (PHA) Production.

Acinetobacter	*Gamphosphaeria*	*Photobacterium*
Actinomyceles	*Haemophilus*	*Pseudomonas*
Alcaligenes	*Halobacterium*	*Rhizobium*
Aphanothece	*Hyphomicrobium*	*Rhodobacter*
Aquaspirllum	*Lampocystis*	*Rhodospirillum*
Azospirilum	*Lampropedia*	*Sphaerotilus*
Azotobacter	*Leptothrix*	*Spirulina*
Bacillus	*Methylobacterium*	*Spirulina*
Beggiatoa	*Methylocystis*	*Streptomyces*
Beijerinckia	*Methylosinus*	*Syntrophomonas*
Caulobacter	*Micrococcus*	*Thiobacillus*
Chlorofrexeus	*Microcoleus*	*Thiocapsa*
Chlorogloea	*Micricystis*	*Thiocystis*
Chromatium	*Moraxella*	*Thiodictyon*
Chromobacterium	*Mycoplana*	*Thiopedia*
Clostridium	*Nitrobacter*	*Thiosphaera*
Derxia	*Nitrococcus*	*Vibrio*
Ectothrothodospira	*Nocardia*	*Xanthobacter*
Escherichia	*Oceanospirithum*	*Zoogloea*
Ferrobacillus	*Paracoccus*	

The overall structure of the PHA and formulation of PHAs is shown in Figure 3.1. As can be seen, in terms of the structural composition, PHA can be of many types. Some of the most important compounds are given in Table 3.2 as combination of PHAs. Although more than 80 different types of PHA in cells of microorganisms have been identified,[71] usually PHA extracted from the cells of microorganisms, mainly PHB and poly-beta-hydroxy valeric acid (PHV) were formed.

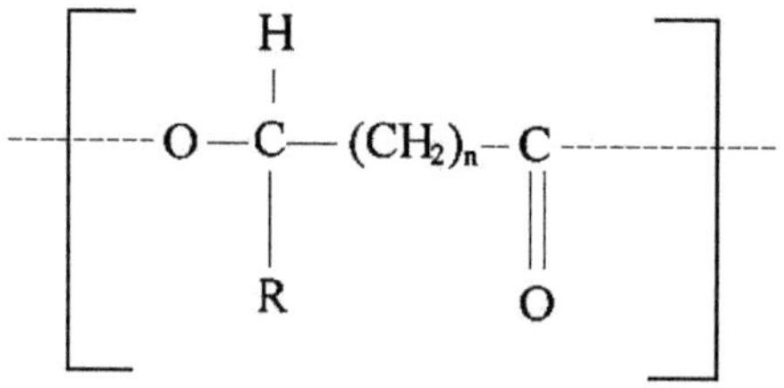

$$\left[-O-\underset{\displaystyle R}{\overset{\displaystyle H}{C}}-(CH_2)_n-\underset{\displaystyle O}{\underset{\|}{C}}- \right]$$

FIGURE 3.1 Formulation of polyhydroxyalkanoates (PHA).

TABLE 3.2 Combination of Polyhydroxyalkanoates (PHA).

n	R group	Combination name	No.
1	Methyl group	3-HB, 3-hydroxybutyrate	1
1	Ethyl group	3-HV, 3-hydroxyvalerate	2
1	*n*-Propyl group	3-EC, 3-hydroxycaproate	3
1	*n*-Butyl group	3-HH, 3-hydroxyheptanoate	4
1	*n*-Pentyl group	3-HO, 3-hydroxyoctanoate	5
1	*n*-Hexyl group	3-HN, 3-hydroxynonanoate	6
1	*n*-Heptyl group	3-HD, 3-hydroxydecanoate	7
1	*n*-Octyl group	3-HUD, 3-hydroxyudecanoate	8
1	*n*-Nonyl group	3-HDD, 3-hydroxydodecanoate	9
2	Methyl group	4-HV, 4-hydroxyvalerate	10
2	Ethyl group	4-HC, 4-hydroxycaproate	11

In today's world, the homopolymer PHB and copolymer are the two types of PHAs 3HB–3HV produced for industrial applications. In Figures 3.2 and 3.3, respectively, 3HB–3HV and 3HB–4HB copolymer structure are shown. Lafferty et al.[69] cited in a study of asymmetric growth conditions such as limits on nutrients (nitrogen, sulfate compounds, and phosphorus), low levels of oxygen, and high ratio of carbon to nitrogen in the substrate as factors enhancing the productivity of the polymer.

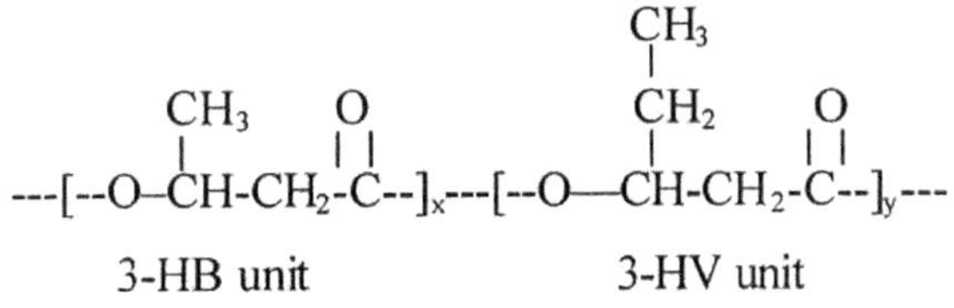

```
                                      CH3
                                       |
            CH3      O                CH2       O
             |       ||                |        ||
---[--O--CH-CH2-C--]x---[--O--CH-CH2-C--]y---
       3-HB unit                 3-HV unit
```

FIGURE 3.2 Formulation of copolymer called PHBV (poly (3-hydroxybutyrate-*co*-3-hydroxyvalerate).

In another studies by Sasykala and Ramana,[106] in addition to the limitations caused by nitrogen, phosphorus, sulfate, and oxygen limitation in iron, magnesium, manganese, potassium, and sodium explained the influence of these elements on PHA production process

In the binary study,[35] if the development is uneven because of the concentration of NADH, acetyl-CoA can obtain more energy for cells in acid cycle (TCA) according to Dawes.[30] The high concentration of NADH inhibits the synthesis of an enzyme citrate (one of the key enzymes cycle TCA) and the amount of acetyl-CoA is subsequently increased. The acetyl-CoA as a substrate for PHA production in three consecutive enzymatic reactions used is shown in Figure 3.4.

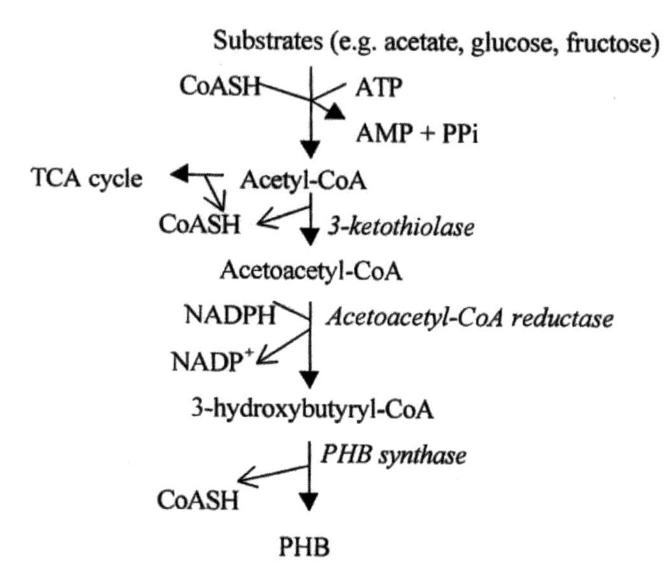

FIGURE 3.3 Synthesis map of polyhydroxyalkanoates (PHA).

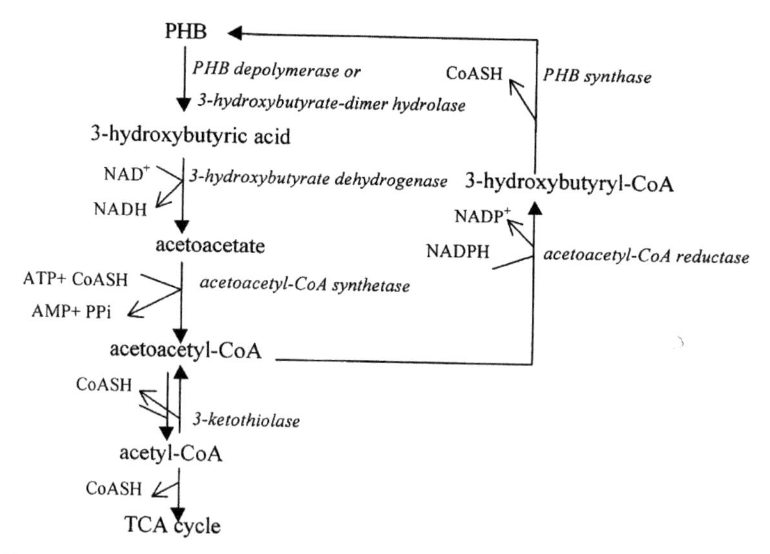

FIGURE 3.4 PHB biosynthesis using microorganisms.

The high concentration of intracellular CoA enzyme prevents the 3-keto-thiolase. One of the three PHA biosynthetic enzymes are effective against the intrusion if there is no-entry barrier acetyl-CoA in the TCA cycle. It is possible that the CoA free as acetyl moiety of the synthesis of citrate release. For example, the acetyl-CoA is used, as long as the concentration of intracellular CoA is increased as a result of PHA production stops.

PHA during the period of food restriction can act as a source of energy or carbon to microorganisms.

According to Lee,[71] four different biological synthesis routes for the production and storage of PHAs, which are outlined below, have been determined.

3.1.2 PATHWAYS FOR BIOSYNTHESIS OF PHA

As mentioned earlier, four biological synthesis routes for the production of PHA are defined in this section with a brief description of each of them.

First, *the biosynthesis of PHA using R. eutropha*: Most microorganisms are capable of producing PHA using the route of which production by *R. eutropha*, *Zoogloen ramigera*, and *Azotobacter beijerinckii* are noted.[35] In this way, the substrate is converted to acetyl-CoA. Then, 2 mol of acetyl-CoA is used and converted to 1 mol of the PHB. In this way, acetyl-CoA to PHB production to three successive enzymatic reactions is considered. Figure 3.4 shows PHB biosynthesis using microorganisms.[4,35]

In this process, propionic acid alone is used as a substrate copolymer PHB–PHV. Acetyl-CoA is formed by eliminating the carbonyl carbon of propionyl-CoA, and then, 2 mol of acetyl-CoA produce one unit of the used HB HV copolymer, while a single reaction and propionyl-CoA and acetyl-CoA is also formed. Figure 3.5 shows the biosynthesis of copolymer 3HB–3HV using *R. eutropha*.[35]

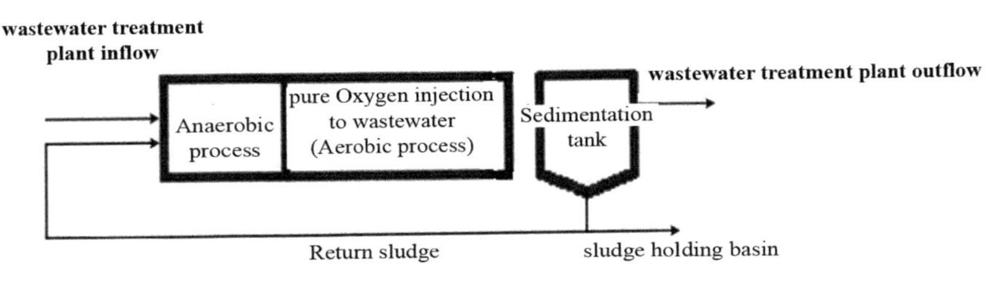

FIGURE 3.5 Batch-flow diagram of activated sludge system.

According to studies[35] concurrently with PHA synthesis by *R. eutropha*, if possible limitation of nitrogen loss (degradation) by PHA, there is the phenomenon of the natural cycle of metabolism called PHA.

For example, about 56% PHA biosynthesis occurs when exposed to butyric acid as the substrate and placed under nitrogen limitation. PHA synthesis within cells is significantly modified so that after about 48 h of PHV production, it resulted in a decrease from 56% to 19%. These findings of synthesis and metabolism, concomitant use of PHA suggest that PHA is the natural cycle of the states. In Figure 3.6, natural cycle of PHA metabolism is shown.

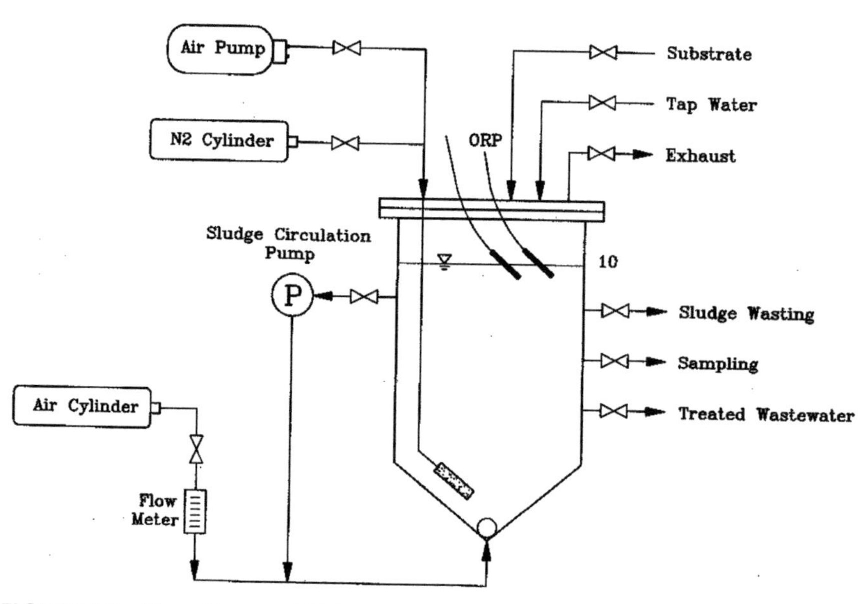

FIGURE 3.6 Flow diagram of SBR Batch for production of PHB.

Second, *biosynthesis of PHA using Rhodospivillum rubrum*: In the same way using *R. eutropha*, PHA biosynthesis occurs with the only difference as involvement of two enoyl catalyzing-CoA in the second stage in the conversion of L-3-hydroxybutyryl-CoA to D-3-hydroxybutyryl-CoA in the vicinity of crotonyl-CoA.[4,35,71] The simple diagram of the biological pathway can be considered in the following:

Acetate acetyl CoA—acetoacetyl CoA—L-3-hydroxybutyryl-CoA—crotonyl CoA—D-3-hydroxybutyryl-CoA—PHB

Third, *biosynthesis of PHA by Pseudomonas oleovorans*: The biosynthesis of *P. oleovorans* and more *Pseudomonas* first group of similar rRNA (rRNA homology group I) is true.[71] These organisms PHA with average (C6–C9) from the average molecular chain alkanes (medium-chain length alkanes) and alcohol produce. According to studies,[35] short-chain length PHA production like copolymer and homopolymer PHB PHB–PHV can also be the result of the action of this group of microorganisms. But in this case, the production rate decreased by less than 5.1%.

Fourth, *biosynthesis of PHA using Pseudomonas aeruginosa*: More *Pseudomonas* of (rRNA homology Group I) except *Pseudomonas oleororans* using this method may produce medium-chain PHA. Austin Bocelli[120] concluded that medium-chain PHA synthesis by this route is unrelated to acetate derived.

3.1.3 FACTORS AFFECTING PHB PRODUCTION AND COMPOSITION

As previously mentioned, the percentage of PHB and PHV type of substrate used in the polymer will be effective. According to the studies,[35] when the polymeric composition is produced by *R. eutropha*, with carbon pair homopolymer PHB used as a substrate. If the number is odd, "Keep Healthy Carbs" acid copolymer chains PHB–PHV will be produced.

In this case, fructose or glucose as a substrate is used to produce the PHB homopolymer.

According to studies, substrate concentration used in the production of polymers has been reported. For example, when the propionate as substrate for PHA production from *R. eutropha* is used, propionate concentration of 14–2 g/l decreased from 56% to 12%.

As mentioned by Binary and colleagues,[34] copolymer composition is produced using a combination of acetate and propionate H16 *R. eutropha* when sodium was used as a substrate. In this study, the higher concentration of sodium propionate percent of PHV in the copolymer is increased. Also when Azprvpyvnat sodium alone was used as a substrate, 3HB–3HV copolymer was produced.

In this study, mole percent of PHV in the copolymer-produced compounds used depends on the substrate in such a way that when the pentanoic acid and butyric acid was used, the carbon source of PHV increased to more than 95%. The copolymer composition, physical properties, and thermal behavior of the producer can be controlled by changing the substrate used.[36] Further experiments using different carbon sources for PHA production using *R. eurtropha* were done. In these experiments, use of different combinations of different carbon sources to produce PHA is presented in Table 3.3.

TABLE 3.3 Kinds of Polyhydroxyalkanoates (PHA) Used by the Excess Supply of Carbon Sources.

PHA produced	Carbon sources	No.
3HB–3HP	Hydroxy propionic acid	1
3HB–3HP	1,5-Pentanediol	2
3HB–3HP	1,7-Heptanediol	3
3HB–3HP	1,9-Nonandiol	4
3HB–4Hb	4-Hydroxybutyric acid	5
3HB–4HB	δ-Butyrolactone	6
3HB–4HB	1,4-Butanediol	7
3HB–4HB	1,6-Hexanediol	8
3HB–4HB	1,8-Octanediol	9
3HB–4HB	1,10-Decanediol	10
3HB–4HB	1,12-Dodecanediol	11
3HB–3HV	Propionic acid	12
3HB–3HV	Propionate	13

In this study, the biodegradation of a thin film of PHA production (weight of 4–8 mg and 10 × 10 dimensions of 0.3–0.6 mm thick) was investigated. According to speed tests, 3HB–4HB copolymer was higher than the other samples tested have been reported. In this experiment, reducing the rate of decomposition of 3HV copolymers is mentioned. Shymyz et al.[115] PHA production from H16 *R. eutropha* using valeric and butyric acid was also studied, and optimum conditions for PHB production by these bacteria by butyric acid at a concentration of 3 g/l and a pH of 8 was also reported. This study also provided information about the association of PHV–PHA. In these conditions, the PHB stored inside microbes to 75% have been reported. In this experiment, the concentration of butyric acid also decreased production of PHB (0.3 concentrations used in this study, 3.0 and 10 g/l to produce butyric acid, respectively, 44%, 55%, and 63% were PHA). Following the test at a concentration of 3 g/l of butyric acid at pH 5.7 and 9.6 was repeated, in which case, the amount of PHA production, respectively, of 58% and 53% have been reported. The test also increases the pH above 4.8, PHB production was stopped. When the propionic and butyric acid was used as a source of carbon, 3HB–3HV copolymer was obtained, but the amount of polymer in comparison with 3HB using butyric acid alone showed decrease in production.

In another study,[119] PHB–PHV copolymer produced using *R. eutropha* strain R3 is studied under conditions of nitrogen limitation. In this study, using fructose, gluconate, acetate, succinate, and lactate in a proportion of 47%, 7/35%, 5/29%, and 2/43%, respectively, polymers (PHA) were produced. In this experiment, the PHV in the copolymer produced in the range of 7–4% have been reported. Also, when the experiment was restricted in magnesium, and sulfur and fructose was used as a carbon supply, production of PHA were 45% and 47% and PHV in the copolymer were 7% and 6%, respectively. To benefit from a cheaper carbon source, Burke et al.[10] studied the PHA production by microorganisms appointed by the 118-methyletrophic on methanol. In this study, using a mixture of methanol and Valerie, and using microorganisms *Methylobacterium extorquens*, significant amounts of PHA were produced. In this study, the rate of PHA production in the range of 60–70%, and about 20%, has reported the presence of PHV in the copolymer.

In another research by Burke et al.,[11] PHA production with the help of *M. extorquens* ATCC 55366 and methanol as carbon and energy source in a fermentation system were examined. In this system, the level of PHB production in the range of 40–46% by weight of dry sludge has been reported. The production of biomass and growth rate of microorganisms is affected by mineral compounds in the system, thus, no ammonium sulfate or manganese sulfate and the absence of a combination of calcium chloride, sulfate bivalent iron, magnesium, and zinc biological mass production were reduced. In this study, high concentrations of ammonium sulfate were introduced as toxicity to the system, while the concentration of $MgSO_4$, $FeSO_4$, and mixture of micronutrients (trace element) increased the rate of microbial growth. In other words, the maximum cell density of *M. extorquens* was noted, when an appropriate proportion of inorganic compounds has been used.

In a study, Suzuki et al.[122] obtained the highest PHB production of 66% dry cell weight using SP.K *Pseudomonas* in methanol, used as the only source of carbon. In this study, to achieve high concentrations of PHB, substrate was tested with different combinations. (It should be noted that in this study, low concentrations of phosphate and ammonium were used.) In this study, nitrogen PHB was reported as the most important factor for accelerating storage failure. The limitations of the concentration of dissolved oxygen reduces the growth rate of the microbial mass and thus the PHB production was decreased, which is inconsistent with results reported by others.

In another study[69] on *R. eutropha* and *A. beijerinckii*, the PHB production under limited oxygen is limited in the amount of oxygen accelerates PHB production.

In the study mentioned above, PHB with 176°C melting point and molecular weight of *Pseudomonas* 105.3 was reached.

Daniel et al.[27] observed that in a medium, feed (methanol) was added to using *Pseudomonas* 135 and experienced the limitation of 55% ammonium polymer (PHB). Then, the microbes in environment limited the level of PHB production to 5/42% and 5/34% for Mg^{2+} and PO_4^{3-}, respectively. The PHB produced by this group of microorganisms has a melting point of about 173°C. Average molecular weight of the polymer produced for each of the states limit for NH^{4+}, Mg^{2+}, and PO_4^{3-} of 1057, 1055/2, and 1051, respectively, was reported..

Bayrvm[18], of the Institute of ICI with his experiences in relation to the industrial production of PHA, has stated that

- *Ralstonia* produced high molecular weight PHA and showed that PHA production can easily do it, In PHA *production* by *Azotobacter* studied methylotrophs production efficiency and low molecular weig*ht polyme*r which has been produced and difficulty to isolate the polymer. *Azotobacter* is also not particularly interesting due to the fact that when it was used carbon source, the *organism p*roduced polysaccharides instead of PHA production.
- *R. eutropha* able to produce 70–80% polymer in phosphate ions is limited. In this case, the mixture of glucose and propionate was used as the carbon source of microorganisms capable of producing the copolymer. In another study,[11] the main problem in the use of different types of inefficient consumption of *R. eutropha* propionate has been reported. For example, only about one-third of the material in the production of HV copolymer decreased. In another research mutant strain, Bs-1 has been reported in more propionate consumption and, in this case, it was reported that more than 80% of propionate production reached in the HV copolymer.

Based on the studies, *Alcaligenes latus* is able under normal growth conditions with a cell storage capacity of 80% PHA. The one-step process of PHA production can be used by the organism. The binary study[35] fed the two-step process with full.

The technique is used to produce the maximum PHB and high concentrations of cells. This method is also called the first stage of the development phase by using the right combination of substrate, producing the maximum amount of biological mass (biomass). The restrictions in a storage nutrient PHA production are in the second phase, and thus the phase of the polymer is accelerated.

In 1996, Yemen et al.[129] studied *A. latus* PHA production by using sucrose as a carbon source. In this case, a high concentration of cells, 142 g/l in a short time (18 h) of PHB concentration of 50%, was registered at the time. Therefore, they concluded that innoculum size reduced planting time. They also need time to process PHB production using glucose as the carbon source, using the same conditions as the test performed by *R. eutropha*. In this experiment, the time required to achieve a cell concentration of 122 g/l is about 30 h, and during this period, the concentration of PHB in the cell increased to about 65%. Average molecular weight of PHB produced in this study is 1056.

Copolymer produced by various bacteria of the third group of 13-family rRNA (rRNA superfamily III) was examined by Renner et al.[100] in 1996. They concluded that different bacteria produce PHAs with the same conditions will be different proportions of PHB–PHV. Copolymer with groups Norcadia, Rhodococcus, and Corynebacterium were studied by Anderson and colleagues[5] in 1990; they showed that polymer in the same condition in storage and its compounds.

So, in summary, we can say that the level and composition of PHAs is mainly influenced by the type of microorganisms, the carbon source (substrate) used, the concentration of the substrate, and environmental growth conditions. PHA production by microorganisms under different growth conditions is provided in Table 3.4.

TABLE 3.4 PHB Material Properties Comparison with Polypropylene (PP) Material Properties.

Properties	PHB	PP
Crystalline melting point (°C)	175	176
Crystallinity (%)	80	70
Molecular weight (Da)	5 × 10	2 × 10
Glass transiting temperature (°C)	–4	–10
Density (g/cm^3)	1.250	0.905
Flexural modulus (GPa)	4.0	1.7
Tensile strength (MPa)	40	38
Extension to break (%)	6	400
Ultraviolet resistance	Good	Poor
Solvent resistance	Poor	Good

3.1.4 PHYSICAL PROPERTIES OF POLYHYDROXYLALKANOATES

As previously mentioned, various types of microorganisms, polymers, mainly as a bridge of PHAs are known as carbon and energy reserves in the produced cells. The reason for this name is b situations are different in alkyl groups. PHB is also one of the polymers of this group which have similar properties of PP. Table 3.5 was tabulated to compare between properties of these polymers. Three features of the original PHB, thermoplastic performance, resistance to water, and potential degradability 100% (100% biodegradability), have led this compound to be considered as more environment friendly now.

TABLE 3.5 Feed Composition to SBR Batch (Aerobic–Anaerobic) Reactor for Production of PHB.

No.	Feed composition	Value (ppm)
1	$(NH_4)_2SO$	72
2	KCl	70
3	K_2HPO_4	60
4	$CaCl_2$	17
5	$MgSO_4$	89
6	Pepton	100
7	Sodium propionate	130
8	Yeast extract	33
9	Acetic acid	117

In 1994, Bouma et al.[9] showed that PHB as an aliphatic homopolymer with a melting point 179°C and with high transparency (80%) and at a temperature above the melting point of the analyzes, are cited.

According to study[31], in the case of 190°C PHB for 1 h at room temperature, molecular weight is reduced to approximately half of the initial value.

PHA can sometimes vary depending on their composition and physical properties. PHB physical properties, such as tensile capacity expansion crystallized and capabilities on the molecular weight, which is influenced by the power of microorganisms used, growth conditions, and the purity of the sample are obtained. In 1990, Bayrvm[17] has two main advantages over homopolymer PHB PHB–PHV copolymer which are summarized as follows:

First, the copolymer has a lower melting point. The HV copolymer molecular weight loss, and particularly temperature does not happen. This important feature allows a larger range of temperatures of copolymer prepared.

Second, the PHB–PHV copolymer grade increases the flexibility and durability and less crystallized of copolymer. (However, one disadvantage of low-power process is that it is crystallized for a longer time.) In 1994, Bouma et al.[9] suggested that it can be used in combination with the right combination to obtain the PHV. Based on this research, composite materials top PHV is solid and enduring. Compounds that have lower rates are PHV hard and brittle. (The product of the company ICI under the name Biopol, PHB–PHV PHV, in the copolymer is in the range of 0% to 30%.)

In studies conducted by Lee in 1996,[71] PHA molecular weight in the range of 1052 –1063 was reported. The study also pointed out that higher molecular weight of PHB is more suitable for industrial applications.

In 1995, Burke and colleagues[11] found that the molecular weight of the polymer to produce the PHB can be reduced. In addition, biomass extraction process may also reduce the molecular weight of PHB.[67]

Research also shows that when the concentration of butyric acid increases, the molecular weight of PHB decreases.[115] The pH effect of PHB molecular weight is nonsignificant alterations. In this study, the highest molecular weight of 1063/3 of PHB in concentration 13 g l of butyric acid is obtained, while the molecular weight 1062 optimum condition for PHA 13 (g l butyric acid concentration at pH = 8) is achieved. Molecular weights in the range of 1066/1–1063/3 have been reported.

3.1.5 APPLICATIONS OF PHAs

Research conducted on PHAs, nature biodegradable, compatibility with critical systems, the nature and properties of thermoplastic polyester polymer is shown. In addition, the PHAs produced from renewable sources explains that these factors justify the application of polyhydroxylalkanoates.[12]

The polymer properties for chemical synthesis of optically active substances only in the form of a special space with biological activity (such as some medications) have been used. In chromatography, separation of optical isomers can be used in the polymer. In general, applications of PHAs in three main areas—agriculture, medicine and pharmacy, and packaging products—are concentrated.

PHAs use in agriculture: In agriculture, to protect plants from soil pests, PHB is an appropriate insecticide that is used. In this case, along with Bazrafshani in autumn Granvlhayy of PHB-containing insect planted, PHB degradation is due to bacterial activity and thereby results in slow release of insecticide seed in the soil pest-free environment sprout. The onset of winter is reduced and bacterial activity decreases for PHB degradation. Therefore, less insecticide is released. It is ideal for low pest activity at the same time. In contrast, in the spring, pest activity is higher; thus, more free insecticide is released and therefore PHB-containing insects planted protect plants from pests.[111] *PHAs use in medicine and pharmacy*: Biodegradability properties of PHAs and their compatibility with biological systems made from this polymer controlled release of drugs in the pharmaceutical and veterinary use.[35] The biodegradable polymers are also used as carriers for long-term doses of drugs within the body, surgical pins, sutures, wound-cleansing cotton, wound covers, pages, and bone substitute, replacing blood vessels and dynamic growth, and bone healing.

Bone and PHB has similar properties. Electrical stimulation resulted in strengthening of bone. The fractures occur while using PHB page can also stimulate bone growth. On the other hand, since PHB is biodegradable, slowly absorbed, repeated surgery to remove the grafts are not necessary.[12]

PHAs use in packaging and production: One of the simplest applications (HB–CO–HV) P used in packaging, prepared food containers, beverage bottles, plastic film, and bags of all kinds. By using this polymer in these cases, the accumulation of waste in the environment is prevented.[12,41] Table 3.6 shows a number of important applications of PHAs.

TABLE 3.6 Dissolved Salts in Water.

Sea water			Brackish water	Water	Water temp. (°C)
Dissolved salts in water (mg/l)					
20,000	**15,000**	**10,000**	**5000**	**0**	
11.32	12.14	12.97	13.74	14.62	0
10.01	10.7	11.39	12.09	12.8	5
8.98	9.55	10.13	10.73	11.33	10
8.14	8.63	9.14	9.65	10.15	15
7.42	7.86	8.3	8.73	9.17	20
6.74	7.15	7.56	7.96	8.38	25
6.13	6.49	6.86	7.25	7.63	30

3.2 MATERIALS AND METHODS

This work includes study on behavior of the wastewater fluids flow state and its effect on pure oxygen penetration in wastewater flow. One of the important factor in wastewater flow are polyhydroxyalkanoates or PHAs.

PHAs are linear polyesters produced in nature by bacterial fermentation of sugar or lipids. The polyester is extracted and purified from the bacteria by optimizing the conditions of microbial fermentation of sugar or glucose. The simplest and most commonly occurring form of PHA is the fermentative production of PHB (P3HB), which consists of 1000–30,000 hydroxy fatty acid monomers. The biosynthesis of PHA is usually caused by certain deficiency conditions (e.g., lack of macro-elements such as phosphorus, nitrogen, trace elements, or lack of oxygen) and the excess supply of carbon sources. The main approach in this research was to decrease BOD5 amount equal to 80–90% due to the lack of oxygen and industrial production of PHA. In this study, to perform the required tests, three different reactors are used. In the first reactor, anaerobic and aerobic activated sludge system is an act of sludge for use in the later stages of implementation. The action of microorganisms in the second reactor to produce polymers in a SBR system is considered. In the third reactor, polymer enrichment of microorganisms from a noncontinuous system (BATCH) is done. After commissioning of these systems, their adaptation to the conditions of the effect of various parameters on the amount and type of polymer produced in each reactor need to be discussed.

Here, in connection with the characteristics of the system used, and how to set up and implementation of microbes, the material is given.

3.2.1 SYSTEMS USED

As previously noted in this study, three different systems will be used for testing. The first system to go online (continues) acts only to provide sludge used in other reactors after the commissioning of the system and ensured that the proper functioning of SBR and polymerization reactor was practically unusable, given the importance of this sector in relation to each of these systems and how they are described.

3.2.1.1 ANAEROBIC AND AEROBIC ACTIVATED SLUDGE REACTOR TO SUPPLY THE REQUIRED RAW SLUDGE

In this study, an activated sludge system (anaerobic and aerobic), which continuously acts as a source of primary sludge, needs to be used. Schematic view of the system includes an anaerobic tank, an aerobic reactor, and a sedimentation tank as shown in Figure 3.7.

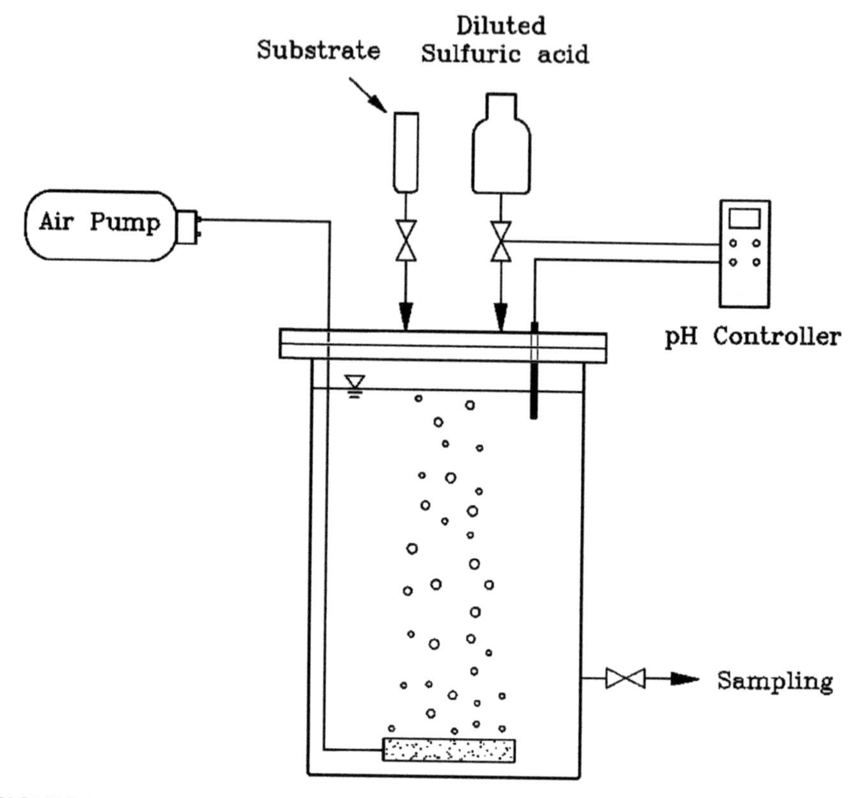

FIGURE 3.7 The reactor for production of polymer.

In this system, wastewater inputs are injected artificially in the laboratory and are prepared for the continuous and uniform rate of 1 l/h of anaerobic tank. A liter of effluent entering the system includes compounds in Table 3.7.

TABLE 3.7 Wastewater Treatment Method Comparison.

Method	Pollut.	Sludge	Solid weight	Return sludge	Air vol. (m^3)	Effic. (%)
Extended aeration	0.1–0.4	0.5–0.15	3–6	50–150	90–125	75–95
Conven	0.3–0.6	0.2–0.4	1.5–3	15–50	45–90	85–95
Taper. aeration	0.3–0.6	0.2–0.4	1.5–3	15–50	45–90	85–95
Step aeration	0.6–1	0.2–0.4	2–3.5	20–75	45–90	85–95
Contact Stab. AS	0–0.2	0.2–0.4	2–5	25–100	45–90	80–90
Two stage	1.6–6	0.4–1.5	3–6	100–500	25–90	75–90

In this system, the tank overflows into the anaerobic–aerobic reactor, and after taking soluble organic material by microorganisms, the separate suspended solids settles in the treated wastewater.. Finally, the percentage of treated wastewater and sludge discharge (80%) is returned to the anaerobic tank.

3.2.1.2 SBR SYSTEM FOR MICROORGANISMS TO PRODUCE POLYMER

In this study, microorganisms produce a system polymer SBR (anaerobic and aerobic). The reactor continuously will be considered and used in different cycles. The reactor is a schematic view as presented in Figure 3.8 showing that all steps are controlled by an electric system precisely. The reactor intensity aerobic and anaerobic oxidation reactions and reduction process using a controller will be ORP measurement and control. SBR system used in this study is shown in Figure 3.9.

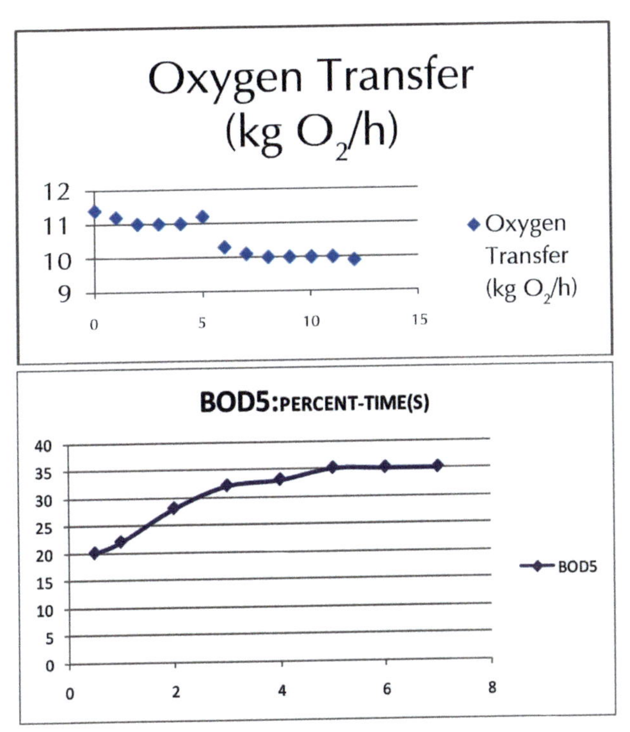

FIGURE 3.8 Amount of the biochemical oxygen demand in 5 days (BOD5) in water treatment process.

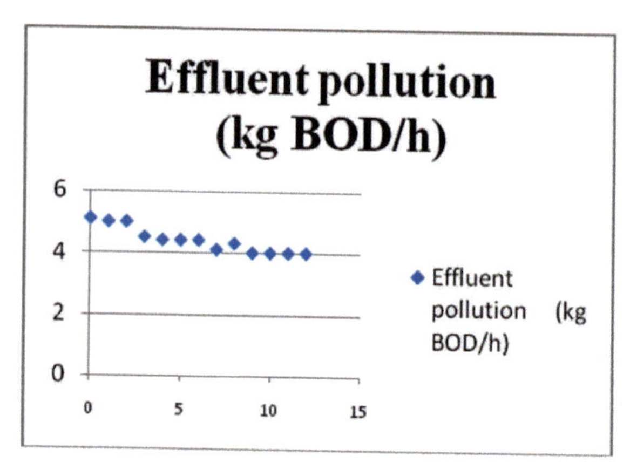

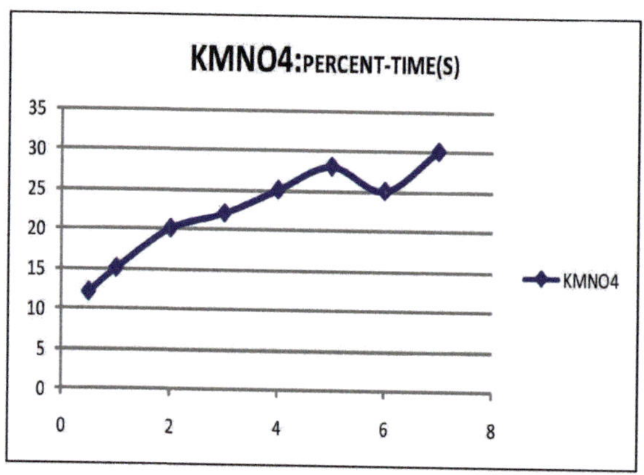

FIGURE 3.9 Oxygen required for pollution decreasing due to detention time.

3.2.1.3 POLYMER REACTOR

To enrich microorganisms, polymer system of a batch with a volume of 5.1 ml is used. The system is completely aerobic and SBR system will be used at the same temperature. In this system, the schematic view in Figure 3.5 shows the amount of excess sludge in SBR system with a certain amount of carbon source containing VFAs and then mixed for 24 h in an aerobic process. The process samples were taken from the system parameters including: MLSS, MLVSS, COD, 3HB, and 3HV as measured. At all stages of the reactor, feed will be added automatically, using an electronic control system, to control the pH of a device with an accuracy of 1.0 pH unit controller that is in the path (online) installed on the system used. A view of the reactor used in Figure 3.6 is shown.

According to surveys conducted, in all parts of the world today, especially in countries with hydrocarbon reserves rich in oil and gas, chemical production cost of polymers compared with biodegradable polymers is much lower and therefore a world of biodegradable polymers to replace conventional polymers like PP is observed. Extensive research is ongoing to reduce production costs and improve the quality of this type of polymer production, especially in developing countries.

In other words, PHA production processes sludge from sewage treatment plants to mix the sludge harmlessly to the environment (stabilized sludge) is the amount of polymer produced that has also a variety of applications.

At first glance, the municipal sewage sludge disposal cost of conventional methods such as incineration, landfill, and composting looks like less expensive than polymer production process. Taking into account operating costs, such as stabilization and sludge dewatering as well weather conditions in the northern cities of Rasht and Anzali, especially with the relatively high annual rainfall, as well as the cost of land, public opinion, the sludge polymer production (PHA) from municipal sewage treatment plants can be justified..

Due to lack of proper incinerator at the provincial level as well as the considerable cost of these units, burning the sludge in the region is not possible.

Carry that with regard to the means of communication (mountain region) and also public opinion in practice, this is avoided.

It also will be difficult.

So taking into consideration, all aspects of polymer-manufacturing process of sludge treatment to seek to reduce the volume of sludge can be considered as the best in the field.

PHAs are linear polyesters produced in nature by bacterial fermentation of sugar or lipids. They are produced by the bacteria to store carbon and energy. More than 150 different monomers can be combined within this family to give materials with extremely different properties.[1] These plastics are biodegradable and are used in the production of bioplastics. They can be either thermoplastic or elastomeric materials, with melting points ranging from 40 to 180°C. The mechanical and biocompatibility of PHA can also be changed by blending, modifying the surface, or combining PHA with other polymers, enzymes, and inorganic materials, making it possible for a wider range of applications.[2]

Irreversibility of wastewater fluid dynamics phenomenon is an important feature for PHA produced by microorganisms. In this research, miscible wastewater interpenetration happened when they move themselves in the separated pipes toward the common joint and pipe in aeration tank. Fluid condition, e.g., velocity, pressure, temperature, and the other properties in the pipes were homogeneous. In this work, pure oxygen was introduced in wastewater in a 40-l pure O_2 container. By submerged porous diffusers and air nozzles, pure O_2 was entered to aeration tank by transient flow. Microorganisms received by pure O_2 grow quickly as illustrated in Figure 3.7. This chapter presents the computational performance of numerical methods for modeling of pure oxygen diffusion in wastewater transient flow. This model was defined by method of the Eulerian based expressed in a method of characteristics (MOC) based on finite difference form. Pure oxygen has been diffused in the activated sludge-wastewater treatment process (pipeline and aeration tank). This method needs to low detention time and low structural space for wastewater treatment plant.[1] In model studies and analysis of prototype problems, similarity law for flow in pumping is generally valid for one-phase flow. Present guidelines and standards for equivalent model and prototype analysis accept that similarity of flow in model and prototype turbomachines (pump stations) exist before the critical cavitation coefficients are reached.

This work presents the application of computational performance of a numerical method by a dynamic model. The model has been presented by the Eulerian method expressed in a MOC. It has been defined by finite difference form for heterogeneous model with varying state in the system. The present work offered MOC as a computational approach from theory to practice in numerical analysis modeling. Therefore, it is computationally efficient for transient flow of irreversibility prediction in a practical case. The difference or improvement of the methods and analysis study is based upon the physical conservation laws of mass, momentum, and energy. The mathematical statements of these laws may be written in either integral or differential form. The integral form is useful for large-scale analyses and provides answers that are sometimes very good and sometimes not, but that are always useful, particularly for engineering applications. The differential form of the equations is used for small-scale analyses. In principle, the differential forms may be used for any problem, but exact solutions can be found only for a small number of specialized flows. Solutions for most problems must be obtained by using numerical techniques, and these are limited by the computer's inability to model small-scale processes. Water hammer (hydrodynamics instability) is caused by a pressure wave or shock wave that travels faster than the sound through the pipes. It is resulted by a sudden stop in the velocity of the water, or a change in the direction. It is also described as a rumbling, shaking vibration in the pipes. Various methods have been developed to solve transient flow in pipes. These ranges of methods are included by approximate equations to numerical solutions of the nonlinear Navier–Stokes equations. In this work, a case study with experimental and computational approach on hydrodynamics instability for a wastewater pipeline has been presented. Present work used the MOC to solve virtually any hydraulic transient problems of wastewater flow in conventional activated sludge system. *Dateline for field tests and lab*: Model data collection was at 10:00 am, 10/02/08–09/02/09. The MOC is based on a finite difference technique, where pressures are computed along the pipe for each time step. Two cases are considered for modeling:

First, the combined elasticity of both the wastewater and the pipe walls is characterized by the pressure wave speed (arithmetic method combination of Joukowski (3.1) formula and Allievi (3.2) formula)[3]:

$$H_2 - H_1 = \left(\frac{C}{g}\right)\left(v_2 - v_1\right) = \rho C\left(v_2 - v_1\right), \tag{3.1}$$

$$c = \frac{1}{\left[\rho\left((1/k)+(d\cdot C_1/E\cdot e)\right)\right]1/2}, \tag{3.2}$$

with combination of Joukowski and Allievi formula (3.3) and (3.4):

$$\begin{aligned} &\lambda\begin{bmatrix}(\partial v/\partial t)+(1/\rho)(\partial p/\partial s) \\ +g(dz/ds)+(f/2D)v|v|\end{bmatrix} \\ &+C^2(\partial v/\partial s)+(1/p)(\partial p/\partial t)=0, \\ &\lambda={}^{+}c \text{ and } \lambda={}^{-}c. \end{aligned} \tag{3.3}$$

Hence, water hammer pressure or surge pressure (ΔH) is a function of independent variables (X), such as

$$\Delta H \approx \rho, K, d, C1, fe, V, g. \tag{3.4}$$

Second, the MOC based on a finite difference technique where pressures ((3.5) and (3.6)) are computed along the pipe for each time step,

$$H_P = \frac{1}{2}\begin{pmatrix} C/g\left(V_{Le} - V_{ri}\right) + \left(H_{Le} + H_{ri}\right) \\ -C/g(f\,\Delta t/2D)\left(V_{Le}|V_{Le}| - V_{ri}|V_{ri}|\right)\end{pmatrix}, \tag{3.5}$$

$$V_P = 1/2\begin{pmatrix}\left(V_{Le} + V_{ri}\right) + (g/c)\left(H_{Le} - H_{ri}\right) \\ -(f\,\Delta t/2D)\left(V_{Le}\left|V_{Le}\right| + V_{ri}\left|V_{ri}\right|\right)\end{pmatrix}. \quad (3.6)$$

Transient analysis results that are not comparable with actual system measurements are generally caused by inappropriate system data (especially boundary conditions) and inappropriate assumptions.[4]

Behavior of the wastewater fluids flow state as a combination of the diffusing process and remixing process have been studied. In this process, high-speed treatment has been achieved. Also, CO_2 is released and O_2 is utilized by microorganisms as result of their activity. In this work, the formulation of process in wastewater transmission line and aeration tank are as following[5]:

$$V = \frac{\text{LO}}{\text{BV}}, \quad (3.7)$$

$$\text{BV} = \text{MLSS} \times \text{MLVSS}, \quad (3.8)$$

$$R_t = \frac{V}{Q}, \quad (3.9)$$

In the case of low F/M, microorganisms feed by organic material in wastewater or feed by other microorganisms.

To save balancing condition, we need to high return sludge and high MLSS.

On the base of European standard, for days without rain,[4] the ratio of RS per max influent become 100% (3.10) and can be achieved from the flowing relations:

$$\frac{\text{RS}}{Q_{\max}} = \left[\frac{\text{MLSS}}{(\text{SMLSS} - \text{MLSS})}\right] \times 100, \quad (3.10)$$

There was a problem in the treatment process, when *sludge volume index* (*SVI*) became more than 200. In this condition, sedimentation was failed. By decreasing *SVI*, aeration time was decreased.

3.2.1.3.1 Sludge Volume Index—SVI

The amount of surplus sludge (3.11) that must be removed from the settling tank was related to biochemical oxygen demand "*BOD5*" of influent entrance to aeration tank and settling tank output.

Sludge age

$$\text{Sludge age} = \frac{(\text{MLSS}) \times (V)}{(\text{SS}_e \times Q + \text{SLS})}. \quad (3.11)$$

3.2.1.3.2 Oxygenation Calculation

A portion of absorbed pure oxygen spent informs the energy consumption and multiplying of bacteria.

The other part of O_2 are spent for oxidation (3.12) of organic carbonate and organic nitrogenous material[5]

$$Q_V = A_Y B_V + B(\text{MLSS}) + 3 - 4(\text{ON}). \quad (3.12)$$

3.2.1.3.3 Process Design

$$O_C = \left[\frac{C_S}{A(F)}\left(C_S - C_X\right)\right] \times \left[0.5\text{YB}_V + 0.1\text{MLSS} + 3.4\text{ON}\right]. \quad (3.13)$$

3.2.1.3.4 Design Criteria

In the air phase, it is suggested to incorporate the air-phase flow component to the mixed flow model.

Influent production per person = 200 l/day

Max influent factor = 1.71 (per 14 h)

BV = 0.5 (kg · BOD5/M3DAY)

BV / MLVSS = F / M = 0.15 & BODS 60g/p-day = 0.06kg/p-day,

Wastewater temperature = 20°C.

3.3 RESULTS AND DISCUSSION

In this work, pure oxygen was introduced in to the wastewater from a 40-l pure O_2 container. By submerged porous diffusers and by air nozzles, pure O_2 was entered to aeration tank. Then, microorganisms by receiving pure O_2 grew quickly.

3.3.1 RESEARCH APPROACH

Pollution decreasing is illustrated in Figure 3.3 as 20 mg/l for BOD 5

- Influent pollution calculation:
- Fixed population = 700
- Office workers = 500
- Total population = 1200

$$Q_m^d = (700)(0.2) + 500(0.05) = 165\,\text{m}^3/\text{day},$$

Max influent per hour is illustrated in Figure 3.10

$$Q_m^d = \frac{165}{14} = 12\text{m}^3/\text{h} = 3.3\,\text{l/s},$$

Total amount of influent pollution: 0.06 (1200) = 72kg / p-day

Average amount of influent pollution: 72/165 = 0.45kg/m^3 = 450mg/1

Due to pollution-detention curves, decrease of influent pollution relation with detention time is detention time:

$$R_t = \frac{V}{Q} = \text{Tank volume/influent flow rate}$$

$$R_t + \frac{200}{12} = 17\,\text{h}$$

From pollution decreasing percent = 0.35

450 (1–0.35) = 293mg/1

Average amount of influent pollution (per 24 h) = 0.293 (165) = 48.5 kg BOD/day

3.3.2 OXYGEN REQUIRED CALCULATION

Figure 3.11:

Cs = 9.17 and A = 0.9 and Y = 0.925 & F = 0.85; MLSS = 3.3kg/m^3 and 3.4 (ON) = 0.23

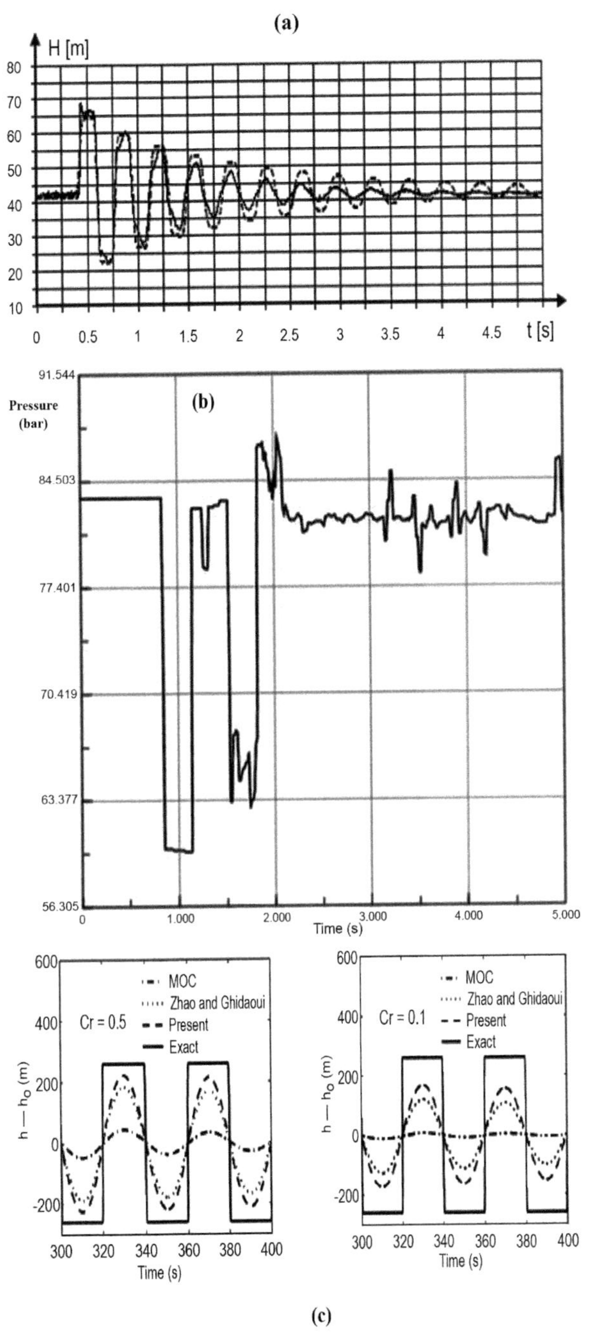

FIGURE 3.10 Present work comparison with other experts: (a) Kodura and Weinerowska research,[8] (b) present research, and (c) (Mohamed and Ghidaoui).[9]

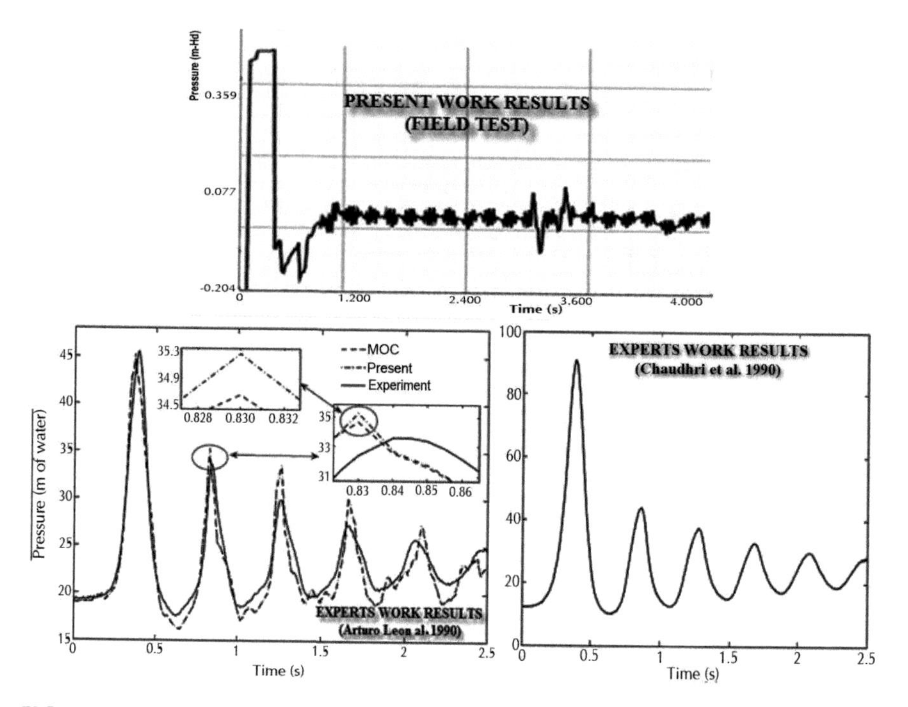

FIGURE 3.11 Wastewater depth versus time for comparison of present work with other experts works.[136,137]

$$O_C = \left[\frac{C_S}{A(F)}\left(C_S - C_X\right)\right] \times \begin{bmatrix} 0.5(Y)B_V + \\ 0.1(\text{MLSS}) + 3.4\text{ON} \end{bmatrix}, \tag{3.14}$$

$$O_C = [9.17/0.9(0.85)(9.17 - 1.5)]$$

$$\times [0.5(0.925)0.5 + 0.1(3.3) + 3.4(\text{ON})],$$

$O_C = 1.2366\,\text{kgO}_2/m^3$ day, $\Sigma O_C = [200(1.2366)]/24 = 10.3(kgO2/h)$; detention time in aeration tank (h).

3.3.3 RESEARCH'S RESULTS

Aeration tank volume (real value) = 200 M^3

Detention time (real value):

$$R_t = \frac{V}{Q} = \frac{200}{12} = 17\,\text{h}$$

Theoretical required volume (existence data)

$$V = \frac{LQ}{BV} = \frac{72}{0.5} = 144\,M^3$$

Theoretical detention time: $R_t = \frac{V}{Q} = \frac{144}{12} = 12\,h$

3.3.4 AIR(PURE O_2) ENTRANCE APPROACHES

Modeling of air or pure O_2 injection in wastewater at present work influenced on hydraulic similarity.

Two different types of air content models have been proposed in the literature in predicting the transient pressure behavior: the concentrated vaporous cavity model (Brown 1968 and Provoost 1976) and the discrete air release model.[2] The concentrated vaporous cavity model produces satisfactory results in slow transients but produces unstable solutions for rapid transients, such as pump's stoppage with reflux valve closure. The discrete air-release model produces satisfactory results in pump shut down cases but is susceptible to long-term numerical damping (Ewing 1980 and Jonsson 1985).

Typically, in the discrete air release model, the wave-speed distribution along a pipeline (with node points $i = 0, 1, \ldots N$) was given by Lee.[2] In this work, at first step, it was a simulated transient pressure in the system due to an emergency power failure without any protective equipment in service. After a careful examination of results, protective equipment was selected and simulated the system again using modeling to assess the effectiveness of the devices which selected to control transient pressures.

The formulated models can accurately describe complex flow features such as wastewater vacuum system-pressurized flow interfaces, interface reversals, and open-pressurized pipeline surges. Much of the complex dynamics in unsteady sewer flow is due to the suggestion to incorporate the air-phase flow component to the mixed flow model. Transient-state pressures were monitored by high-frequency response pressure detectors at two locations. By comparing the simulated results using the *MOC* and the proposed scheme, it can be seen that the pressure traces computed using the *MOC* are lower than the proposed scheme for all the simulations. This means that the *MOC* is more dissipative than the proposed scheme.[6] They can be verified numerically or logically in Figure 3.10. However, the *MOC* agrees with the experiments slightly better than the proposed scheme, when the physical dissipation is estimated using only a steady friction formulation (as used in present work). The important aim in this research was to decrease of *BOD5* to amount of 80–90%. Comparison between laboratory observations on fluid streamlines in pipes for three cases including: laminar flow-transient flow and turbulent flow showed the little error in the prediction of fluid flow software analysis results.[6,7]

Between the zone of incipient cavitation and critical cavitation, similarity is considered to be satisfactory for the analyses of internal flow in pumping. However, for flow in the downstream of the pump station and pipe system, a small quantity of oxygen gas bubbles will change system wave speed characteristics and the similarity for the model. The study of hydraulic transient and hydraulic vibration problems may be difficult to satisfy. When pump stations operate in the zone of cavitation, air bubbles or oxygen will flow through the pump resulting in two-phase flow in the downstream of the fluid system. Though de-aeration devices may be used to minimize the air content, some air will still remain in the fluid system. Since it is almost impossible to predict the quantity of air getting through the pump and remain downstream of the system, systematic analysis has to be carried out. It is assumed with various amount of oxygen content in the model analysis of the transient fluid flow problem. The first study on the effects of air content on wave speed in a transient fluid system was conducted by Whiteman and Pearsall (1959). Detail survey on effects of air on wave speed in fluid systems was given by Lee (1991).[2] Oxygen diffusion in wastewater first was conducted by Union Carbide Corporation (1974). This work is the first research on the effects of oxygen diffusion in wastewater transient flow.

3.4 CONCLUSIONS

As it was mentioned in the introduction, a robust and efficient numerical model able to reproduce unsteady gravity flows, unsteady pressurized flows and the simultaneous occurrence of gravity and pressurized flows in sewers was developed by the authors. The MOC approach transforms the water hammer partial differential equations into the ordinary differential equations along the characteristic lines defined as the continuity equation and the momentum equation are needed to determine V and P in a one-dimensional flow system. Solving these two equations produces a theoretical result that usually corresponds quite closely to actual system measurements if the data and assumptions used to build the numerical model are valid. The test procedure was as follows: A steady state flow of an O_2–wastewater mixture was established in the wastewater pipe by controlling the exit valves and the pressure of the injected O_2 at the inlet. The flow velocity of the O_2–wastewater mixture was maintained at a high enough rate so that slug flow could be avoided by limiting the rate of O_2 injection. Transient flow was created by a rapid control valve closure at the downstream end of the wastewater pipe. Transient-state pressures were monitored by high-frequency-response pressure detectors at two locations. By comparing the simulated results using the MOC and the proposed scheme, it can be seen that the pressure traces computed using the MOC are lower than the proposed scheme for all the simulations. This means that the MOC is more dissipative than the proposed scheme. However, the MOC agrees with the experiments slightly better than the proposed scheme, when the physical dissipation is estimated using only a steady friction formulation (as used in present work). The important aim in this research was to decrease of BOD5 to amount of 80–90%. An activated sludge system (anaerobic and aerobic) continuously acts as a source of primary sludge. Anaerobic and aerobic activated sludge reactor to supply the required raw sludge. The important results in this work were to decrease BOD5 amount equal to 80–90% and the industrial production of PHA to production of plastics which are biodegradable and are used in the production of bioplastics.

ACKNOWLEDGMENTS

The author thanks to all specialists for their valuable observations and advice, and the referees for recommendations that improved the quality of this chapter.

KEYWORDS

- **polyhydroxyalkanoates**
- **pure oxygen**
- **wastewater treatment**
- **activated sludge**
- **microorganism**

REFERENCES

1. Chua, A. S. M.; Satoh, H. T. H.; Mino, T. Production of Polyhydroxyalkanoayes (PHA) by Activated Sludge Treating Municipal Wastewater: Effect of pH, Sludge Retention Time (SRT), and Acetate Concentration in Influent. *Wat. Res.* **2003,** *37*, 3602–3611.
2. Akiyama, M.; Taima, Y. M.; Doi, Y. Production of PHB by a Bacterium of the Genus *Alcaligesesis* Utilizing Long Chain Fatty Acids. *Appl. Microbiol. Biotechnol.* **1992,** *37*, 689–701.
3. Akiyama, M.; Doi, Y. Production of PHB from Alkanedioic Acids and Hydroxylated Fatty Acids by *Alcaligesesis* sp. *Biotechnol. Lett.* **1993,** *15* (2), 163–168.
4. Anderson, A. J.; Dawes, E. A. Occurrence, Metabolism, Metabolic Role, and Industrial Uses of Bacterial Polyhydroxyalkanoates. *Microbiol. Rev.* **1990,** *54* (4),450–472.
5. Anderson, A. J.; Haywood, G. W.; Williams, D. R.; Dawes, E. A. The Production of Polyhydroxyalkanoates from Unrelated Carbon Sources. In *Novel Biodegradable Microbial Polymers*; Dawes, E. A., Ed.; Kluwer Academic Publishers: Dordrecht, The Netherlands, 1990; pp 119–129.
6. Anderson, A. J.; Williams, D. R.; Taidi, B.; Dawes, E. A.; Ewing, D. F. Studies in Copolyester Synthesis by *Rhodococcus ruber* and Factors Influencing the Molecular Mass of Polyhydroxybutyrate Accumulated by *Methylobacterium extorquens* and *Alcaligenes eutrophus*. *FEMS Microbiol. Rev.* **1992,** *130*, 93–102.
7. Ayorinde, F. O.; Saeed, K. A.; Eribo, E.; Morrow, A.; Collis, W. E.; Mclnnis, F.; Polack, S. K.; Eribo, B. Production of PHB from Saponified *Veronica galamesis* Oil by *Alcaligenesis eutrophus*. *J. Ind. Microbiol. Biotechnol.* **1998,** *21*, 46–50.
8. Berger, E.; Ramsay, B. A.; Ramsay, J. A.; Chavarie, C. PHB Recovery by Hypochlorite Digestion of Non-PHB. *Biomass Biotechnol. Tech.* **1989,** *3* (4), 227–232.
9. Booma, M.; Selke, S. E.; Giacin, J. R. Degradable Plastics. *J. Elastom. Plast.* **1994,** *26*, 104–142.
10. Bourque, Ouellette, B.; Andre, G.; Groleau, D. Production of Poly-β-Hydroxybutyrate from Methanol: Characterization of a New Isolate of *Methylobacterium extorquens*. *Appl. Microbiol. Biotechnol.* **1992,** *37*, 7–12.
11. Bourque, D.; Pomerleau, Y.; Groleau, D. High Cell Density Production of Poly-β-Hydroxdybutyrate (PHB) Form Methanol by *Methylobacterium extorquens*: Production of High-Molecular-Mass PHB. *Appl. Microbiol. Biotechonol.* **1995,** *44*, 367–376.
12. Brandl, H.; Gross, R. A.; Lenz, R. W.; Fuller, R. C. Plastics Form Bacteria and for Bacteria: Poly(β-Hydroxybutyrate) as Natural, Biodegradable Polyesters. *Adv. Biochem. Eng.* **1990,** *41*, 77–93.
13. Brandl, H.; Bachofen, B.; Mayer, J.; Wintermantel, E. Degradation and Applications of Polyhydroxyalkanoates. *Can. J. Microbiol.* **1995,** *41* (Suppl. 1), 143–153.
14. Braunegg, G.; Sonnleitner, B.; Lafferty, R. M. A Rapid Gas Chromatographic Method for the Determination of PHB in Microbial Biomass. *Eur. J. Appl. Microb. Biotechnol.* **1978,** *6*, 905–910.
15. Braunegg, G.; Lefebvere, G.; Genser, K. F. Polyalkanoates, Biopolyester from Renewable Resources: Physiological and Engineering Aspects. *J. Biotechnol.* **1998,** *65*, 127–161.
16. Byrom, D. Polymer Synthesis by Microorganisms: Technology and Economics. *TIBTECH* **1987,** *5*, 246–250.
17. Byrom, D. Industrial Production of Copolymer from *Alcaligenes eutrophus*. In *Novel Biodegradable Microbial Polymers*; Dawes, E. A., Ed.; Kluwer Academic Pubishers: Dordrecht, The Netherlands, 1990; pp 113–117.

18. Byrom, D. Production of Poly-β-Hydroxybutyrate:Poly-β-Hydroxyvalerate Copolymers. *FEMS Microbiol. Rev.* **1992,** *103*, 247–250.
19. Castella, J. M.; Urmenta, J.; Lafuente, R.; Navarrte, A.; Guerrero, R. Biodegradation of PHAs in Aerobic Sediments. *Int. Biodeter. Biodegrad.* **1995,** *35*, 155–174.
20. Chen, Y.; Chen, J.; Yu, C.; Du, G.; Lun, S. Recovery of Poly-3-Hydroxybutyrate from *Alcaligenes eutrophus* by Surfactant-Chelate Aqueous System. *Process Biochem.* **1999,** *34* (2), 153–157.
21. Chen G, Page WJ. The Effect of Substrate on the Molecular Weight of PHB Production by *Azotobacter vinelandii* UWD. *Biotechnol. Lett.* **1994,** *16* (2), 155–160.
22. Cho, G. D.; Yoo, J. Y.; Oh, J. T.; Kim, W. S. Study on the Biosynthesis of PHB with *Alcaligenesis latus*. *Hwahak Konghak* **1997,** *35*, 412–418.
23. Chua, H.; Yu, P. H. F.; Ho, L. Y. Recovery of Biodegradable Polymers from Food-Processing Wastewater Activated Sludge System. *J. IES* **1997,** *37* (2), 9–13.
24. Chua, H.; Yu, P. H. F. Production of Biodegradable Plastics Form Chemical Wastewater—A Novel Method to Reduce Excess Activated Sludge Generated from Industrial Wastewater Treatment. *Wat. Sci. Technol.* **1999,** *39* (10–11), 273–280.
25. Chuang, S. H.; Ouyang, C. F.; Yuang, H. C.; You, S. J. Evaluation of Phosphorus Removal in Anaerobic–Anoxic–Aerobic System via Polyhydroxyalkanoates Measurements. *Wat. Sci. Technol.* **1998,** *38* (1), 107–114.
26. Chuang, S. H.; Ouyang, C. F.; Yuang, H. C.; You, S. J. Effects of SRT and DO on Nutrient Removal in a Combined AS-Biofilm Process. *Wat. Sci. Technol.* **1997,** *36* (12), 19–27.
27. Daniel, M.; Choi, J. H.; Kim, J. H.; Lebeault, J. M. Effect of Nutrient Deficiency on Accumulation and Relative Molecular Weight of PHB by Methylotrophic Bacterium *Pseudomonas* 135. Appl. Microbiol. Biotechnol **1992,** *37*, 702–706.
28. Dawes, E. A.; Senior, P. J. The Role and Regulation of Energy Reserve Polymers in Micro-organisms. *Adv. Microb. Physiol.* **1973,** *10*, 135–266.
29. Dawes, E. A. Starvation, Survival and Energy Reserves. In *Bacteria in their Natural Environment*, Fletcher, M., Floodgate, G. D., Eds.; Academic Press, Miami, FL, 1985; pp 43–79.
30. Dawes, E. A. Novel Microbial Polymers: An Introductory Overview. In *Novel Biodegradable Microbial Polymers*; Dawes, E. A., Ed.; Kluwer Academic Publishers: Dordrecht, The Netherlands, 1990; pp 3–16.
31. De Koning, G. Physical Properties of Bacterial Poly(R)-3-Hydroxyalkanoates). *Can. J. Microbiol.* **1995,** *41* (Suppl. 1), 303–309.
32. Dionisi, D.; Majone, M.; Ramadori, R.; Beccari, M. The Storage of Acetate under Anoxic Conditions. *Water Res.* **2001,** *35* (11), 2661–2668.
33. Doi, Y.; Kunioka, M.; Nakamura, Y.; Soga, K. Nuclear Magnetic Resonance Studies on Poly(β-Hydroxybutyrate) and a Copolymer of -Hydroxybutyrate and β-Hydroxyvalerate Isolated from *Alcaligenes eutrophus* H16. *Macromolecules* **1986,** *19*, 2860–2864.
34. Doi, Y.; Kunioka, M.; Nakamura, Y.; Soga, K. Biosynthesis of Copolyesters in *Alcaligenes eutrophus* H16 from ^{13}C-Labeled Acetate and Propionate. *Macromolecules* **1987,** *20*, 2988–2991.
35. Doi, Y. Microbial Polyesters. VCH Publishers, Inc.: New York, NY, 1990.
36. Doi, Y.; Segawa, A.; Nakamura, S.; Kunioka, M. Production of Biodegradable Copolymers by *Alcaligenes eutrophus*. In *Novel Biodegradable Microbial Polymers*; Dawes, E. A. Ed.; Kluwer Academic Pubishers: Dordrecht, The Netherlands, 1990; pp 37–48.

37. Fidler, S.; Dennis, D. Polyhydroxyalkanoate Production in Recombinant *Escherichia coli*. *FEMS Microbiol. Rev.* **1992,** *103*, 231–236.
38. Fuchtenbusch, B.; Steinbuchel, A. Biosynthesis of PHAs from Low Rank Coal Liquefaction Products by *Pseudomonas oleovorance* and *Rhodococcus ruber*. *Appl. Microbiol. Biotechnol.* **1999,** *52*, 91–95.
39. Fuchtenbusch, B.; Wullbrandt, D.; Steinbuchel, A. Production of PHAs by *Ralstonia eutroph* and *Pseudomonas oleovorance* from Oil Remaining from Biotechnological Rhamnose Production. *Appl. Microbiol. Biotechnol.* **2000,** *53*, 167–172.
40. Gorenflo, V.; Steinbuchel, A.; Marose, S.; Rieseberg, M.; Scheper, T. Quantification of Bacterial Polyhydroxyalkanoic Acids by Nile Red Staining. *Appl. Microbiol. Biotechnol.* **1999,** *51*, 765–772.
41. Griffin, G. J. L. *Chemistry and Technology of Biodegradable Polymers*, first ed. Chapman & Hall: London, 1994.
42. Gross, R. A.; DeMello, C.; Lenz, R. W.; Brandl, H.; Fuller, C. Biosynthesis and Characterization of PHAs Produced by *Pseudomnas oleovorance*. *Macromolecules* **1989,** *22*, 1106–1115.
43. Grothe, E.; Chisti, Y. PHB Thermoplastic Production by *Alcaligenesis latus*: Behavior of Fed-Batch Culture. *Bioprocess Eng.* **2000,** *22*, 441–449.
44. Hahn, S. K.; Chang, Y. L.; Kim, B. S.; Chang, H. N. Optimization of microbial Poly(3-Hydroxybutyrate) Recovery Using Dispersions of Sodium Hypochlorite Solution and Chloroform. *Biotechnol. Bioeng.* **1994,** *44*, 250–262.
45. Harrison, S.; Dennis, J. S.; Chase, S. A. The Effect of Culture History on the Disruption of *Alcaligenes eutrophus* by High Pressure Homogenization. In *Separation for Biotechnology II*, first ed.; Pyle, D. L., Ed.; Elsevier: London, 1990.
46. Hassan, M. A.; et al. The Production of Polyhydroxyalkanoate Form Anaerobically Treated Palm Oil Mill Effluent by *Rhodobacter sphaeroides*. *J. Rerment. Bioeng.* **1997,** *83* (5), 485–488.
47. Haywood, G. W.; Anderson, A. J.; Dawes, E. A. A Survey of the Accumulation of Novel Polyhydroxyalkanoates by Bacteria. *Biotechnol. Lett.* **1989,** *11* (7), 471–476.
48. Haywood, G. W.; Anderson, A. J.; Williams, D. R.; Dawes, E. A. Accumulation of a Poly(Hydroxyalkanoate) Copolymer Containing Primarily 3-Hydroxyvarate from Simple Carbohydrate Substrates by *Rhodococcus* sp. NCIMB 40126. *Int. J. Biol. Macromol.* **1991,** *13*, 83–88.
49. Hejazi, P.; Vasheghani-Farahani, E.; Yamini, Y. Supercritical Fluid Disruption of *Ralstonia eutropha* for Poly(β-hydroxybutirate) Recovery. *Biotechnol. Prog.* **2003,** *19*, 1519–1523.
50. Hezayen, F. F.; Rehm, B. H. A.; Eberhardt, R. *Polymer Production by 2 Newly Isolation Extremely Halophilic Archaea: Application of a Novel Corrosion Resistance Bioreactor*, 2000.
51. Holmes, P. A. Application of PHB—A Microbially Produced Biodegradable Thermoplastics. *Phys. Technol.* **1985,** *16*, 32–36.
52. Hood, C. R.; Randall, A. A. A Biochemical Hypothesis Explaining the Response of Enhanced Biological Phosphorus Removal Biomass to Organic Substrates. *Wat. Res.* **2001,** *35*, 2758–2766.
53. Horib, K, Marsudi, S.; Unno, H. Simultaneous Production of PHAs and Rhamnolipids by *Pseudomonas aeroginosa*. *Biotechnol. Bioeng.* **2002,** *78* (6), 699–707.
54. Hrabak, O. Industrial Production of Poly-β-Hydroxybutyrate. *FEMS Microbiol. Rev.* **1992,** *103*, 251–256.

55. Huang, J.; Shetty, A. S.; Wang, M. Biodegradable Plastics: A Review. *Adv. Polym. Technol.* **1990,** *10* (1), 23–30.
56. Ishihara, Y.; Shimizu, H.; Shioya, S. Mole Fraction of PHB in Fed-Batch Culture of *A. eutrophus*. Biotechnol. Bioeng. **1994,** *81* (5), 422–428.
57. Jan, S.; Roblot, C.; Courtois, B.; Babotin, J. N.; Seguin, J. P. H NMR Spectroscopic Determination of Poly-β-Hydroxybutyrate Extracted from Microbial Biomass. *Enzym. Microb. Technol.* **1996,** *18*, 195–201.
58. Yu, J.; Si, Y.; Keung, W.; Wong, R. Kinetics Modeling of Inhibition and Utilization of Mixed Volatile Fatty Acids in the Formation of Polyhydroxyalkanoates by *Ralstonia eutropha. Process Biochem.* **2002,** *37*, 731–738.
59. Jung, K.; Hazenberg, W.; Prieto, M. W. Two Stage Continuous Process Development for the Production of Medium Chain Length PHAs. *Biotechnol. Bioeng.* **2001,** *72* (1), 19–24.
60. Juttner, P. R; Lafferty, R. M.; Knackmuss, H. Y. A Simple Method for the Determination of Poly-β-Hydroxybutyric Acid in Microbial Biomass. *Eur. J. Appl. Microbiol.* **1975,** *1*, 233–237.
61. Kato, N.; Konishi, H.; Shimo, M.; Sakazawa, C. Production of PHB trimmer by *Bacillus megaterium* B124. *J. Fermen. Bioeng.* **1992,** *73* (3), 246–247.
62. Kellerhals, M. B.; Hazenberg, W.; Witholt, B. High Cell Density Fermentation of *Pseudomonas oleovorance* for the Production of mcl-PHA in Two Liquid Phase Media. *Enzym. Microb. Technol.* **1999,** *24*, 111–116.
63. Khatipov, E.; Miyake, M.; Miyake, J.; Asads, Y. Accumulation of PHB by *Rhodobacter sphearoides* on Various Carbon Source and Nitrogen Substrates. *FEMS Microbiol. Lett.* **1998,** *1652*, 39–45.
64. Khosravi-Darani, K.; Vasheghani-Farahani, E.; Shojaosadati, S. A.; Yamini, Y. Effect of Process Variables on Supercritical fluid Disruption of *Ralstonia eutropha* Cells for poly(*R*-hydroxybutirate) Recovery. *Biotechnol. Prog.* **2004,** *20*, 1757–1765.
65. Kim, B. S.; Lee, Sc.; Lee, S. Y.; Chang, H. N. Production of PHB by Fed-Batch Culture of *Ralstonia eutropha* with Glucose Concentration Control. *Biotecnol. Bioeng.* **1994,** *43*, 892–898.
66. Kimura, H.; Yoshida, Y.; Doi, Y. Production of PHB by *Pseudomonas acidovorance. Biotechnol. Lett.* **1992,** *14* (6), 149–158.
67. Kulaev, I. S.; Vagabov, V. M. Polyphosphate Metabolism in Microorganisms. *Adv. Microbial Physiol.* **1983,** *24*, 83–171.
68. Kumagai, Y. Enzymatic Degradation of Binary Blends of Microbial Poly(3-Hydroxy Butyrate) with Enzymatically Active Polymers. *Polym. Degrad. Stab.* **1992,** *37*, 253–256.
69. Lafferty, R. M.; Korsatko, B.; Korsatko, W. Microbial Production of Poly-β-Hydroxybutyric Acid. In *Biotechnology*; Rehm, H. J.; Reed, G., Eds.; VCH Publishers: New York, 1988; pp 135–176.
70. Law, J.; Slepecky, R. Assay of Poly-β-Hydroxybutyric Acid. *J. Bacteriol.* **1961,** *82*, 33–36.
71. Lee, S. Y. Review Bacterial Polyhydroxyalkanoates. *Biotechnol., Bioeng.* **1996,** *49*, 1–14.
72. Lee, S.; Yu, J. Production of Biodegradable Thermoplastics from Municipal Sludge by a Two-Stage Bioprocess. *Resour., Conserv. Recycl.* **1997,** *19*, 151–164.
73. Lee, S.; Yup, C. J. Production and Degradation of Polyhydroxyalkanoates in Waste Environment. *Waste Manage.* **1999,** *19* (2), 133–139.
74. Lemos, P. C.; et al. Effect of Carbon Source on the Formation of Polyhydroxyalkanoates (PHA) by a Phosphate-Accumulation Mixed Culture. *Enzym. Microb. Techonol.* **1998,** *22*, 662–671.

75. Libergesel, M.; Husted, E.; Timm, A.; Steinbuchel, A.; Fuller, R. C.; Len, Z. R. W.; Schlegel, H. G. Formation of PHA by Phototrophic and Chemulithotrophic Bacteria. *Arch. Microbial.* **1991,** *155*, 415–421.
76. Ling, Y.; Wong, H. H.; Thomas, C. J.; Williams, D. R. G.; Middelberg, A. P. J. Pilot-Scale Extraction of PHB from Recombinant *E. coli* by Homogenization and Centrifugation. *Biosepatation* **1997,** *7*, 9–15.
77. Linko, S.; Vaheri, H.; Seppala, J. Production of PHB by *Alcaligenesis eutrophus* on Different Carbon Sources. *Appl. Microbiol. Biotechnol.* **1993,** *39*, 11–15.
78. Madigan, M. T.; Martinko, J. M.; Parker, J. *Brock Biology of Microorganisms*. Princeton Hall: Upper Saddle River, NJ, 1997.
79. Maranogoni C, Furigo A, Glancia MF. Production of PHB by *Ralstonia eutropha* in Whey and Inverted Sugar with Propionic Acid Feeding. *Process Biochem.* **2002,** *38*, 137–141.
80. Masow, T.; Babel, W. Calorimetrically Recognized Maximum Yield of PHB Continuously Synthesized from Toxic Substrate. *J. Biotechnol.* *77*, 247–253.
81. McCool, G. J.; Fernandez, T.; Li, N.; Cannon, M. Polyhydroxyaklanoate Inclusion-Body Growth and Prokiferation in *Bacillus megaterium. FEMS Microbial. Lett.* **1996,** *138*, 41–48.
82. Mergeay, M.; Houba, C.; Gerits, J. Extra Chromosomal Inheritance Controlling Resistance to Cadmium and Zinc Ions: Evidence from Curing in a *Pseudomonas. Arch. Int. Physiol. Biochem.* **1978,** *86*, 440–441.
83. Mino, T.; Kawakmi, T.; Matsuo, T. Location of Phosphorus in Activated Sludge and Fraction of Intracellular Polyphosphate in Biological Phosphorus Removal Process. *Wat. Sci. Technol.* **1985,** *17* (2/3), 93–106.
84. Mino, T.; Kawakami, T.; Matsuo, T. Behavior of Intracellular Polyphosphate in Biological Phosphorus Removal Process. *Wat. Sci. Technol.* **1985,** *17* (11,12), 11–21.
85. Ohi, K.; Takaida, N.; Komemushi, S.; Okazaki, M.; Muira, Y. A New Species of Hydrogen-Oxidizing Bacterium. Appl. Microbiol. **1979,** *25*, 53–58.
86. Page, W. J. Production of PHB by *Azotobacter vinelandii* UWD during Growth on Molasses and Other Complex Carbon Sources. *Appl. Microbiol. Biotechnol.* **1989,** *31*, 329–333.
87. Page, W. J. Production of Polyhydroxyalkanoates by *Azotobacter vinelandii* Strain UWD in Beet Molasses Culture. *FEMS Microbiol. Rev.* **1992,** *103*, 149–158.
88. Page, W. J.; Cornish, A. Growth of *Azotobacter vinelandii* in Fish Peptone Medium and Simplified Extraction of PHB. *Appl. Environ. Microbiol.* **1993,** *59*, 4236–4244.
89. Page, W. J. Bacterial Polyhydroxyalkanoates, Natural Biodegradable Plastics with a Great Future. *Can J. Microbiol.* **1995,** *41* (Suppl. 1), 1–3.
90. Page, W. J.; Manchak, J. The Role of β-Oxidation of Short-Chain Alkanoates in Polyhydroxyalkanoate Copolymer Synthesis in *Azotobacter vinelandii* UWD. *Can. J. Microbiol.* **1995,** *41* (Suppl. 1), 106–115.
91. Pelissero, A. Update on Biodegradable Plastics Materials. Imballaggio **1987,** *38*, 54.
92. http://www.env.tu-tokyo.ac.jp//7Esato/PHAinfo/PHAinfo.html.
93. Preiss, J. Bacterial Glycogen Synthesis and Its Regulation. *Annu. Rev. Microbiol.* **1984,** *38*, 419–458.
94. Preusting, H.; Kingama, J.; Witholt, B. Physiology and Polyester Formation of *Pseudomonas oleovorance* in Continuous Two Liquid Phase Cultures. *Enzym. Microbial Technol.* **1991,** *13*, 770–780.

95. Preusting H, Hazenberg, W. Continuous Production of PHB by *Pesudomonas oleovorance* in a High Cell Density, Two Liquid Phases Chemostat. *Enzym. Microbial Technol.* **1993,** *15*, 311–316.
96. Ramsay, J. A.; Berger, E.; Voyer, C.; Chavarie, C.; Ramsay, B. A. Extraction of Poly-3-Hydroxybutyrate Using Chlorinated Solvents. *Biotechnol. Tech.* **1990,** *8* (8), 589–594.
97. Ramsay, J. A.; Berger, E.; Ramsay, B. A.; Chavarie, C. Recovery of Poly-3-Hydroxyalkanoic Acid Granules by a Surfactant-Hypochlorite Treatment. *Biotechnol. Tech.* **1994,** *9* (10), 709–712.
98. Ramsay, J. A.; Berger, E.; Voyer, R.; Chavarie, C.; Ramsay, B. A. Extraction of Poly-3-Hydroxybutyrate Using Chlorinated Solvents. *Biotechnol. Tech.* **1994,** *8* (8), 589–594.
99. Ramsay, J. A.; Berger, E.; Ramsay, B. A.; Chavarie, C. Recovery of Poly-β-Hydroxyalkanoic Acid Granules by a Surfactant–Hypochlorite Treatment. *Biotechnol. Tech.* **1990,** *4* (4), 221–226.
100. Renner, G.; Haage, G.; Braunegg, G. Production of Short-Side-Chain Polyhydroxyalkanoates by Various Bacteria from the rRNA Superfamily III. *Appl. Microbiol. Biotechnol.* **1996,** *46*, 268–272.
101. Resch, S.; Gruber, K.; Wanner, G.; Slater, S.; Dennnis, D.; Lubitz, W. Aqueous Release and Purification of Poly(β-Hydroxybutyrate) from *Escherichia coli*. *J. Biotechnol.* **1998,** *65*, 173–182.
102. Roh, K. S.; Yeom, S. H.; Yoo, Y. J. The Effects of Sodium Bisulfate in Extraction of PHB by Hypochlorite. *Biotechnol. Tech.* **1995,** *4* (4), 221–226.
103. Roy, R. K. *A Primer on the Tguchi Method*, first ed. Van Nostrand Reinhold: New York, 1990.
104. Saad, B.; Neuenschwander, P.; Uhlschmid, G. K.; Suter, U. W. New Versatile Elastomeric Degradable Polymer Materials for Medicine. *Int. J. Biol. Macromol.* **2001,** 158–163.
105. Lee, S.; Yu, J. Production of Biodegradable Thermoplastics from Municipal Sludge by a Two Stage Bioprocess. *Resour., Conserv. Recycl.* **1997,** *19*, 151–164.
106. Sasikala, C. H.; Ramana, C. H. V. Biodegradable Polyesters. In *Advances in Applied Microbiology*; Neidleman, S. L., Laskin, A. I., Eds.; Academic Press: California, 1996; vol. 42, pp 97–218.
107. Satoh, H.; Mino, T.; Matsuo, T. Uptake of Organic Substrates and Accumulation of Polyhydroxyalkanoates Liked with Glycolysis of Intracellular Carbohydrates under Anaerobic Conditions in the Biological Excess Phosphorus Removal Processes. *Wat. Sci. Technol.* **1992,** *26* (5–6), 933–942.
108. Satoh, H.; Iwamoto, Y.; Mino, T.; Matsuo, T. Activated Sludge as a Possible Source of Biodegradable Plastic. In *Proceedings, Water Quality International 1998, Book 3 Wastewater Treatments. IAWQ 19th Biennial International Conference*, Vancouver, BC, Canada, 21–26 June 1998; pp 304–311.
109. Schwien, U.; Schmidt, E. Improved Degradation of Monochlorophenol by Constructed Strain. *Appl. Environ. Microbiol.* **1982,** *44*, 33–39.
110. Senior, P. J.; Beech, G. A.; Ritchie, G. A. F.; Dawes, E. A. The Role of Oxygen Limitation in the Formation of Poly-β-Hydroxybutyrate during Batch and Continuous Culture of *Azotobacter beijerinckii*. *Biochem. J.* **1972,** *128*, 1193–1201.
111. Shahhosseini, S. A Fed-Batch Model for PHA Production Using *Alcaligenes eutrophus*. M.S. Thesis, Department of Chemical Engineering, The University of Queensland, Australia, 1994.

112. Shimao, M. Biodegradation of Plastics. *Curr. Opin. Biotechnol.* **2001,** *12*, 242–247.
113. Shimizu, H.; Tamira, S.; Shioya, S.; Suga, K. Kinetics Study of PHB production and its Molecular Weight Distribution Control in Fed-Batch Culture of *A. eutrophus. J. Ferment. Bioeng.* **1993,** *76* (6), 465–469.
114. Chen, G. Q.; Wu, Q. The Application of Polyhydroxyalkanoates as Tissue Engineering Materials. *Biomaterials* **2005,** *26* (33), 6565–6578. doi:10.1016/j.biomaterials.2005.04.036.
115. Shimizu, H.; Tamura, S.; Ishihara, Y.; Shioya, S.; Suga, K. Control of Molecular Weight Distribution and Mole Fraction in Poly(-D-(–)-3-hydroxyalkanoates) (PHA) Production by *Alcaligenes eutrohpus*. In *Biodegradable Plastics and Polymers*; Doi, Y., Fukuda, K., Eds.; Elsevier Science BV: New York, NY, 1994; pp 365–372.
116. Shimizu H, Kozaki Y, kodama H, Shioya, H. Maximum Production Strategy for Biodegradable Copolymer PHBV in Fed-Batch Culture of *Alcaligenesis eutrophus, Biotechnol. Bioeng.* **1998,** *62* (5), 518–525.
117. Smolders, G. J. F.; van der Meij, J.; van Loosdrecht, M. C. M. Model of the Anaerobic Metabolism of the Biological Phosphorus Removal Process: Stoichiometry and pH Influence. *Biotechnol. Bioeng.* **1994,** *43*, 461–470.
118. Sonneleitner, B.; Heinzle, G.; Braunegg, G.; Lafferty, R. M. Formal Kinetics of PHB Production in *Alcaligenesis eutrophus* H16 and *Mycoplana rubra* to the Dissolved Oxygen Tension in Ammonium-Limited Batch Cultures. *Eur. J. Appl. Microbiol.* **1979,** *7*, 1–10.
119. Steinbuchel, A.; Pieper, U. Production of a Copolyester of 3-Hydroxybutyric Acid and 3-Hydroxyvaleric Acid from Single Unrelated Carbon Sources by a Mutant of *Alcaligenenes eutrophus*. *Appl. Microbiol. Biotechnol.* **1992,** *37*, 1–6.
120. Steinbuchel, A. PHB and Other Polyhydroxyalkanoic Acids. In *Biotechnology*; Rehm, H. J., Reed, G., Eds.; VCH: New York, NY, 1986; pp 403–464.
121. Steinbuchel, A.; Fuchtenbusch, B. Bacterial and Other Biological Systems for Polyester Production, *TIBTECH*, **1998,** *16*, 419–427.
122. Suzuki, T.; Yamane, T.; Shimizu, S. Mass Production of Poly-β-Hydroxybutyric Acid by Fully Automatic Fed-Batch Culture of Methylotroph. *Appl. Microbiol. Biotechnol.* **1986,** *23*, 322–329.
123. Suzuki, T.; Yamane, T.; Shimizu, S. Mass Production of PHB by Fully Automatic Fed-Batch Culture of Methylotroph. *Appl. Microbiol. Biotechnol.* **1989,** 23, 322–329.
124. Takabatake, H.; Satoh, H.; Mino, T.; Matsuo, T. Recovery of Biodegradable Plastics from Activated Sludge Process. *Wat. Sci. Technol.* **2000,** *42* (3–4), 351–356.
125. Tsuchikura, K. BIOPOL Properties and Processing. In *Biodegradable Plastics and Polymers*; Doi, Y., Fukuda, K., Eds. Elsevier Science BV: New York, NY, 1994; pp 362–364.
126. Van Groenestijn, J. W.; Deinema, M. H.; Zehnder, A. J. B. ATP Production from Polyphosphate in *Acinetobacter* Strain 210A. *Arch. Microbial.* **1987,** *148*, 14–19.
127. Varma, A. K.; Peck, H. D. Utilization of Short and Long-Chain Polyphosphate as Energy Sources for the Anaerobic Growth of Bacteria. *FEMS Microbial. Lett.* **1983,** *16*, 281–285.
128. Yamane, T. Yield of poly-D(–)-3-Hydroxybutyrate from Various Carbon Sources: A Theoretical Study. *Biotechnol. Bioeng.* **1993,** *41* (1), 165–170.
129. Yamane, T.; Fukunage, M.; Lee, Y. W. Increase PHB Productivity by High-Cell-Density Fed-Batch Culture of *Alcaligenes latus*, a Growth-Associated PHB Producer. *Biotechnol. Bioeng.* **1996,** *50*, 197–202.

130. Yu, P. H. F.; Chua, H.; Huang, A. L.; Lo, W. H.; Ho, K. P. Transformation of Industrial Food Wastes into Polyhydroxyalkanoates. *Wat. Sci. Technol.* **1999,** *40* (1), 365–370.
131. Hariri Asli, K.; Nagiyev, F. B.; Haghi, A. K. Interpenetration of Two Fluids at Parallel Between Plates and Turbulent Moving in Pipe, In *Computational Methods in Applied Science and Engineering*, Nova Science Publisher: USA, 2009; Chapter 7, pp 115–128.
132. Lee, T. S.; Pejovic, S. Air Influence on Similarity of Hydraulic Transients and Vibrations. *ASME J. Fluid Eng.* **1996,** *118* (4), 706–709.
133. Hariri Asli, K.; Nagiyev, F. B.; Haghi, A. K. Some Aspects of Physical and Numerical Modeling of Water Hammer in Pipelines. *Int. J. Nonlin. Dyn. Chaos Eng. Syst.* **2010,** *60* (4), 677–701. ISSN: 1573-269X (Electronic Version), ISSN: 0924-090X (Print Version).
134. Hariri Asli, K.; Nagiyev, F. B.; Haghi, A. K. Physical Modeling of Fluid Movement in Pipelines. *Nanomaterials Yearbook*, USA, 2009.
135. Monzavi, M. T. *Wastewater Treatment*; Tehran, Iran, 1991; pp 7–162.
136. Leon Arturo, S.; Mohamed, S.; Ghidaoui, M.; Arthur, R.; Schmidt, M.; Marcelo, H.; Garc, A. M. *Godunov Type Solutions for Transient Flows in Sewers*. Illinois USA, 2005; pp 20–44.
137. Leon Arturo, S. Improved Modeling of Unsteady Free Surface, Pressurized and Mixed Flows in Storm-Sewer Systems. Submitted in Partial Fulfillment of the Requirements for the Degree of Doctor of Philosophy in Civil Engineering in the Graduate College of the University of Illinois at Urbana-Champaign, 2007; pp 57–58.
138. Kodura, A.; Weinerowska, K. The Influence of the Local Pipeline Leak on Water Hammer Properties. In *Materials of the II Polish Congress of Environmental Engineering*, Lublin, 2005; Doi, Y.; Steinbuchel, A. *Biopolymers*. Wiley-VCH: Weinheim, Germany, 2002.
139. http://www.google.com/images?q=+biocycle&hl=en&lr=&start=140&sa=N.
140. Jacquel, N.; et al. Isolation and Purification of Bacterial Poly(3-hydroxyalkanoates). *Biochem. Eng. J.* **2008,** *39* (1), 15–27. doi:10.1016/j.bej.2007.11.029.
141. Rudnik, E. Compostable Polymer Materials. Elsevier, 2008; p 21. ISBN 978-0-08-045371-2 (retrieved 10 July 2012).
142. Lamonica, M. Micromidas to Test Sludge-To-Plastic Tech. CNET, 2010 (retrieved 23 October 2015).
143. Egerton-Read, S. A New Way to Make Plastic. *Circulate* **2015** (retrieved 23 October 2015).
144. Jacquel, N.; et al. Solubility of Polyhydroxyalkanoates by Experiment and Thermodynamic Correlations. *AiChE J.* **2007,** *53* (10), 2704–2714. doi:10.1002/aic.11274.

CHAPTER 4

COMPLEX CALCULATION OF A CRITICAL PATH OF MOTION OF A CORPUSCLE TAKING INTO ACCOUNT A REGIME AND DESIGN OF THE APPARATUS

REGINA RAVILEVNA USMANOVA[1] and
GENNADY EFREMOVICH ZAIKOV[2*]

[1]*Ufa State Technical University of Aviation, Ufa 450000, Bashkortostan, Russia*

[2]*N. M. Emanuel Institute of Biochemical Physics, Russian Academy of Sciences, Moscow 119991, Russia*

[*]*Corresponding author. E-mail: GEZaikov@yahoo.com*

CONTENTS

ABSTRACT

In this chapter, flow hydrodynamics at different values of routine-design parameters are examined and discussed.

4.1 INTRODUCTION

To study vortex devices, in many research works, extensive experimental data are reported. However, many important problems of analysis and design of vortex devices have not yet found a systematic review.

Existing research in this area shows a strong sensitivity of the output characteristics of the regime and design of the device. This indicates a qualitatively different flow hydrodynamics at different values of routine-design parameters.

Thus, it is important to consider the efficiency of fluid flow and vortex devices, receipt, and compilation dependencies between regime-design parameters of the machine. Creating effective designs is the actual problem. The method of forecasting of efficiency of a gas cleaning on the basis of the analysis of hydrodynamics of gas streams is developed. Calculation of a mechanical trajectory of corpuscles gives the chance to define effect of major factors on gas-cleaning process. In the capacity of key parameter, the dimensionless group (the separation factor) is installed. The factor gives the chance to size up a critical way of a corpuscle and considers regime-design data the apparatus.

4.2 DERIVE THE EQUATION OF MOTION OF A PARTICLE

Centrifugal machines are considered as one of the most common devices for dust cleaning, due to their widespread use simplicity of design, reliability, and low capital cost.

Consider that the mechanism of dust–gas cleaning scrubber is something dynamic[1] and capture dust scrubber is based on the use of centrifugal force. Dust particle flowing at a high velocity tangentially enters the cylindrical part of the body and makes a downward spiral. The centrifugal force by the rotational motion flow causes the dust particles to move to the sides of the device (Fig. 4.1).

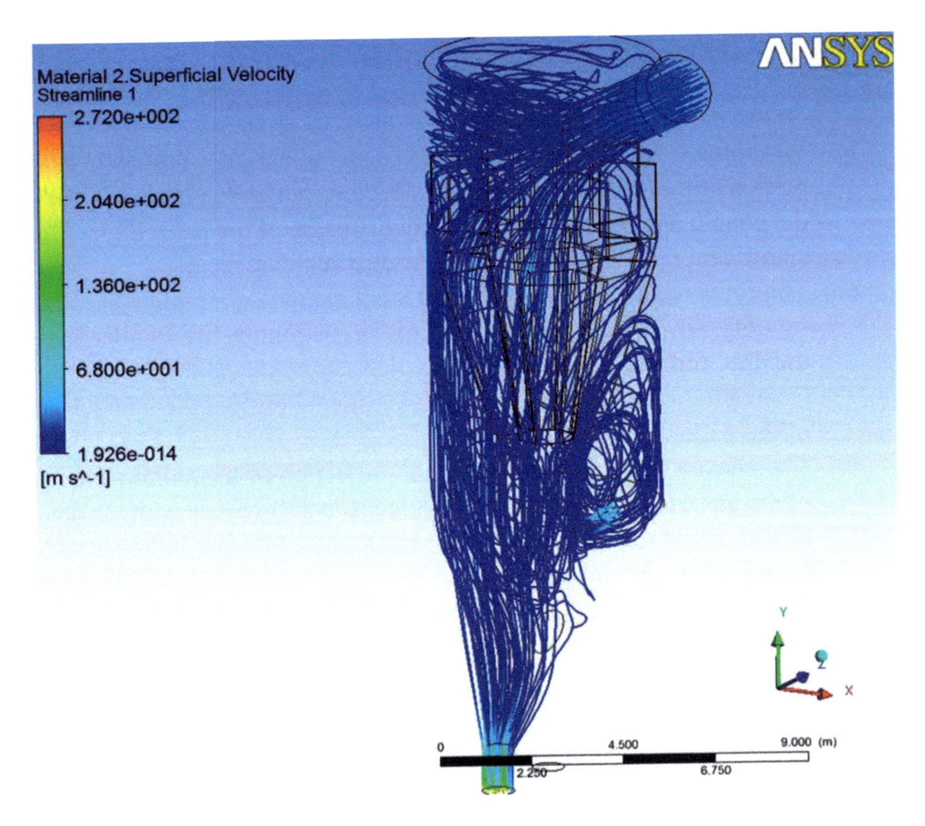

FIGURE 4.1 The trajectory of the particles in a dynamic scrubber.

When moving in a rotating curved gas flow, the dust will act under the influence of centrifugal force and resistance.

Analysis of the swirling dust and gas flow in the scrubber will be carried out under the following assumptions:

1. Gas is considered ideal incompressible fluid and, therefore, its potential movement.
2. Gas flow is axisymmetric and stationary.
3. Circumferential component of the velocity of the gas changes in law

$$w_\phi = \text{const} \cdot \sqrt{r}$$

This law is observed in the experiments[2,3] and will provide a simple solution that is convenient for the quantitative analysis of the particles.

4. Particle does not change its shape over time and diameter, nor does any crushing or coagulation happens. Deviation of the particle shape on the field is taken into account as the coefficient *K*.
5. Wrapping a strong flow of gas is viscous in nature. Turbulent fluctuations of gas are not taken into account, which is consistent with the conclusion of Ref. [4]: turbulent diffusion of the particles has no significant impact on the process of dust removal.
6. The forces that are not considered are Zhukovsky, Archimedes, and severity, since these forces by orders of magnitude are smaller than the drag force and centrifugal force.[3,5,6]
7. Concentration of dust is small, so we cannot consider the interaction of the particles.
8. The uneven distribution of the axial projection of the radial velocity of the gas can also be neglected, which is in accordance with the data of Ref. [7]. Axial component of the velocity of the particles changes little on the tube radius.

The rotation of the purified stream scrubber creates a field of inertial forces, which leads to the separation of a mixture of gases and particles. Therefore, to calculate the trajectories of the particles, it is necessary to know their equations of motion and aerodynamics of the gas flow. In accordance with the assumption of a low concentration of dust particles, the influence of them on the gas flow can be neglected. Consequently, we can consider the motion of a single particle in the velocity field of the gas flow. Therefore, the task to determine the trajectories of particles in the scrubber is decomposed into two parts:

- determination of the velocity field of the gas flow; and
- integration of the equations of motion of a particle for a calculation of the velocity field of the gas.

The assumption of axial symmetry of the problem (with the exception of the mouth) allows for the consideration of the motion of the particles using a cylindrical coordinate system.

The greatest difficulty is to capture fine dust, for which the strength of the resistance with sufficient accuracy is given by Stokes. By increasing the ratio, dust cleaning of the machine grows,[8] so the calculation parameters scrubber with low dust content (by assumption) guarantees a minimum efficiency.

4.3 CALCULATE THE TRAJECTORIES OF THE PARTICLE

To calculate the trajectories of the particles, we need to know their equations of motion. Such a problem for some particular case is solved by the author of Ref. [9].

We introduce a system of coordinates *OXYZ*. Its axis is directed along the *OZ* axis of the symmetry scrubber (Fig. 4.2). Law of motion of dust particles in the fixed coordinate system *OXYZ* can be written as follows:

$$m\frac{d\vec{v}_p}{dt} = \vec{F}_{st} \tag{4.1}$$

where m is the mass of the particle; $\vec{v}_p$ is the velocity of the particle; $\vec{F}_{st}$ is the aerodynamic force.

For the calculations necessary to present the vector equation of (4.1) motion is in scalar form. Position of the particle will be given by its cylindrical coordinates (r, φ, z). Velocity of a particle is defined by three components: U_p—tangential, V_p—radial, and W_p—axial velocity.

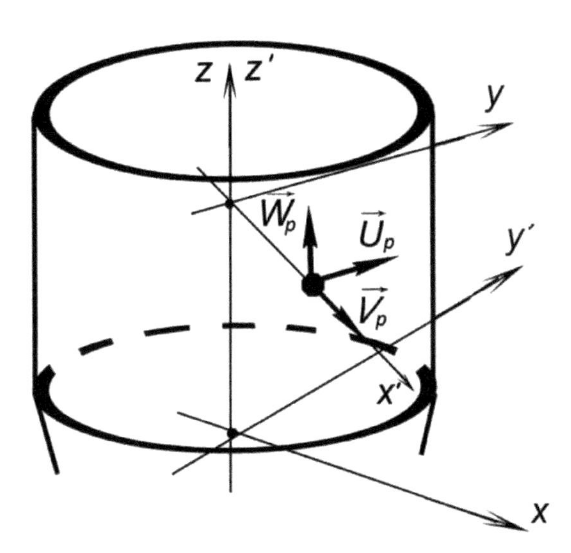

FIGURE 4.2 The velocity vector of the particle.

We take a coordinate system *O'X'Y'Z'*; let *O'X'* pass axis through the particle itself, and the axis *O'Z'* lies on the axis *OZ*. Adopted reference system moves forward along the axis *OZ* W_p speed and rotates around an angular velocity:

$$\omega(t) = \frac{U_p}{r_p} \tag{4.2}$$

The equation of motion of a particle of mass, $m = (1/6)\,\pi\rho_p d_p^3$ coordinate system $O'X'Y'Z'$ becomes:

$$m\frac{\mathrm{d}\vec{v}'_{\mathrm{p}}}{\mathrm{d}t} = \vec{F}_{\mathrm{st}} - m\vec{a}'_0 + m\left[\vec{r}'_{\mathrm{p}} \cdot \vec{\omega}\right] + m\left[\vec{r}' \cdot \vec{\omega}\right]$$
$$+m\left[\vec{\omega}\cdot\left[\vec{r}' \cdot \vec{\omega}\right]\right] + 2m\left[\vec{v}'_{\mathrm{p}} \cdot \vec{\omega}\right]$$

or

$$\frac{\mathrm{d}\vec{v}_{\mathrm{p}}}{\mathrm{d}t} = \frac{1}{m}\vec{F}_{\mathrm{st}} - \vec{a}'_0 + \left[\vec{r}'_{\mathrm{p}} \cdot \vec{\omega}\right] + \left[\vec{\omega}\cdot\left[\vec{r}'_{\mathrm{p}} \cdot \vec{\omega}\right]\right] + 2\left[\vec{v}_{\mathrm{p}} \cdot \vec{\omega}\right] \tag{4.3}$$

where $\vec{a}'_0$ is the translational acceleration vector of the reference frame; $\mathrm{d}\vec{v}_{\mathrm{p}}$ is the velocity of the particle; r'_{p} is the radius vector of the particle; $\left[\vec{r}'_{\mathrm{p}} \cdot \vec{\omega}\right]$ is the acceleration due to unevenness of rotation; $\left[\vec{\omega}\cdot\left[\vec{r}'_{\mathrm{p}} \cdot \vec{\omega}\right]\right]$ is the centrifugal acceleration; and $2\left[\vec{v}_{\mathrm{p}} \cdot \vec{\omega}\right]$ is the Coriolis acceleration.

The first term on the right-hand side of eq 4.3 is the force acting *c* gas flow on the particle and is given by Stokes:

$$F_{\mathrm{st}} = 3\pi\mu_{\mathrm{g}} d_{\mathrm{p}}\left[\vec{v}_{\mathrm{g}} - \vec{v}_{\mathrm{p}}\right] \tag{4.4}$$

μ_g is the dynamic viscosity of the gas.

The second term in (4.3) is defined as

$$\frac{\mathrm{d}W_{\mathrm{p}}}{\mathrm{d}t}\vec{e}_z = \frac{\mathrm{d}W_{\mathrm{p}}}{\mathrm{d}t}\vec{e}_{z'}$$

Convert the remaining terms:

$$\left[\vec{r}_{\mathrm{p}} \cdot \vec{\omega}\right] = \left[\vec{r}'_{\mathrm{p}} \cdot \frac{\mathrm{d}\vec{\omega}}{\mathrm{d}t}\right] = \left[\vec{r}'_{\mathrm{p}} \cdot \frac{\mathrm{d}}{\mathrm{d}t}\left(\frac{U_{\mathrm{p}}}{r_{\mathrm{p}}}\vec{e}_{z'}\right)\right]$$
$$= -r_{\mathrm{p}}\left(\frac{1}{r_{\mathrm{p}}}\frac{\mathrm{d}U_{\mathrm{p}}}{\mathrm{d}t} - \frac{U_{\mathrm{p}}}{r_{\mathrm{p}}^2}V_x\right)\vec{e}_{y'} = \left(-\frac{\mathrm{d}U_{\mathrm{p}}}{\mathrm{d}t} + \frac{U_{\mathrm{p}}V_{\mathrm{p}}}{r_{\mathrm{p}}}\right)e_y$$

$$\left[\vec{\omega}\cdot\left[r_{\mathrm{p}}^{-1} \cdot \vec{\omega}\right]\right] = \frac{U_{\mathrm{p}}^2}{r_x}\left[\vec{e}_z \cdot \left[\vec{e}_x \cdot \vec{e}_z\right]\right] = -\frac{U_{\mathrm{p}}^2}{r_{\mathrm{p}}}\left[\vec{e}_z \cdot \vec{e}_v\right] = \frac{U_{\mathrm{p}}^2}{r_{\mathrm{p}}}e_x$$

$$2\left[\vec{v}_{\mathrm{p}} \cdot \vec{\omega}\right] = 2v_{x'}\left[\vec{e}_{x'} \cdot \vec{\omega}\right] = 2v_{x'}\frac{U_x}{r_{\mathrm{p}}}\left[\vec{e}_{x'} \cdot \vec{e}_{z'}\right] = \left(-2\frac{U_{\mathrm{p}}V_{\mathrm{p}}}{r_{\mathrm{p}}}\right)e_{y'}$$

$\vec{e}_y, \vec{e}_v, \vec{e}_z$ are the vectors of the reference frame and used the fact that $\vec{r}_p = \vec{e}_x \cdot r_x \cdot v = V_p$

Substituting these expressions in the equation of motion (4.3), we get

$$m\frac{d\vec{v}'_p}{dt} = \vec{F}_{st} - m\vec{a}'_0 + m\left[\vec{r}'_p \cdot \vec{\omega}\right] + m\left[\vec{r}'_p \cdot \vec{\omega}\right] + m\left[\vec{\omega} \cdot \left[\vec{r}'_p \cdot \vec{\omega}\right]\right] + 2m\left[\vec{v}_p \cdot \vec{\omega}\right]$$

or

$$\frac{d\vec{v}_p}{dt} = \frac{1}{m}\vec{F}_{st} - \vec{a}'_0 + \left[\vec{r}'_p \cdot \vec{\omega}\right] + \left[\vec{\omega} \cdot \left[\vec{r}'_p \cdot \vec{\omega}\right]\right] + 2\left[\vec{v}_p \cdot \vec{\omega}\right]$$

We write this equation in the projections on the axes of the coordinate system $O'X'Y'Z'$:

$$\begin{cases} \dfrac{dV_{x'}}{dt} = \dfrac{1}{m}F_{stx'} + \dfrac{U_p^2}{r_p} \\ 0 = \dfrac{1}{m}F_{sty} - \dfrac{dU_p}{dt} - \dfrac{U_p V_p}{r_p} \\ 0 = \dfrac{1}{m}F_{stz} - \dfrac{dW_p}{dt} \end{cases}$$

$$\begin{cases} \dfrac{dV_p}{dt} = \dfrac{1}{m}F_{stx} + \dfrac{U_p^2}{r_p} \\ \dfrac{dU_p}{dt} = \dfrac{1}{m}F_{sty} - \dfrac{U_p V_p}{r_p} \\ \dfrac{dW_p}{dt} = \dfrac{1}{m}F_{stz} \end{cases} \quad (4.5)$$

We have the equation of motion of a particle in a rotating gas flow projected on the axis of the cylindrical coordinate system.

Substituting (4.2) and (4.4) in (4.5), we obtain the system of equations of motion of the particle:

$$\begin{cases} \dfrac{dV_p}{dt} = \dfrac{18\mu}{\rho_p d_p^2}\left(V_g - V_p\right) + \dfrac{U_p^2}{r_p} \\ \dfrac{dU_p}{dt} = \dfrac{18\mu}{\rho_{\div} d_{\div}^2}\left(U_g - U_p\right)\dfrac{U_p V_p}{r_p} \\ \dfrac{dW_p}{dt} = \dfrac{18\mu}{\rho_p d_p^2}\left(W_g - W_p\right) \end{cases} \quad (4.6)$$

4.4 OUTPUT RELATIONSHIP BETWEEN THE GEOMETRICAL AND OPERATIONAL PARAMETERS

Formal analysis of relationships that define the motion of gas and solids in the scrubber shows that the strict observance of similarity of movements in the devices of different sizes requires the preservation of four dimensionless complexes, such as

$$\mathrm{Re}_{\mathrm{d}} = \frac{wD}{\nu};\quad F_r = \frac{w^2}{Dg};\quad A_{\mathrm{r}} = \frac{\delta\rho_2}{D\rho_1};\quad \mathrm{Re}_{\delta} = \frac{v\delta}{\nu}$$

Not all of these systems are affecting the motion of dust. Experimentally found that the influence of the Froude number F_r is negligible [9] and can be neglected. It is also clear that the effect of the Reynolds number for large values is also insignificant. However, even if unchanged, the remaining two complexes still introduces significant difficulties in modeling devices.

On the other hand, there is no need for strict observance of the similarity in the trajectory of the particle in the apparatus. What is important is the end result—providing the necessary efficiency unit. To estimate the parameters that characterize the removal of particles of a given diameter, consider the approximate solution of the problem of the motion of a solid particle in a scrubber. A complete solution for a special case considered in the literature,[10] this solution can be used to obtain the dependence of the simplified model of the flow.

For the three coordinates—radial, tangential, and vertical equations of motion of a particle at a constant resistance can be written in the following form:

$$\frac{\mathrm{d}w}{\mathrm{d}t} - \frac{w_\phi^2}{r} = -\alpha(w_r - v_r)$$

$$\frac{\mathrm{d}w_z}{\mathrm{d}t} \cong -\alpha(w_z - v_z)$$

$$\frac{\mathrm{d}w_\phi}{\mathrm{d}t} + 2w_r\frac{w_\phi}{r} - \alpha(w_\phi - v_\phi)$$

where α is the factor resistance to motion of a particle divided by its mass.

$$\alpha = \frac{\mu}{K\rho\delta^2}$$

where K is the factor, which takes into account the effect of particle shape (take $K = 2$).

The axial component of the velocity of gas and particles are the same as follows from the equations of motion by neglecting gravity. Indeed, if $\mathrm{d}w_z/\mathrm{d}t = \alpha(w_z - v_z)$, then taking $w_z - v_z = \ddot{A}w_z$, we get

$$\frac{\mathrm{d}\Delta w_z}{\mathrm{d}t} = -\alpha\Delta w_z; \quad \Delta W_z = \Delta w_{z_0} e^{-\mathrm{d}t}$$

If initially

$$w_z = v_z\left(\Delta W_z = 0\right); \qquad \Delta W_t = 0, \quad W_t = \mathrm{const}$$

projection speed:

$$w_\phi = \mathrm{const}\cdot\sqrt{r}$$

Valid law $w_\varphi(r)$ may differ markedly from the accepted, but this is not essential. In this case, it only makes us enter into the calculation of average

$$\left(\frac{w_\phi^2}{r}\right)$$

Under these simplifications, the first of the equations of motion is solved in quadrature. Indeed, for now

$$\frac{\mathrm{d}w_r}{\mathrm{d}t} + aw_r = \frac{w_\phi^2}{r}$$

then with the obvious boundary condition $t = 0$, $v_r = 0$, we have

$$w_r \cong \frac{1}{\alpha}\left(\frac{w_\phi^2}{r}\right)av\left(1 - e^{-\mathrm{d}t}\right)$$

The time during which the flow passes from the blade to the swirler exit from the apparatus as well

$$t_1 = \frac{l}{w_{zav}}$$

On the other hand, knowing the law of the radial velocity, we can find the time during which the particle travels a distance of r_1 (the maximal distance from the wall) to the vessel wall (r_2).

$$r_2 - r_1 = \int_0 w_r \mathrm{d}t = \frac{w_\phi}{d_r} \int_0 \left(1 - e^{-dt}\right) \mathrm{d}t$$

$$r_2 - r_1 = \frac{v_\phi}{\alpha r_{av}} \left[t_1^l + \frac{1}{\alpha}\left(e^{-dt} - 1\right) \right]$$

Substituting in this equation as the limiting value of $t_1 = t_t$, we get

$$\frac{\alpha r_{av}\left(r_2 - r_1\right)}{v_{\phi av}^2} \geq \frac{l}{w_z} + \frac{1}{\alpha}\left(e^{\left(\mu/\rho\delta^2\right)\times\left(l/w_z\right)} - 1\right)$$

or

$$\frac{\mu}{K\rho\delta^2} \cdot \frac{r_1^2 - r_2^2}{2V_{\phi\, av}^2} \cdot \frac{w_z}{l} \geq 1 + \frac{K\rho\delta^2}{\mu} \cdot \frac{w_z}{l}\left(e^{-\left(\mu/\rho\delta^2 K\right)\times\left(l/w_t\right)} - 1\right) \tag{4.7}$$

The presented approach is based on the known dependence and model of the flow; it's different in a number of studies approach is only in the details. However, there are two sets; one of which characterizes the geometry of the device, and the other characterizes the operating data. The use of these systems simplifies the calculation and, most importantly, takes into account the influence of some key factors to the desired gas velocity and height of the apparatus $W_{Zav.}$

Dependence structure (4.7) shows the feasibility of introducing two sets, one of which

$$A = \frac{\mu l}{2\rho\delta^2 W_t K}$$

characterizes the effect of the flow regime and the particle diameter, and the other is a geometric characteristic of the device.

$$A_r = \frac{r_2}{l}\sqrt{\left(1 - r_1^{-2}\right)} ctg\,\beta \tag{4.8}$$

In (4.8), through r_1 marked relative internal radius apparatus:

$$r_1 = \frac{r_1}{r_2}$$

and β_1 is the average angle of the flow at the exit of the guide apparatus

$$tg\,\beta_1 = \frac{V_\phi}{V_z}.$$

Then (4.7) takes the simple form $A_r \geq f(A)$, where

$$f(A) = \sqrt{\frac{1}{A} + \frac{1}{2A^2}\left(l^{-2A} - 1\right)} \tag{4.9}$$

Expressed graphically in Figure 4.3, this dependence allows to determine the minimum value of the mode parameter $A_{\min}$ for the scrubber with geometrical parameters of A_p. And must take $A > A_{\min}$

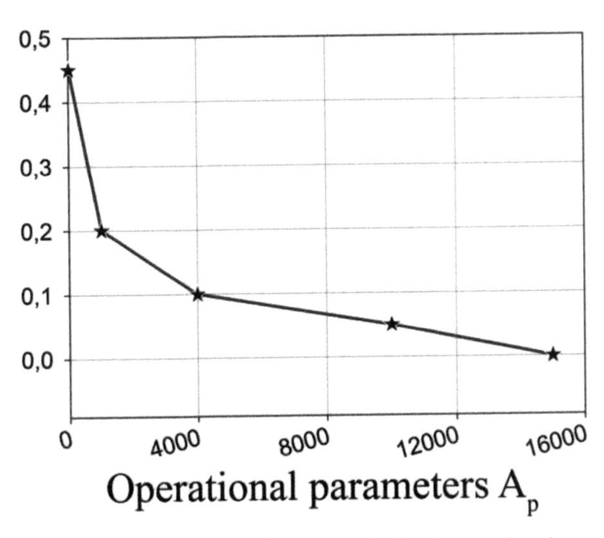

FIGURE 4.3 The relationship between the geometrical and operational complexes.

One of the important consequences of the resulting function is the relationship between the diameter of the dust particles and the axial velocity of the gas. With this machine, with data $A_{\min}$ = const and therefore

$$w_z \delta^2 = \text{const}$$

This means that reducing the particle size axis (expenditure), the rate should be increased according to the dependence

$$\frac{w_z}{w_{z_0}} = \left(\frac{\delta_0}{\delta}\right)^2$$

Unfortunately, a significant increase W_z permitted as this may lead to the capture of dust from the walls and ash. You can also change the twist angle

β and the height of the flow system, without changing the axial velocity. If the reduction of the particle diameter δ or increasing the size of the unit has increased the value of A, it must be modified accordingly A_g (Fig. 4.3), and the new value of A_r to find an angle β:

$$\frac{ctg\,\beta}{ctg\,\beta_0} = \left(\frac{A_g}{A_{g_0}}\right)$$

4.5 CONCLUSIONS

1. Creating a mathematical model of the motion of a particle of dust in the swirling flow allowed us to estimate the influence of various factors on the collection efficiency of dust in the apparatuses of the centrifugal type and create a methodology to assess the effectiveness of scrubber.
2. Identified settlement complexes, one of which characterizes the geometry of the scrubber and the other operational parameters. The use of these systems simplifies the calculation and takes into account the influence of several key factors.
3. The developed method can be used in the calculation and design of gas-cleaning devices, as constituent relations define the relationship between the technological characteristics of the dust collectors and their geometrical and operational parameters.

KEYWORDS

- **separation factor**
- **movement equation**
- **geometrical complex**
- **regime complex**
- **trajectory of particles**

REFERENCES

1. Usmanova, R. R. *Dynamic Gas Washer*: 2339435, November 2008.
2. Barahtenko, G. M.; Idelchik, I. E. *Industrial and Sanitation Gas*. Chemistry: Moscow, 1984.
3. Straus, V. *Industrial cleaning gases*. Chemistry: Moscow, 1981.
4. Shilyaev, M. I. *Aerodynamics and Heat and Mass Transfer of Gas-Dispersion Flow: Studies. Allowance*. Industrial Engineering: Tomsk, 2003.
5. Lagutkin, D. A. *Supersonic Two-Phase Currents*. The Higher School: Minsk, 2004.
6. Deutsch, M. E.; Filippov, G. A. *Gas Dynamics of Two-Phase Media*. Energy: Moscow, 1988.
7. Starchenko, A. V.; Bells, A. M. *Thermophys. Aeromech.* **1999,** *1*, 59–70.
8. Gupta, A.; Lilly, N. In *Sayred Swirling Flow Trans*; Krashennikov, S., Ed.; Mir: Moscow, 1987.
9. Tarasova, L. A. Hydraulic Calculation of the Resistance of the Vortex Unit. *Chem. Petrol. Mech. Eng.* **2004,** *2*, 11–12.
10. Bulgakov, V. K. *Finite Element Scheme of the High-Order for the Navier–Stokes equations. Modified by the SUPG-Method*; vol. 1, 2003; pp 129–132.

CHAPTER 5

THE MODERN APPROACH TO MODELING AND CALCULATION OF EFFICIENCY OF PROCESS OF A GAS CLEANING

R. R. USMANOVA[1] and G. E. ZAIKOV[2*]

[1]*Ufa State Technical University of Aviation, Ufa 450000, Bashkortostan, Russia*

[2]*N. M. Emanuel Institute of Biochemical Physics, Russian Academy of Sciences, Moscow 119991, Russia*

[*]*Corresponding author. E-mail: GEZaikov@yahoo.com*

CONTENTS

ABSTRACT

In this chapter, a new approach for modeling and calculation of efficiency of gas cleaning process is discussed.

5.1 INTRODUCTION

A wide heading of new powers in industrial production, and also the intensification of processes, as a rule involves sharp growth of the flying emissions and loading raise on air-cleaning constructions that causes one of the basic problems of an intensification of process of clearing of gases.

Working out of new mathematical model approaches for calculation of turbulent twirled currents is an important step for creation of adequate methods of calculation of the inertia apparatuses for the purpose of their optimization, technological and design data, and exclusion of expensive experimental researches.[1] Now, there were considerable changes in the areas of mathematical modeling connected with application of computing production engineering and software packages that gives the chance to predict integrated characteristics of apparatuses already on a design stage. It is possible to provide such constructive solutions of separate knots of the apparatus which will allow to raise efficiency of a gas cleaning considerably.

Mathematical models of a current of multiphase medium should predict as much as possible precisely; on the one hand, gas cleaning parameters at inoculation of any parameter, and on the other hand, to show possible ways of an intensification of process of separation. For this purpose, the model should provide characteristics of all prominent aspects of a current (boundary conditions, physical parameters of multiphase medium, turbulence, and geometrical characteristics) with possibility of the solution of such equations.

Modeling of a current of a dispersoid in the inertia apparatuses becomes complicated determinirovanno-stochastic character traffic of corpuscles in the turbulent twirled stream which becomes complicated interacting of corpuscles with each other and with apparatus walls, complexity of the task of entrance conditions, and modification of corpuscles as a result of crushing and concretion.

There are good enough bases for the use of methods of modeling in the research of streams of suspended matters when corpuscles are so small that they traffic the time bulk which is defined by the Stokes law.

Contrary to it for a case of too large corpuscles perspective methods in this direction is absent. Last, refers to those cases, when concentration of corpuscles large and is great a collision of frequency of corpuscles. Agglomeration and concretion of corpuscles are also especially difficult factor for modeling.

Calculation of the turbulent twirled currents at creation of adequate methods of calculation of the inertia apparatuses should yield enough results in a wide range of variables and combine it with simple and inexpensive laboratory researches of characteristics of the dust, which results can be used in the capacity of an input information. The best method of calculation can be chosen only after accumulation of considerable experience; noncoincidence of results of the first numerical calculations with equipment characteristics is thus inevitable.

5.2 APPROACH SAMPLING TO CALCULATION OF MULTIPHASE STREAMS

The boundary conditions, fulfilling to transport equations in areas $R > r$, should consider taking into account the equality of particle fluxes on radius r and also equalities of concentration of corpuscles on radius R.

The analytical approach of the majority of researchers[2] to the description of hydrodynamics of the inertia apparatuses is based on system of the equations of the Navier–Stokes added with continuity equations of the installed axisymmetric twirled gas stream.

$$\frac{\partial}{\partial t}(\rho \upsilon_i) + \frac{\partial}{\partial q_j}(\rho \upsilon_i \upsilon_j) = -\frac{\partial}{\partial q_i} + \frac{\partial}{\partial q_j}\left[\mu\left(\frac{\partial \upsilon_i}{\partial q} + \frac{\partial \upsilon_j}{\partial q_j}\right)\right]$$
$$\frac{\partial \rho}{\partial t} + \frac{\partial}{\partial q_j}(\rho \upsilon_i) = 0 \tag{5.1}$$

The solution of system of the equations of Navier–Stokes is difficult that explains the necessity of adoption of various, not absolutely, correct assumptions that reduce adequacy of offered analytical descriptions to a real flow pattern in the inertia apparatuses and, finally, lead to essential divergences of results of scalings with empirical data.

The great interest is represented by working out of effective numerical methods of the solution of the multidimensional equations of purely hyperbolic type or the parabolic equations containing a hyperbolic part. Such mathematical models many present are space-nonstationary problems of mechanics of multiphase currents. Construction of a computational algorithm for the specified sort of problems represents rather a challenge which usually dares stage by stage.

General provisions of mathematical production engineering and construction that such circuit designs consist are follows:

1. *A hyperbola in the equations.* The hyperbolic part of the parabolic equations is most difficult in the computing plot as it is a source of origination of big gradients in narrow zones (in purely hyperbolic problems, there are ruptures of solutions). Effective methods of the solution of the hyperbolic equations have, thus, wide sphere of application. Really, splitting methods on physical processes allow to induct formally enough and effectively practically any earlier developed (for the solution of the hyperbolic equations) method into the general algorithm of the solution of the parabolic problem containing a hyperbolic part.
2. *Multidimensionality.* The most universal and effective approaches of the solution of the multidimensional parabolic and hyperbolic equations (obvious and implicit circuit designs) are various methods of splitting on the space variables. Janenko, Marchuka, and Godunova's known circuit designs are referred here. Using similar approaches, it is possible to generalize naturally almost any one-dimensional numerical algorithm on a multidimensional case. The initial problem, thus, becomes essentially simpler and reduced to search of comprehensible one-dimensional circuit designs.
3. *A method of uncertain factors.* Use of a method of uncertain factors (with introduction of linear spaces of these factors) at an analysis stage of different circuit designs for the elementary transport equations is rather constructive. This approach allows to build for any grid templates an assimilation of all different circuit designs with positive approximation (monotonous or majorant circuit designs), playing the important role in calculus mathematics. In most general cases, it is possible to prove the absence of different circuit designs with positive approximation and with a higher order of accuracy on solutions of the initial equations. On the basis of the same approach for the most used grid templates (both obvious and implicit), the Setochno-characteristic circuit designs of the second and third order of accuracy closest in space, injected into consideration of factors to circuit designs with positive approximation, are built. In particular,

new, more effective modifications of different circuit designs widely applied in computing practice have been gained. The given approach is rather a perspective and at construction of so-called hybrid circuits for numerical solution.

4. *Verification of results of numerical modeling.* This problem gets an especial urgency, so far, as concerned with direct numerical modeling of the difficult (complex) phenomena or processes. In this case detection, recognition and identification of new effects appear as the essential moment. For such problems, it is difficult enough to build formal theories and to apply classical mathematical methods. However, now principles of rational numerical modeling allow to promote essentially in the field of construction of the systems simulating such phenomena that gives a basis for progress in creation of simulators and calculation of currents' multiphase.

5.3 WE DEFINE CONDITIONS ON BOUNDARY LINES

The boundary conditions, fulfilling to transport equations in areas $R > r$, should register taking into account equality of particle fluxes on radius r, and also equalities of concentration of corpuscles on radius R. Near to a bounding surface, tangential speeds of gas decrease and accept a zero value on a surface. A centrifugal force acting on small corpuscles also decreases and accepts a zero value on the wall. Corpuscles near to a wall take great interest in turbulent pulsations and depart from a wall, and by a centrifugal force are refunded to a wall. Thus, near to a motionless surface of a corpuscle are in a dynamic equilibrium, on boundary line a stream—a firm wall, carrying over of corpuscles is absent, and the total particle flux at the expense of a centrifugal force and turbulent carrying over should be equal to null. On a gas bottle axis, owing to symmetry of a current, the derivative on radius from concentration of corpuscles is equal to null. In a settlement grid, the firm wall is represented by boundary lines Γ_1, Γ_3, Γ_4, Γ_5 (Fig. 5.1).

As lines Γ_1–Γ_8 and Γ_3–Γ_4 are stream lines on these walls, current function ψ can accept any constant value. On character of a current for Γ_1, Γ_5 current function $\psi = 0$; for Γ_3, Γ_3, current function $\psi = \psi_{max}$. Boundary lines Γ_4 and Γ_6 represent target cross-section or a penetrable wall. If on Γ_1, Γ_3, Γ_8, the sticking condition is satisfied: if $\upsilon_z = \upsilon_\varphi = \upsilon_r$ on Γ_4 and Γ_6 change of speed can be set as some function $f(r)$, then $\upsilon_z = \upsilon_z(r)$.

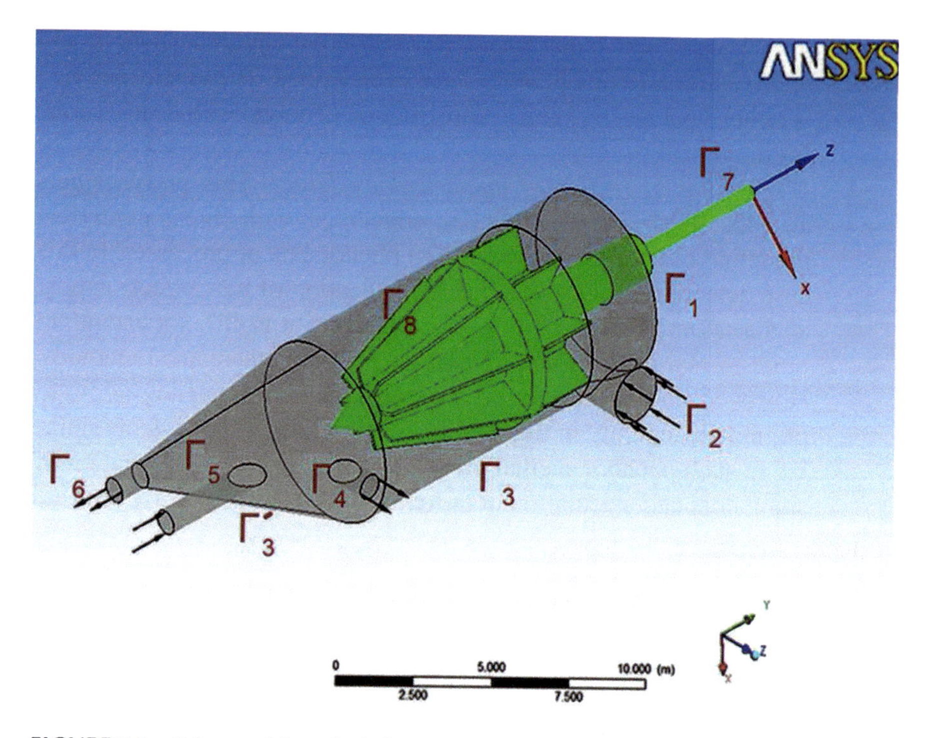

FIGURE 5.1 Scheme of the calculation of the boundary conditions.

Boundary line Γ_7 represents a symmetry axis. For Γ_7, it is $\upsilon_r, \upsilon_\varphi = 0$; therefore $(d\upsilon_r/dz) = 0$. Account speed is symmetric concerning an axis $(d\upsilon_z/dz) = 0$, when $\omega_z = 0$. On target boundary line Γ_4, the most reliable way of the task of boundary conditions is full definiteness of values ψ, ω, υ and statement of boundary conditions of Neumann is applied:

$$\frac{d\psi}{dz} = 0; \quad \frac{d\omega}{dz} = 0; \quad \psi_\Gamma = \psi_{\Gamma-1}; \quad \omega_\Gamma = \omega_{\Gamma-1}$$
$$\frac{d\upsilon_r}{dz} = \frac{d\omega}{dz} = 0; \quad \frac{d^2\psi}{dz^2} = 0 \tag{5.2}$$

These conditions have the second order of accuracy. The numerical solution of model has been spent for one mid-flight pass from entrance cross-section of a swept volume to the day off by means of integration to neighborhoods of each knot is the final-difference grid by which all space of a gas bottle (Fig. 5.2), unknown values of speed and pressure are found in knots of this grid.

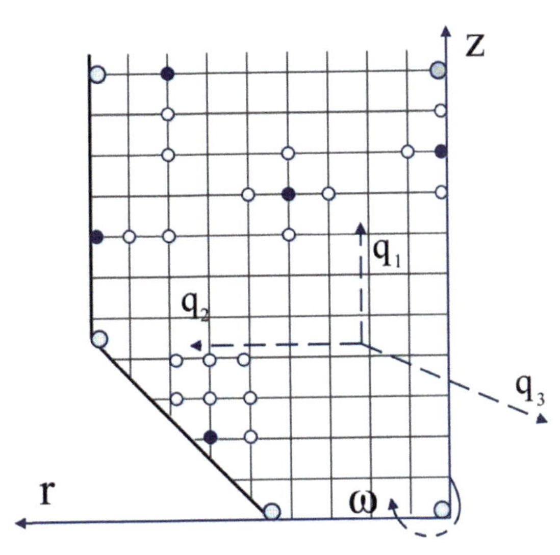

FIGURE 5.2 Computational grid.

Each knot of a grid is defined by values of projections of speed of a stream: the radial υ_r, tangential υ_φ, and axial υ_z. Transitions between knots are carried out in steps by replacement of one value of speed with another or a finding of intermediate values between knots by means of interpolation. At such statement of a regional problem, the sticking condition was realized on each time step and, is analogous to conditions for functions ψ and ω, was put on different boundary lines. This results from the fact that use of conditions of sticking on the same boundary line changes a modeling problem, and at its numerical solution, there can be an accuracy decrease.

On boundary lines of settlement area for each knot of a grid, it is possible to write down:

$$\frac{d}{dr}\left(r\frac{d\phi}{dr}\right)+\frac{d}{dr}\left(r\frac{d\phi}{dz}\right)=0 \tag{5.3}$$

For the function addressing in a zero on boundary line of a grid, we will compute a scalar product and norms:

$$(y,v)=\sum_{i,j=1}^{N} h_r h_z y_i, v_{i,j}, \quad \|y\|=\sqrt{(y,y)}$$

Let's execute approximation of eq 5.3 on a grid step h, having made replacement of derivatives with the following function:

$$\frac{i+1/2}{h_r}\cdot\phi_{i+I,j}+\frac{i\cdot h_r}{h_z^2}\cdot\phi_{i,j+I}-\left(\frac{2\cdot i}{h_r}+\frac{2\cdot i\cdot h_r}{h_z^2}\right)\cdot\phi_{i,j}$$
$$+\frac{i-1/2}{h_r}\cdot\phi_{i-I,j}+\frac{i\cdot h_r}{h_z^2}\cdot\phi_{i,j-I}=0 \tag{5.4}$$

Let's inject a designation:

$$(Ay)_{ij}=-\frac{i+\frac{1}{2}}{h_r}y_{i+I,j}-\frac{i\cdot h_r}{h_z^2}y_{i,j+I}+\left(\frac{2i}{h_r}+\frac{2i\cdot h_r}{h_z^2}\right)\cdot y_{i,j}$$
$$-\frac{i-\frac{1}{2}}{h_r}y_{i-I,j}-\frac{i\cdot h_r}{h_z^2}y_{i,j-I} \tag{5.5}$$

Then, eq 5.3 will register as

$$Ay=f;\quad (Ay,y)\geq 0;\quad (Ay,v)=(y,Av)$$

Over the range sizes of $0\leq r_{\min}\leq r\leq r_{\max}$, boundary line of parameter A will be in limits $\gamma_1\leq A\leq\gamma_2$, forming system of the linear equations for each knot of a grid

$$\gamma_1=2\cdot r_{\min}\left(\frac{4}{h_z^2}\sin^2\frac{\pi\cdot h_r}{2l_r}+\frac{4}{h_z^2}\sin^2\frac{\pi\cdot h_r}{2l_z}\right),$$
$$\gamma_2=2\cdot r_{\max}\left(\frac{4}{h_z^2}\cos^2\frac{\pi\cdot h_r}{2l_r}+\frac{4}{h_z^2}\cos^2\frac{\pi\cdot h_z}{2l_z}\right) \tag{5.6}$$

At $r_{\min}=h_r$, $r_{\max}=l_r$, where l is the length of settlement area and h is a grid step.

The system of the linear is final-difference equations which solution allows to define value of potential in grid knots is gained. By results of scalings, pictures of stream-lines and profiles of speed in various cross-sections of a stream were under construction.

It is installed that at increase in a Reynolds number, the current structure changes from layered to complicated by the developed secondary whirlwinds. Three types of a current are qualitatively discriminated: a layered current, a current with peripheral a whirlwind, a current with peripheral and the peripheral bound vortexes. At considerable intensity of process,

$Re = 6 \times 10^4$, the forming has big tangential speeds that leads to the considerable pressure gradients calling a reverse flow along an axis, reducing efficiency of separation. Character and intensity of return currents depends on intensity and character of falling twistings. This leading-out is necessary for considering in practice and appropriate way to organize hydrodynamics of streams in the apparatus.

5.4 VISUALIZATION OF RESULTS OF CALCULATION

The analysis of the gained velocity profiles on (Fig. 5.3) allows to reveal three characteristic areas on an apparatus axis: area of formation of a gas stream, area of a stable stream, and damping area.

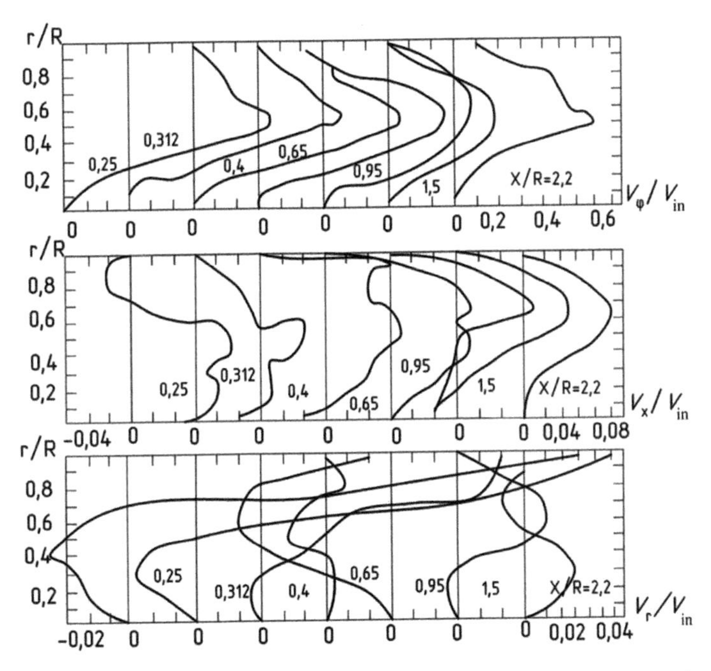

FIGURE 5.3 Projections of the tangential, axial, and radial velocities along the machine sections. $X/R = 0.25, 0.4, 0.65, 0.95, 1.5, 2.2$; the parameter values: $V_r/V_{in} = 0.05$; $V_\varphi/V_{in} = 1.8$; $Re = 6 \times 10^4$

In the first area, making speeds are stable; there is their formation. In stable area, profiles of speed are similar on apparatus altitude. Practically in all cross-sections, it is possible to gate out two zones: peripheral (a zone of a free whirlwind) and axial (a zone of the forced whirlwind). For tangential speed drift of a maximum from periphery to the center and abbreviation of a zone of the forced whirlwind is characteristic. Tangential speed much more axial in peripheral and quasipotential zones, and in the field of an axis practically one order with it. The axial component practically does not change the profile, its maximum is near to an apparatus wall.

Characteristic pictures of fields of speed for settlement currents are presented in Figure 5.4.

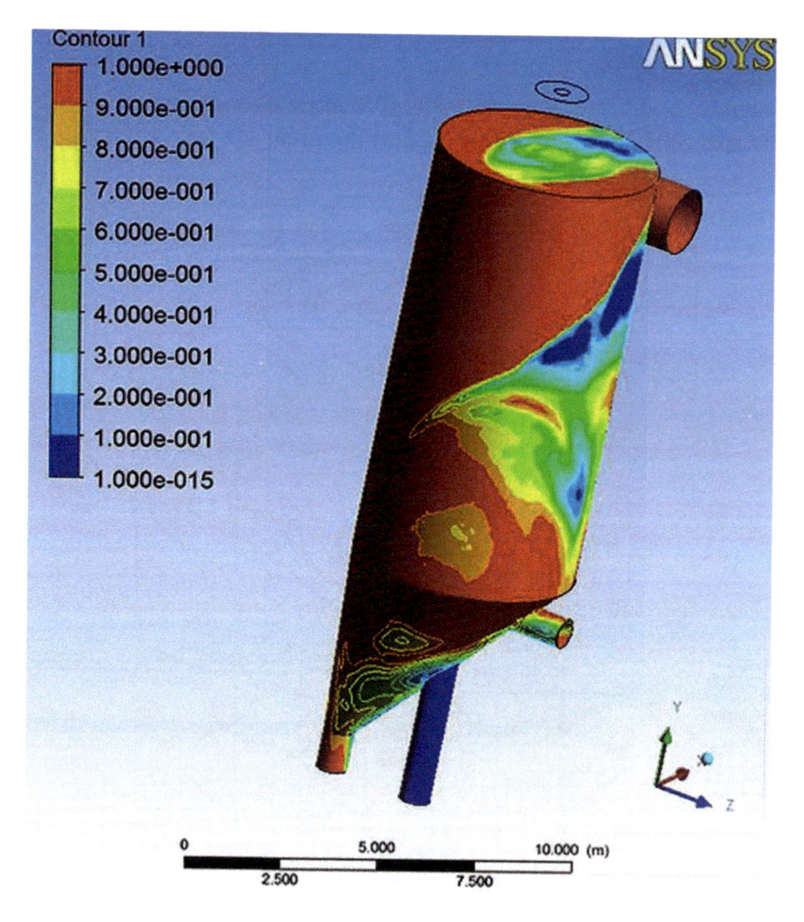

FIGURE 5.4 The velocity field of gas.

The weak twisting $\omega_0 = 15\ s^{-1}$ slightly influences character of a current which is close to a usual tangential twisting of a stream; the increase in a twisting to $\omega_0 = 50\ s^{-1}$ leads to formation of zones of recirculation near to twirled vortex generator; fields of speed form a bend. Extent of such zone considerably increases with increase in a Reynolds number and at the further increase in a twisting to $\omega_0 = 100\ s^{-1}$ leads to center displacement recirculation to area and an elongation of a zone of a counter-flow in axial and radial directions.

The qualitative picture gained theoretically will well be coordinated with known experimental data, a divergence, depending on a flow, fluctuates over the range from 4% to 7%. The gained results are used in the equation of traffic of corpuscles for forecasting of efficiency of process of a gas cleaning.

5.5 CALCULATION OF TRAFFIC OF CORPUSCLES IN THE TWIRLED STREAM

For calculation of traffic of corpuscles in the twirled stream, the mathematical model of process of separation of dispersion particles on drops of an irrigating liquid has been made. The following forces acting on a corpuscle were considered: a gravity, force of Koriolis, force of an aerodynamic resistance of medium, and a centrifugal force[4]:

$$\begin{cases} \dfrac{dU_x}{dt} = \dfrac{3}{4} \cdot \dfrac{\rho_g}{\rho_p} \cdot \dfrac{\xi_p}{d_p} \cdot U_0 \cdot \left(V_x - U_x\right) + g; \\ U_x = \dfrac{dx}{dt} \\ \dfrac{dU_\phi}{dt} = \dfrac{3}{4} \cdot \dfrac{\rho_g}{\rho_p} \cdot \dfrac{\xi_p}{d_p} \cdot U_0 \cdot \left(V_\phi - U_\phi\right) + \dfrac{\omega \cdot r}{t}; \\ U_\phi = \omega \cdot r = r\dfrac{d\phi}{dt} \\ \dfrac{dU_r}{dt} = \dfrac{3}{4} \cdot \dfrac{\rho_g}{\rho_p} \cdot \dfrac{\xi_p}{d_p} \cdot U_0 \cdot \left(V_r - U_r\right) - \dfrac{\omega \cdot r}{t} + \omega^2 r; \\ U_r = \dfrac{dr}{dt} \end{cases} \tag{5.7}$$

From the analysis of the gained system of the equations, obviously the corpuscle path depends on following factors: d_p—diameter of corpuscles; ρ_p—corpuscle density; μ_g—dynamic viscosity of gas; r—a characteristic size (geometry) of the apparatus; U_0—initial tangential speed of a stream.

The solution of system (5.7) at initial conditions:

$r'_p\big|_{t=0} = r'_0, z'_p\big|_{t=0} = z'_0, \phi'_p\big|_{t=0} = 0$ for cylindrical coordinates, $U'_x\big|_{t=0} U'_{g_0}, U'_\phi\big|_{t=0} = U'_{g_0}, U'_g\big|_{t=0} = U'_{g_0}$ for definition of speed allow to build mechanical trajectories of corpuscles (Fig. 5.5).

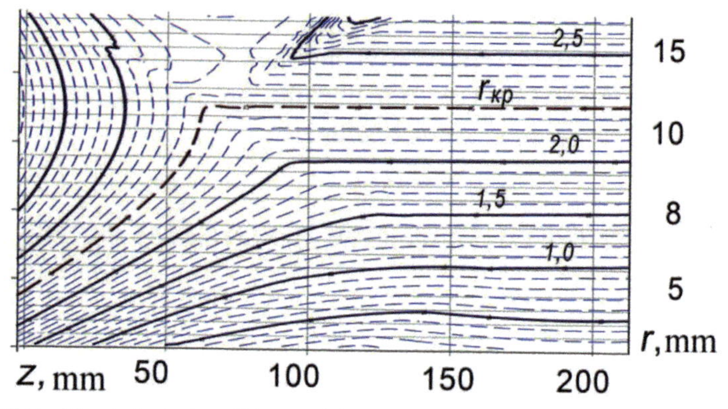

FIGURE 5.5 The trajectories of the particles $d \leq 5$ mic. (---) critical; (—) equilibrium.

The greatest agency of a twisting of a stream on a path of corpuscles appears on distance $r \leq 5D$ from vortex generator, at the further increase of r, the equilibrium path practically does not vary. And, the extent of agency of a twisting for corpuscles in diameter $d \leq 5$ mic essentially depends on their initial rule: the corpuscles which are more close to walls are more poorly subject to twisting agency, and the path of their traffic is close to the critical. Numerical experiment was spent for the most adverse case of an initial arrangement of corpuscles of the set diameter in the upstream end that allows to predict trapping of a corpuscle of larger size. Corpuscles, moving on a critical path, have a cut-off diameter which is possible for conducting of process of separation in a gas bottle with probability of 100%.

5.6 CONCLUSIONS

1. The complex of researches of aerohydrodynamic parameters of the inertia apparatuses with active hydrodynamics which has allowed to size up character of interconnection of the basic aerohydrodynamic parameters from design features of the apparatus is executed.
2. The analytical model of a current of the gas is developed in the medium, allowing to count distributions of all a component of speed U'_φ, U'_r, U'_x, and also current functions $\psi(r,z)$, and to build a

characteristic flow pattern of a current in program complex ANSYS CFX.

3. The calculations of currents defined by a regional problem were spent for values of Reynolds number, $Re = 1 \times 10^2 – 60 \times 10^4$. The analysis of the gained profiles of speed allows to reveal three characteristic areas on an apparatus axis: area of formation of a gas stream, area of a stable stream, and damping area.
4. Mechanical trajectories of corpuscles of the set diameter in the twirled stream are gained. The numerical algorithm is developed for traffic calculation of the gas streams which allows to predict trapping of corpuscles moving on critical paths.

KEYWORDS

- **mathematical model**
- **boundary conditions**
- **the inertial devices**
- **efficiency gas cleaning**
- **computational grid**
- **gas stream**
- **vortex generator**

REFERENCES

1. Varaksin, A. J. *Turbulent Flows of Gas with Firm Corpuscles*. Russian Academy of Sciences: Moscow, 2003.
2. Kochevsky, A. N.; Nenja, V. G. Modern Approach to Modelling and Calculation of Currents of a Liquid in Blade Hydrocars. *Bull. SumGu* **2003,** *13* (59), 195–210.
3. Usmanova, R. R.; Zaikov, G. E. Sampling of Boundary Conditions to Calculation of Parameters of an Eddy Flow Gas-dispersed streams. *Encycl. Eng.* **2015,** *3*, 36–42.
4. Usmanova, R. R. *Dynamic Gas Washer*. 2339435. November 2008.
5. Wilcox, D. C. *Turbulence Modeling for CFD*. DCW Industries: Canada, California, USA, 1993.
6. Menter, F. R.; Esch, T. *Advanced Turbulence Modelling in CFX. CFX Update* **2002,** *20*, 4–5.

CHAPTER 6

NUMERICAL MODELING AND VISUALIZATION OF TRAFFIC OF DISPERSION PARTICLES IN THE APPARATUS

R. R. USMANOVA[1] and G. E. ZAIKOV[2]

[1]*Ufa State Technical University of Aviation, Ufa 450000, Bashkortostan, Russia*

[2]*N. M. Emanuel Institute of Biochemical Physics, Russian Academy of Sciences, Moscow 119991, Russia*

Corresponding author. E-mail: GEZaikov@yahoo.com

CONTENTS

ABSTRACT

In this chapter, numerical modeling and visualization of traffic of dispersion particles in the apparatus, with particular applications in chemical industry, are reported.

6.1 INTRODUCTION

In the modern chemical industry, various devices are applied to conducting of hydrodynamic processes with direct contact of phases.

From modern apparatuses for gas-clearing cyclone separators, separators, scrubbers, whirlwind apparatuses are most popular. Apparatuses differ from each other in the way of formation of the twirled streams and a design of knot of a phase separation. The basic deficiencies of known devices are low efficiency of trapping of finely dispersed corpuscles, secondary ablation of a dispersoid, and a high water resistance.

Gas bottles of new generation in which low power expenses for a gas cleaning and operate reliability combine with high efficiency of process of clearing of gas are necessary.

Now, there were considerable changes in the areas of mathematical modeling connected with application of computing production engineering that gives the chance to predict hydrodynamic characteristics of apparatuses already on a design stage.

6.2 NUMERICAL MODELING OF TRAFFIC OF DISPERSION PARTICLES IN THE APPARATUS

Now essential progress in creation of simulators and calculation of currents of multiphase medium has been attained. Calculation can be carried out with high reliability of the gained results; therefore, the necessary volume of experiment in many cases is reduced to a minimum. Unlike experiment, numerical modeling allows to vary a row of significant parameters of a problem (viscosity, metering characteristics, speed of twirl) which essential impact on formation and behavior of the twirled currents make.

Analytical and numerical methods of calculation can be applied to the description of hydrodynamic characteristics of traffic of streams in working space of a gas bottle.[1]

The analytical approach of the majority of researchers[2,3] to the description of hydrodynamics of the inertia apparatuses is based on system of the equations of the Navier–Stokes added with continuity equations of the installed axisymmetric twirled gas stream. The solution of system of the equations of Navier–Stokes is difficult in a numerical aspect. Assumptions reduce adequacy of a real flow pattern in and lead to essential divergences of results of scalings with empirical data.

The description of traffic of the twirled stream is based on an analytical method on one of following approaches[3]:

1. *Hyperbole of the equations.* The hyperbolic part of the parabolic equations is most difficult in the computing plot as is a source of origination of the big gradients in narrow zones.
2. *Multidimensionality.* The most universal and effective approaches of the solution of the multidimensional parabolic and hyperbolic equations (obvious and implicit circuit designs) are various methods of splitting on the space variables.
3. *A method of uncertain factors.* This approach allows to build for any grid templates all assemblage of difference circuit designs with the positive approximation, playing the important role in calculus mathematics.
4. *Verification of results of numerical modeling.* This problem gets an especial urgency, so far as concerns direct numerical modeling of the difficult (complex) phenomena or processes.

For modeling conducting to solve the specified equations, it is convenient.

Numerical methods (discrete element method) are applied instead of the continuous solution for discrete set of required values in a certain place (a mesh, grid knot) and spaces (at a stationary regime of traffic of streams).

As a result, the mathematical problem of the solution of system of the differential or integrated equations can be reduced to a problem of the solution of system of the algebraic equations.

In practice, various models of the numerical solution of the classical equations of hydrodynamics for turbulent flows, which with that or other success are used in various cases are used, have merits and demerits[3]: direct numerical simulation, Reynolds-averaged Navier–Stokes, large Eddy simulation.

For the task in view, solution in-process used the right-angled, adaptive, locally comminuted is final-volume grid with a size of 1 mesh of an order of 2×10^{-3} mm. The grid circuit design is presented in Figure 6.1.

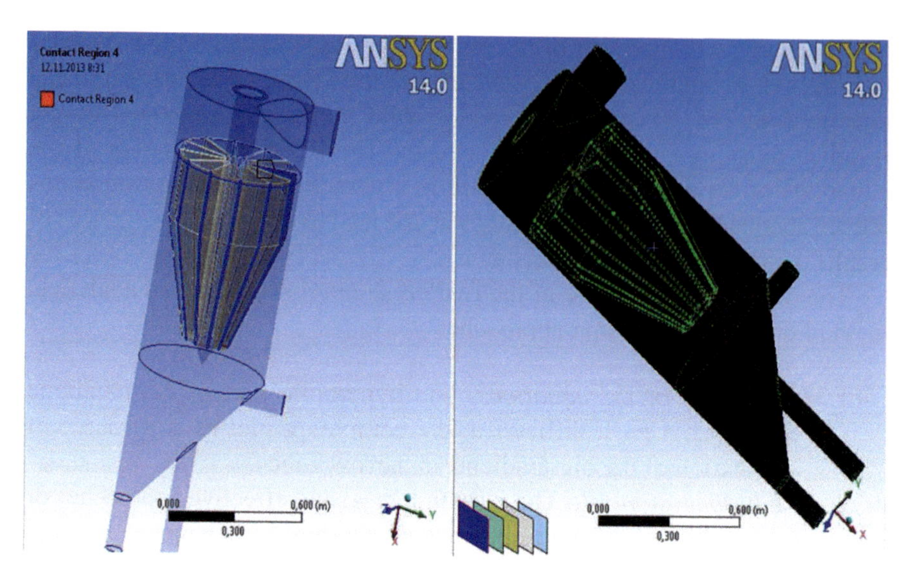

FIGURE 6.1 Typical rated area of a gas bottle.[4]

Each knot of a grid is defined by values of projections of speed of a stream: the radial υ_r, tangential υ_φ, axial υ_z. Transitions between knots are carried out in steps by replacement of one value of speed with another or a finding of intermediate values between knots by means of interpolation.

Adaptation of the first level has been executed on following surfaces: a conic surface of the case of the apparatus, a wall of a tangential connecting pipe of feeding into of gas, entries of an axial flow of a liquid, and a peripheral flow sludge. Adaptation of the second level was spent on surfaces twirled swirl vane in areas of blade passages.

The numerical solution of model has been spent from entrance cross-section of a swept volume to the day off by a path of integration to neighborhoods' of each knot finite difference grid, by which all space of a gas bottle is covered (Fig. 6.2), unknown values of speed and pressure are defined in knots of this grid.

At such statement of a regional problem, the sticking condition was realized on each time step and, is analogous to conditions for functions ψ and ω, was put on different boundary lines. This results from the fact that use of conditions of sticking on the same boundary line changes a modeling problem, and at its numerical solution, there can be an accuracy decrease.

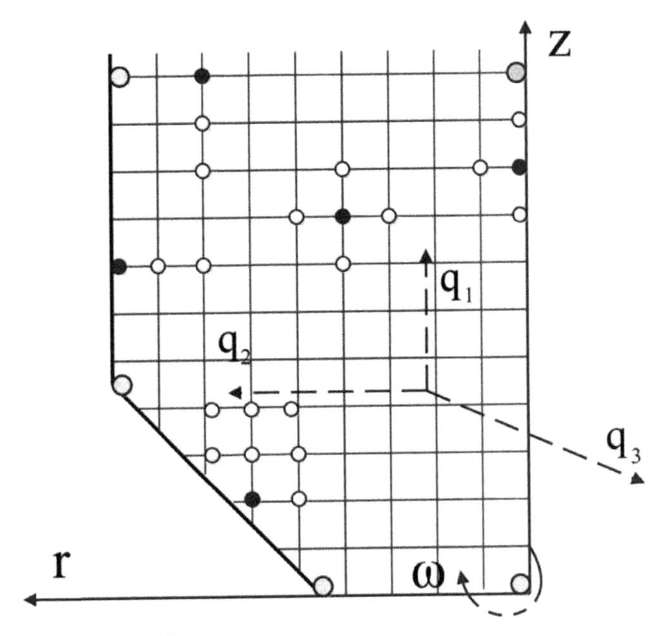

FIGURE 6.2 The rated grid.

On boundary lines of settlement area for each knot of a grid, it is possible to write down:

$$\frac{d}{dr}\left(r\frac{d\phi}{dr}\right)+\frac{d}{dr}\left(r\frac{d\phi}{dz}\right)=0 \tag{6.1}$$

For the function addressing in a zero on boundary line of a grid, we will compute a scalar product and norms:

$$(y,v)=\sum_{i,j=1}^{N} h_r h_z y_i, v_{i,j}, \quad \|y\|=\sqrt{(y,y)}$$

Let's execute approximation of eq 6.1 on a grid step h, having made replacement of derivatives with the following function:

Let's inject a designation:

$$\begin{aligned}&\frac{i+1/2}{h_r}\cdot\phi_{i+I,j}+\frac{i\cdot h_r}{h_z^2}\cdot\phi_{i,j+I}-\left(\frac{2\cdot i}{h_r}+\frac{2\cdot i\cdot h_r}{h_z^2}\right)\cdot\phi_{i,j}\\&+\frac{i-1/2}{h_r}\cdot\phi_{i-I,j}+\frac{i\cdot h_r}{h_z^2}\cdot\phi_{i,j-I}=0\end{aligned} \tag{6.2}$$

register:

$$(Ay)_{ij} = -\frac{i+1/2}{h_r} y_{i+I,j} - \frac{i \cdot h_r}{h_z^2} \cdot y_{i,j+I} + \left(\frac{2 \cdot i}{h_r} + \frac{2 \cdot i \cdot h_r}{h_z^2} \right) \cdot y_{i,j}$$
$$- \frac{i-1/2}{h_r} \cdot y_{i-I,j} - \frac{i \cdot h_r}{h_z^2} \cdot y_{i,j-I} \tag{6.3}$$

Then, eq 6.1 will register:

$$Ay = f; \quad (Ay, y) \geq 0; \quad (Ay, v) = (y, Av)$$

Over the range size $0 \leq r_{min} \leq r \leq r_{max}$, parameter A will be in limits $\gamma_1 \leq A \leq \gamma_2$, forming system of the linear equations for each knot of a grid:

$$\gamma_1 = 2 \cdot r_{min} \left(\frac{4}{h_z^2} \sin^2 \frac{\pi \cdot h_r}{2l_r} + \frac{4}{h_z^2} \sin^2 \frac{\pi \cdot h_r}{2l_z} \right),$$
$$\gamma_2 = 2 \cdot r_{max} \left(\frac{4}{h_z^2} \cos^2 \frac{\pi \cdot h_r}{2l_r} + \frac{4}{h_z^2} \cos^2 \frac{\pi \cdot h_z}{2l_z} \right) \tag{6.4}$$

At $r_{min} = h_r$, $r_{max} = l_r$, where l is the length of settlement area; h is a grid step.

According to Figure 6.3, the system of the linear is finite difference equations which allow to define value of potential in grid knots.

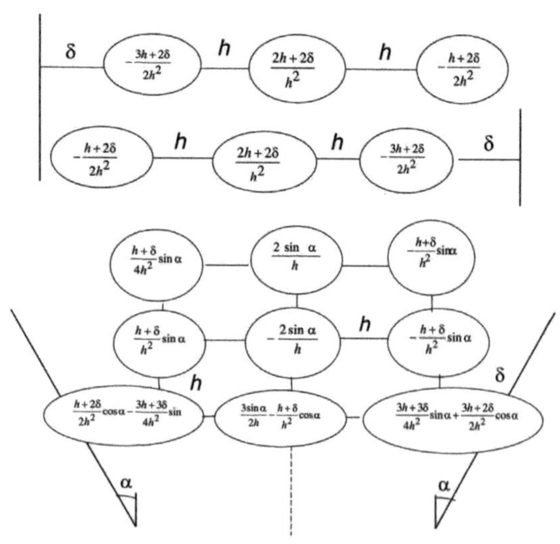

FIGURE 6.3 Knots of a settlement grid to approximation of boundary conditions.

The results of scalings pictures of stream-lines and profiles of speed in various cross-sections of a stream were under construction.

6.3 VISUALIZATION AND THE ANALYSIS OF RESULTS OF CALCULATION

Process of calculation of a current in ANSYS CFX is carried out before achievement of the set convergence criterion. It was visualized that a time history of a picture of a current with conservation of coordinates of all knots of a settlement grid and values of key parameters of a current in these knots. The integrated parameters of calculation typical for dedusters have been also gained: a water resistance, an input, efficiency of clearing, a stream twisting. As a result of numerical experiment, distributions of static pressure of a gas stream in all cross-sections of settlement space that has allowed to size up an apparatus water resistance have been gained. Distribution of static pressure is presented in Figure 6.4.

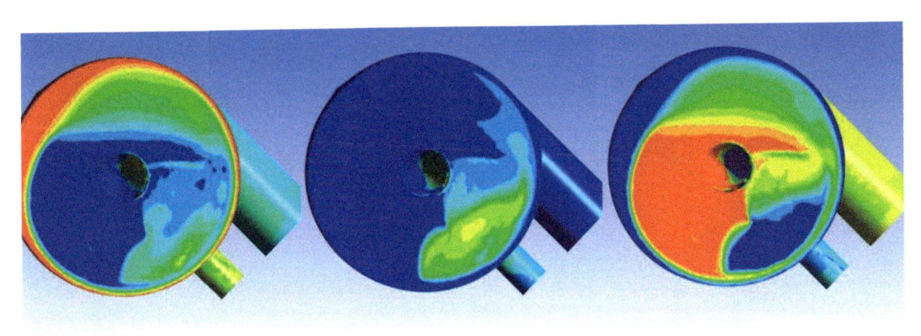

FIGURE 6.4 The static component of pressure: a cross-section 0–1; 0–2; 0–3.

It is installed that in brake-off zones, considerable pressure decreases both in comparison with a main stream, and in a zone of blades, vortex generator is observed. Irregularity of static making pressures in a gas bottle has a reducing effect on efficiency of clearing. By comparison to empirical data on separation efficiency, it is installed that decrease in efficiency of separation does not exceed 3%, though on level of irregularity of a pressure pattern, a difference more considerable. It is possible to explain it that irregularity of pressure is compensated by positive agency of zones of the brake-off promoting separation of small impurity from main current in a zone of a rarefaction and their removal on a helicoid path from working space, and further on walls of a conic part of the apparatus in the sludge remover.

The density and thickness of a wall layer are influenced by speed of gas, a twisting angle vortex generator, and character of feeding into of a stream in the apparatus. High speeds of a gas stream lead to decrease in a thickness of a wall layer, despite a reinforcement, thus magnitudes of a turbulent diffusion.

The component of tangential speed υ_φ essentially varies on gas bottle radius that testifies to presence of differential twirl in which result whirl lines twist on a spiral. For tangential speed drift of a maximum from periphery to the center and abbreviation of a zone of the forced whirlwind is characteristic. Tangential speed much more axial in zones about walls and in the field of an axis practically one order with it. The axial component practically does not change the profile; its maximum is near to an apparatus wall.

The results of modeling presented in Figure 6.5, show that speeds are in regular intervals distributed on a round.

Let's observe distribution axial, tangential, and the radial component of speed in the form of the stream-lines passing along cross-section of the apparatus (Fig. 6.5).

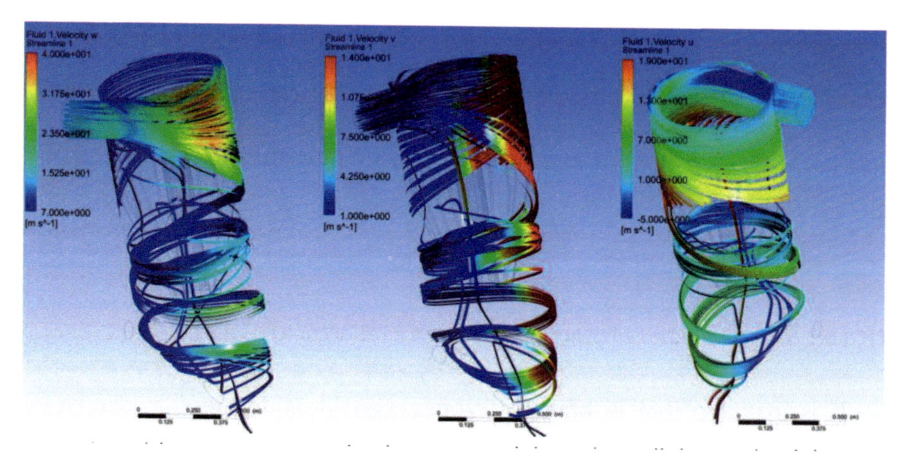

FIGURE 6.5 Speed in a stream: projections tangential υ_φ, the radial υ_r, and axial υ_z speeds in a stream.

From Figure 6.5, it is visible that the radial velocity (υ_r) resistantly to keeps the value on all cross-section of working space, the axial velocity (υ_z) gradually decreases from the center to periphery, whereas tangential, on the contrary, increases (υ_φ). The gained results will be coordinated with the literary data and speak about interacting in the apparatus of two streams—forward and rotational.

6.4 COMPARISON OF RESULTS OF CALCULATION IN ANSYS CFX WITH RESULTS OF EXPERIMENTS

The carried out calculations are compared with results of other authors.

Profiles of an axial velocity along a current, counted at $Re \geq 1000$, are compared in Figure 6.6 to results of Refs. [5–7]. Satisfactory coincidence of results specifies in perfection of a package of the program and serves as an estimation of a method of the solution.

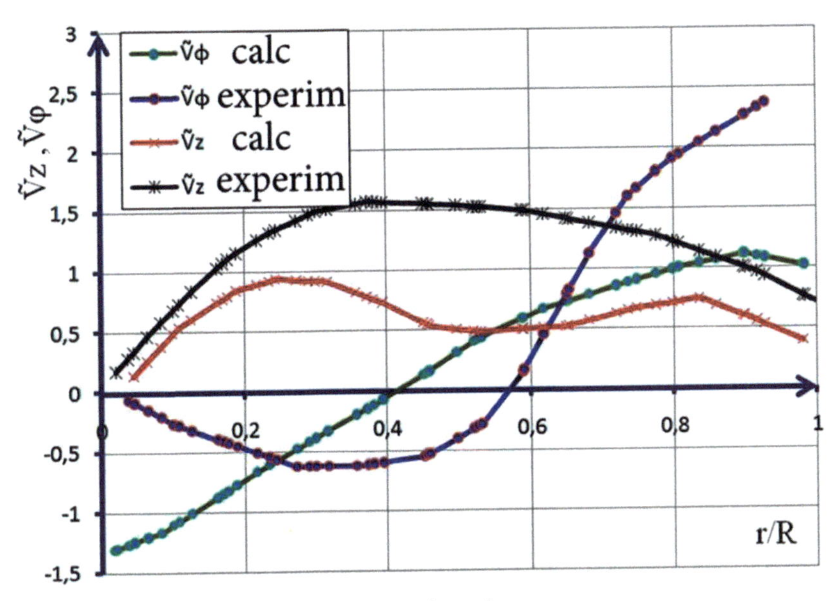

FIGURE 1.6 Comparison of design values of speed.

In the experiments presented for comparison, the solution method is approved in hydrodynamic calculations of streams of viscous incompressible liquid with a primary direction of a current (along an axis).

6.5 CONCLUSIONS

The algorithm of modeling of process of separation of a dispersoid in a gas stream with irrigation by a liquid has been developed. The carried out calculations allow to define potential possibilities of a dynamic gas bottle at its use in the capacity of the apparatus for clearing of gas emissions. Numerical research of work of a gas bottle will allow to analyze its work for the purpose of decrease of power inputs at conservation of quality of a gas cleaning. The developed model helps to simulate sweepingly and visually traffic of a dusty gas stream taking into account brought in to geometry of the apparatus of changes. Thus, the model can be applied to optimization of a design of a dynamic gas bottle.

KEYWORDS

- **the rated grid**
- **mathematical modeling**
- **ANSYS CFX**
- **pressure of gas**
- **speed of a stream**

REFERENCES

1. Zhaofeng, T. *Numerical Modelling of Turbulent Gas–Particle Flow and its Applications.* School of Aerospace, Manufacturing & Mechanical Engineering, RMIT University, 2006.
2. Gidaspow, D. *Multiphase Flow and Fluidization: Continuum and Kinetic Theory Descriptions with Applications*. Academic Press: San Diego, CA, 1994.
3. Neuwirth, A. Vortex Methods for Fluid Simulation in Computer Graphics. *Thesis for the Ph.D. Degree in Computer Science*, University of Ottawa, 2013; pp 114–116.
4. Usmanova, R. R. *Dynamic Gas Washer*. 2339435. November 2008.
5. Ahmadzadeh, A.; Arastoopour, H.; Teymour, F. Numerical Simulation of Gas and Particle Flow in a Rotating Fluidized Bed. *Ind. Eng. Chem. Res.* **2003,** *42*, 2627–2633.
6. Artyukhov, A. E. Computer Simulation of Vortex Flow. *Hydrodyn. J. Manuf. Ind. Eng.* **2013,** *12*, 25–29.
7. Vasquez, S. A.; Ivanov, V. A. A Phase Coupled Method for Solving Multiphase Problems on Unstructured Meshes. In *ASME 2000 Fluids Engineering Division Summer Meeting*, Boston, 2000; pp 247–258.

CHAPTER 7

COMPUTING THE AUGMENTED ECCENTRIC CONNECTIVITY INDICES OF THE NANOSTAR DENDRIMER $D_3[N]$

WEI GAO[1], MOHAMMAD REZA FARAHANI[2*], and MUHAMMAD KAMRAN JAMIL[3]

[1]*School of Information Science and Technology, Yunnan Normal University, Kunming 650500, China*

[2]*Department of Applied Mathematics, Iran University of Science and Technology (IUST), Narmak, Tehran 16844, Iran*

[3]*Abdus Salam School of Mathematical Sciences, Government College University (GCU), Lahore, Pakistan*

[*]*Corresponding author. E-mail: MrFarahani88@gmail.com*

CONTENTS

ABSTRACT

The augmented eccentric connectivity index is defined as ${}^{A}\xi(G)=\sum_{v\in V} M(v)/\varepsilon(v)$, where $M(v)$ and $\varepsilon(v)$ denote the product of degrees of all neighbors of vertex v and the eccentricity of vertex v, respectively. In this chapter, we compute the augmented eccentric connectivity index of nanostar dendrimers $D_3[n]$ ($\forall n \in \mathbb{N}$).

7.1 INTRODUCTION

Let G be a simple connected graph with vertex set $V(G)$ and edge set $E(G)$. A molecular graph is a graph such that its vertices and represents the atoms and the edges, respectively.

If $u,v \in V(G)$, then the distance, $d(x,y)$, between u and v is the length of a minimum path connecting u and v. The *eccentricity* $\varepsilon(u)$ is the largest distance between u and any other vertex u of G, that is, $\varepsilon(u)=\max\{d(u,v)\mid v\in V(G)\}$. The *degree*, $d(v)$, is the number of vertices attached to the vertex v.

In 1997, Sharma, Goswami, and Madan[1] introduced the *eccentric connectivity index* of the molecular graph G, $\xi(G)$ and defined as

$$\xi(G)=\sum_{v\in V(G)} d_v \times \varepsilon(v)$$

where the eccentricity of vertex v of G $\varepsilon(u)$ is the largest distance between v and any other vertex u of G or $e(v)=\max\{d(u;v)\mid \forall u\in V(G)\}$. The radius $R(G)$ and diameter $D(G)$ are defined as the minimum and maximum eccentricity among vertices of G, respectively.[2–10] In other words,

$$D(G)=\max_{v\in V(G)}\{d(u,v)\mid \forall u\in V(G)\}$$
$$R(G)=\min_{v\in V(G)}\{\max\{d(u,v)\mid \forall u\in V(G)\}\}$$

The *augmented eccentric connectivity index* ${}^{A}\xi(G)$ is defined as the summation of the quotients of the product of adjacent vertex degrees and eccentricity of the concerned vertex and expressed as

$${}^{A}\xi(G)=\sum_{v\in V}\frac{M(v)}{\varepsilon(v)}$$

where $M(v)$ denotes the product of degrees of all neighbors of vertex v of G.[11–13] In other words, $M(v)=\prod_{v\in N_G(u)} d_v$ where $N_G(u)=\{v\in V(G)\mid uv\in E(G)\}$.

Nanobiotechonolgy is a promptly advancing area of scientific and technological opportunity that applies the tools and processes of nanofabrication to build devices for studying biosystems. Dendrimers are one of the main objects of this new area of science. A dendrimer is an artificially manufactured or synthesized molecule built up from branched units called monomers using a nanoscale fabrication process. The nanostar dendrimer is a part of a new group of macroparticles that appear to be photon funnels just like artificial antennas. These macromolecules and more precisely those containing phosphorus are used in the formation of nanotubes, micro-, and macrocapsules, nanolatex, colored glasses, chemical sensors, modified electrodes, etc.[14–22]

7.2 MAIN RESULTS

In this section, we compute the augmented eccentric connectivity index ${}^{A}\xi(G)$ of an infinite family of the dendrimers.

Theorem 1. Let $D_3[n]$ be the nth growth of the nanostar dendrimer ($\forall n \in \mathbb{N}$). Then the *augmented eccentric connectivity index* of $D_3[n]$ is

$$
{}^{A}\xi(D_3[n]) = \sum_{i=1}^{n}\left(\frac{3^4 2^{i-1}}{5i+5(n+1)} + \frac{3^2 2^{i+2}}{5i+1+5(n+1)} + \frac{3^2 2^{i+2}}{5i+2+5(n+1)} + \frac{3^2 2^{i+2}}{5i+3+5(n+1)} + \frac{3\left(2^{i+2}\right)}{5i+4+5(n+1)}\right) + \left(\frac{3^2\left(2^{n-1}+3\right)}{5(n+1)}\right)
$$

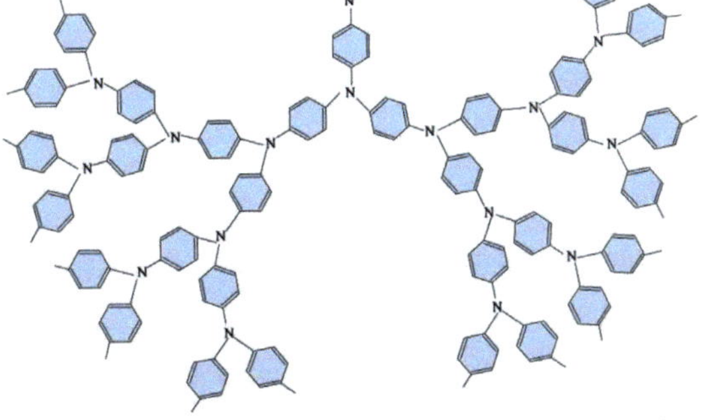

FIGURE 7.1 The two-dimensional of the nth growth of the nanostar dendrimer $D_3[n]$, $\forall n \in \mathbb{N}$.[22]

Proof of Theorem 1. Consider the graph of the nanostar dendrimer $D_3[n]$ ($\forall n \in \mathbb{N}$), with $21(2^{n+1}) - 20$ vertices/atoms ($=|V(D_3[n])|$). According to the structure of Dendrimer Nanostar $D_3[n]$, we see that $D_3[n]$ have $3(2^n)$ hydrogen atoms or vertices with degree 1 and $12(2^{n+1} - 1)$ vertices with degree 2 and $15(2^n) - 8$ vertices with degree 3. Here, we denote the sets of these vertices by V_1, V_2, and V_3, respectively. In other words,

$$|V(D_3[n])| = 21(2^{n+1}) - 20$$
$$V_1 = \{H \in V(D_3[n]) | d_H = 1\} \rightarrow |V_1| = 3(2^n)$$
$$V_2 = \{v \in V(D_3[n]) | d_v = 2\} \rightarrow |V_2| = 12(2^{n+1} - 1)$$
$$V_3 = \{v \in V(D_3[n]) | d_v = 3\} \rightarrow |V_3| = 15(2^n) - 8$$

Thus, the number of edges/bonds in the nanostar dendrimer $D_3[n]$ is

$$\left|E\left(D_3[n]\right)\right| = \frac{1}{2}[V_1 + 2\times V_2 + 3\times V_3]$$
$$= \frac{1}{2}\left[1\times 3\left(2^n\right) + 2\times 12\left(2^{n+1} - 1\right) + 3\times 15\left(2^n\right)\right] = 24(2^{n+1} - 1).$$

According to the structure of the nanostar dendrimer (see Fig. 7.1),[19] we see that an element is added to $D_3[n-1]$ in the nth growth of $D_3[n]$ (see Fig. 7.2), we called this additional element by *Leaf* and denote its vertices by H, $v_1, v_2, \ldots, v_5$, respectively, where the vertices v_1, v_2, and v_5 have degree 3 and the vertices v_3 and v_4 have degree 2, and only hydrogen atoms H have degree 1.

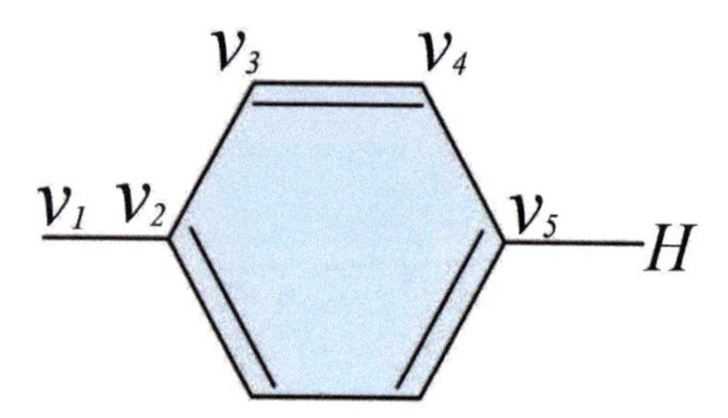

FIGURE 7.2 A leaf is an additional element in the nth growth of Nanostar Dendrimer $D_3[n]$.

Here, from Figure 7.2, one can see that the maximum eccentricity of a leaf of Dendrimer Nanostar $D_3[n]$ is 5; therefore, the eccentricity of previous vertices increase 10.

Now, from Figures 7.1 and 7.2, we can see that all adjacent vertices with v_1 have degree 3 and the product of degrees of all neighbors of vertex v_1, thus $M(v_1) = 3 \times 3 \times 3 = 27$. The vertex v_2 is adjacent with v_1, v_3, v_3' that $v_3', v_3 \in V_2$ and $v_1 \in V_3$, thus $M(v_2) = 3 \times 2 \times 2 = 12$. Also for v_4, v_3 from V_2, we see that $M(v_3) = M(v_4) = 2 \times 3 = 6$. The vertex, $v_5 \in V_3$, is adjacent with $v_4', v_4 \in V_2$ and $H \in V_1$ and $M(v_5) = 1 \times 2 \times 2 = 4$. Obviously, $M(H) = d_{v_5} = 3$.

Here from Figure 7.1 of the nth growth of the nanostar dendrimer $D_3[n]$ and Refs. [17–21], we can see that the eccentricity $\varepsilon(v)$ for all members of a leaf of the nanostar dendrimer $D_3[n]$ in Table 7.1.

Therefore, by using above results and Table 7.1, the augmented eccentric connectivity index ${}^A\xi(G)$ of the nanostar dendrimer $D_3[n]$ is as follow ($\forall n \in \mathbb{N}$):

TABLE 7.1 The Eccentricity for Members of a Leaf in nanostar dendrimer $D_3[n]$.

The members of an addition leaf	The eccentricity of members in ith growth	$M(v)$	The number of members
Center vertex c	$5n + 5$	27	1
v_1	$5i + 5n + 5$	27	$3(2^{n-1})$
v_2	$5i + 5n + 6$	12	$3(2^n)$
V_3	$5i + 5n + 7$	6	$2 \times 3(2^n)$
V_4	$5i + 5n + 8$	6	$2 \times 3(2^n)$
V_5	$5i + 5n + 9$	4	$3(2^n)$
H	$10n + 10$	3	$3(2^n)$

$$
\begin{aligned}
{}^A\xi(D_3[n]) &= \sum_{v \in V(D_3[n])} \frac{M(v)}{\varepsilon(v)} \\
&= \sum_{c} \frac{M(c)}{\varepsilon(c)} + \sum_{v_1} \frac{M(v_1)}{\varepsilon(v_1)} + \sum_{v_2} \frac{M(v_2)}{\varepsilon(v_2)} + \sum_{v_3} \frac{M(v_3)}{\varepsilon(v_3)} + \sum_{v_4} \frac{M(v_4)}{\varepsilon(v_4)} + \sum_{v_5} \frac{M(v_5)}{\varepsilon(v_5)} + \sum_{H} \frac{M(H)}{\varepsilon(H)} \\
&= \frac{3\times3\times3}{5n+5} + \sum_{v_1} \frac{3\times3\times3}{5i+5n+5} + \sum_{v_2} \frac{3\times2\times2}{5i+5n+6} + \sum_{v_3} \frac{3\times2}{5i+5n+7} + \sum_{v_4} \frac{3\times2}{5i+5n+8} + \sum_{v_5} \frac{1\times2\times2}{5i+5n+9} + \sum_{H} \frac{1\times3}{10(n+1)} \\
&= \frac{27}{5(n+1)} + 27\sum_{i=1}^{n} \frac{3(2^{i-1})}{5i+5n+5} + 12\sum_{i=1}^{n} \frac{3(2^i)}{5i+5n+6} + 2\times6\sum_{i=1}^{n} \frac{3(2^i)}{5i+5n+7} \\
&\quad + 2\times6\sum_{i=1}^{n} \frac{3(2^i)}{5i+5n+8} + 4\sum_{i=1}^{n} \frac{3(2^i)}{5i+5n+9} + 3\left(\frac{3(2^n)}{10(n+1)}\right) \\
&= \sum_{i=1}^{n} \left(\frac{81\times2^{i-1}}{5i+5(n+1)} + \frac{36\times2^i}{5i+1+5(n+1)} + \frac{36\times2^i}{5i+2+5(n+1)} + \frac{36\times2^i}{5i+3+5(n+1)} + \frac{12\times2^i}{5i+4+5(n+1)} \right) \\
&\quad + \left(\frac{9\times2^n}{10(n+1)} + \frac{27}{5(n+1)} \right).
\end{aligned}
$$

Here, the augmented eccentric connectivity index of the nanostar dendrimer $D_3[n]$ is

And, this completed the proof of Theorem 1.

Example 1. Let $D_3[n]$ be the nth growth of the nanostar dendrimer ($\forall n \in \mathbb{N}$) depicted in Figure 7.1. Then, some values of the augmented eccentric connectivity index ${}^{A}\xi(D_3[n])$ in Table 7.2 for integer $n = 1,2,3,\ldots,1000$.

TABLE 7.2 Some Values of ${}^{A}\xi(D_3[n])$ for Integer $n = 1,2,3,\ldots,1000$.

*n*th Growth	The number of vertices $\lvert V(D_3[n])\rvert$	The number of edges $\lvert E(D_3[n])\rvert$	Augmented eccentric connectivity index ${}^{A}\xi(D_3[n])$
1	64	72	86.8984520123839
2	148	168	114.031557912353
3	316	360	146.22688704495
4	652	744	194.657438743142
5	1324	1512	274.719234382093
6	2668	3048	412.83433596355
7	5356	6120	656.683819194426
8	10,732	12,264	1093.71316533907
9	21,484	24,552	1885.60973017552
10	42,988	49,128	3333.05404428061
20	44,040,172	50,331,624	1671,623.09464811
30	45,097,156,588	51,539,607,528	1143,322,292.42846
40	46,179,488,366,572	52,776,558,133,224	879,036,865,717.697
50	4.72877960873902E16	5.40431955284459E16	720,610,029,274,822
60	4.84227031934876E19	5.53402322211287E19	6.15216653976531E17
70	4.95848480701313E22	5.66683977944357E22	5.40174294827289E20
80	5.07748844238144E25	5.80284393415022E25	4.84125794634043E23
90	5.1993481649986E28	5.94211218856983E28	4.40754923821703E26
100	5.32413252095856E31	6.0847228810955E31	4.06268841563725E29
200	6.74913978588776E61	7.71330261244315E61	2.57704959949348E59
300	8.55555110060484E91	9.77777268640553E91	2.17844663296216E89
400	1.0845449487965E122	1.23947994148172E122	**≫E100**
500	1.37482405531638E152	1.57122749179015E152	**≫E100**
600	1.74279653893002E182	1.99176747306288E182	**≫E100**
700	2.20925707865032E212	2.52486523274322E212	**≫E100**
800	2.80056606180954E242	3.20064692778233E242	**≫E100**
900	3.55013924923167E272	4.05730199912191E272	**≫E100**
1000	4.50033615018232E302	5.14324131449408E302	**≫E100**

KEYWORDS

- **molecular graphs**
- **eccentric connectivity index**
- **augmented eccentric connectivity index**
- **Nanostar dendrimer**
- **computational approach**

REFERENCES

1. Sharma, V.; Goswami, R.; Madan, A. K. Eccentric Connectivity Index: A Novel Highly Of Descriptor for Structure–Property and Structure–Activity Studies. *J. Chem. Inf. Comput. Sci.* **1997,** *37*, 273–282.
2. Ghorbani, M.; Ghazi, M. Computing Some Topological Indices of Triangular Benzenoid. *Digest. J. Nanomater. Biostruct.* **2010,** *5* (4), 1107–1111.
3. Kumar, V.; Sardana, S.; Madan, A. K. Predicting Anti-HIV Activity of 2,3-Diary l-1,3-Thiazolidin-4-ones: Computational Approaches Using Reformed Eccentric Connectivity Index. *J. Mol. Model.* **2004,** *10*, 399–407.
4. Fischermann, M.; Homann, A.; Rautenbach, D.; Szekely, L. A.; Volkmann, L. Wiener Index versus Maximum Degree in Trees. *Discr. Appl. Math.* **2002,** *122*, 127–137.
5. Gupta, S.; Singh, M.; Madan, A. K. Application of Graph Theory: Relationship of Eccentric Connectivity Index and Wiener's Index with Anti-inflammatory Activity. *J. Math. Anal. Appl.* **2002,** *266*, 259–268.
6. Sardana, S.; Madan, A. K. Application of Graph Theory: Relationship of Molecular Connectivity Index, Wiener's index and Eccentric Connectivity Index with Diuretic Activity. *MATCH Commun. Math. Comput. Chem.* **2001,** *43*, 85–89.
7. Ashrafi, A. R.; Ghorbani, M.; Jalali, M. Eccentric Connectivity Polynomial of an Infinite Family of Fullerenes. *Optoelectron. Adv. Mater.-Rapid Commun.* **2009,** *3*, 823–826.
8. Ashrafi, A. R.; Jalali, M.; Ghorbani, M.; Diudea, M. V. Computing PI and Omega Polynomials of an Infinite Family of Fullerenes. *MATCH. Commun. Math. Comput. Chem.* **2008,** *60*, 905–916.
9. Diudea, M. V.; Vizitiu, A. E.; Gholaminezhad, F.; Ashrafi, A. R. Omega Polynomial in Twisted (4,4)Tori, *MATCH. Commun. Math. Comput. Chem.* **2008,** *60*, 945–953.
10. Ghorbani, M.; Ashrafi, A. R. Counting the Number of Hetero Fullerenes. *J. Comput. Theor. Nanosci.* **2006,** *3*, 803–810.
11. Bajaj, S.; Sami, S. S.; Madan, A. K. Model for Prediction of Anti-HIV Activity of 2-Pyridinone Derivatives Using Novel Topological Descriptor. *QSAR Comb. Sci.* **2006,** *25*, 813–823.
12. Dureja, H.; Madan, A. K. Superaugmented Eccentric Connectivity Indices: Highly Discriminating Topological Descriptors for QSAR/QSPR Modeling. *Med. Chem. Res.* **2007,** *16*, 331–341.

13. YarAhmadi, Z. Eccentric Connectivity and Augmented of N-Branches Phenylacetylenes Nanostar Dendrimers. *Iran. J. Math. Chem.* **2010,** *1* (2), 105–110.
14. Alikhani, S.; Iranmanesh, M. A. Chromatic Polynomials of Some Dendrimers. *J. Comput. Theor. Nanosci.* **2010,** *7* (11) 2314–2316.
15. Alikhani, S.; Iranmanesh, M. A. Chromatic Polynomials of Some Nanostars. Iran. J. Math. Chem. **2010,** *3* (2), 127–135.
16. Ashrafi, A. R.; Mirzargar, M. PI, Szeged and edge Szeged Indices of an Infinite Family of Nanostar Dendrimers. *Indian J. Chem.* **2008,** *47*, 538–541.
17. Farahani, M. R. Fourth Atom-Bond Connectivity Index of an Infinite Class of Nanostar Dendrimer $D_3[n]$. *J. Adv. Chem.* **2013,** *4* (1), 301–305.
18. Farahani, M. R. Computing Fifth Geometric-Arithmetic Index of Dendrimer Nanostars. *Adv. Mater. Corros.* **2013,** *1*, 62–64.
19. Farahani, M. R.; Gao, W.; Kanna, R. The Connective Eccentric Index for an Infinite Family of Dendrimers. *Indian J. Fundam. Appl. Life Sci.* **2015,** *5*, 766–771.
20. Farahani, M. R. Some Connectivity Index of an Infinite Class of Dendrimer Nanostars. *J. Appl. Phys. Sci. Int.* **2015,** *3* (3), 99–105.
21. Farahani, M. R. On Multiple Zagreb Indices of Dendrimer Nanostars. *Int. Lett. Chem., Phys. Astron.* **2015,** *52*, 147–151.
22. Gao, W.; Farahani, M. R. Degree-Based Indices Computation for Special Chemical Molecular Structures Using Edge Dividing Method. *Appl. Math. Nonlinear Sci.* **2015,** *1* (1), 94–117.

CHAPTER 8

HYDRAULIC MODEL CALIBRATION PROCESS

KAVEH HARIRI ASLI[1], SOLTAN ALI OGLI ALIYEV[1], and HOSSEIN HARIRI ASLI[2*]

[1]*Department of Mathematics and Mechanics, National Academy of Science of Azerbaijan "AMEA," Baku, Azerbaijan*

[2]*Civil Engineering Department, Faculty of Engineering, University of Guilan, Rasht, Iran*

**Corresponding author. E-mail: hariri_k@yahoo.com*

CONTENTS

ABSTRACT

This work attends to water distribution network calibration and data collection process. The hydraulic model calibration process adapts it to the current situation. There are more theoretical parameters that are consistent with reality system. The purpose of calibration operation includes the following items: increased reliability model, opportunity to predict the future state of the system, obtain accurate results than the current situation, and the discovery of previously unknown problems in the system. For this purposes, the models must constantly be calibrated according to the changes made. For this purpose, the phase correction model is used. The network analysis and computing include theoretical parameters conformity with reality. It also includes the amount and direction of the input and output streams. This work shows the changes of maximum flow rate and pressure drop in the pipes by hydraulic model comparison with the flow rate and pressure drop related to the control valves operation in the system. The results were compared with any future changes in the network calibration.

8.1 INTRODUCTION

Water distribution network must be seen to be operating efficiently and effectively and must be able to cost justifies their level of leakage and works designed to manage leakage, particularly to their customers who want to see their costs minimized. Leakage is often seen as synonymous with waste, and reducing leakage is seen as a means of saving money. Water lost through leakage has a value and so reducing the level of leakage offers benefits. However, eliminating leakage completely is impracticable and the cost of reducing it to low levels may exceed the cost of producing the water saved. Conversely, when little effort is expended on active leakage control, leakage levels will rise to levels where the cost of the water lost predominates. Water suppliers must therefore strike a balance between the cost of reducing leakage and the value of the water saved. The level of leakage at which it would cost more to make further reductions than to produce the water from another source is what is known as the economic level of leakage (ELL). Operating at model calibration economic levels of leakage means that the total cost to the customer of supplying water is minimized, and suppliers are operating efficiently. This means that leakage reduction should be pursued to the point where the long-run marginal cost of leakage control is equal to the long-run marginal benefit of the water saved. The latter depends on the long-run marginal costs of augmenting supplies by alternative means, including an assessment of the environmental benefits. The ELL is not fixed for all time. It depends on a wide range of factors, which will vary over time. For example, the cost of detecting and repairing leaks will fall as new technology is introduced. This will cause the ELL to fall. Conversely, if total demand falls to a point where there is a large surplus of water, it may not be economic to reduce leakage, unless the water can be sold to other suppliers. Once the economic optimum level is known, this can be compared to the present level of leakage, and the supplier can then set targets for leakage control in conjunction with other corporate policies on customer metering, mains rehabilitation, resource development, and pressure control. To do so, they will need to appraise the investment required for these various different supply and demand management solutions and the benefits which are expected to accrue. Due to the complexity of the issues, it is not possible to generalize to provide standardized formulae for setting leakage targets. There is a need to examine each system to determine the most appropriate method of leakage control and to plan the required capital investment, manpower and revenue resource. However, any supplier who is prepared to commit resources to collecting the required data, and to carry out the analysis and appraisals, will develop a greater understanding of the factors which are important to target setting. They will also be less likely to have unrealistic or uneconomic targets imposed on them from outside, or fall into the trap of setting leakage targets themselves without full consideration of the practicalities of achieving them, or the economic consequences.[1–32]

8.2 MATERIALS AND METHODS

There are many possible ways of setting a leakage target. These can include targets based on minimum night flows, areas with excess pressure, areas with expensive water, or the most urbanized areas. The setting of economic targets, that is, a level of leakage which provides the most economic mix of leakage related costs, is independent of variations in physical factors such as property density, pressure, etc. and can provide clear information upon which sound management decisions may be based. However, it is recognized that there may be social, environmental factors which dictate the target leakage level, as well as economic ones related to the suppliers' own operating environment. This has given rise to a broader concept of "the most appropriate leakage target," being described as "that level of leakage which, over a long term planning horizon, provides the least cost combination of demand management, model calibration and resource development, whilst adequately providing a low risk of security of supply to customers, and not unduly over abstracting water from the environment."

This book will present many theoretical results of water distribution network calibration and data collection process.

Much software will apply to water distribution network calibration including: WATER GEMS8.2 & ArcGIS9-ArcMap9.3 in this book.[33–54]

8.3 RESULTS AND DISCUSSION

Field tests—The field test was included at the end of water transmission line. All of these parts have been tied into existing water networks.

Laboratory model—A scale model have been built to reproduce transients observed in a prototype (real) system, typically for forensic or steam system investigations. This research Lab. model has recorded flow and pressure data. The model is calibrated using one set of data and without changing parameter values (Fig. 8.1).

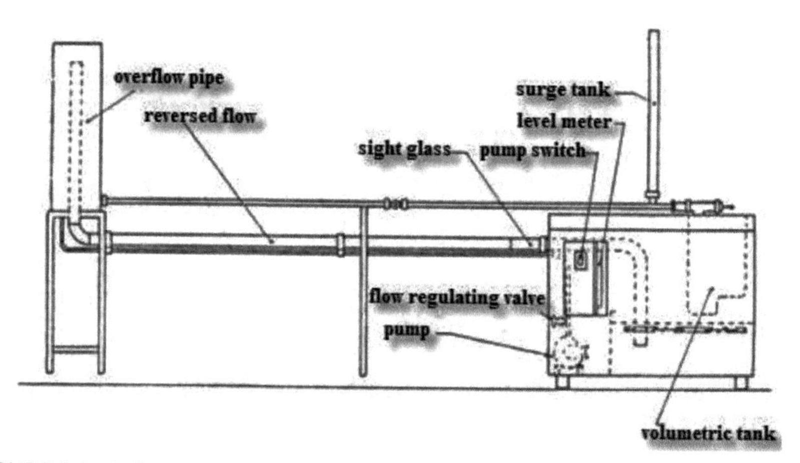

FIGURE 8.1 Laboratory model for experiments with flow and pressure data records.

- **Laboratory model dateline**—The model has been calibrated by the changes of maximum flow rate and pressure in laboratory models.
- Subatmospheric leakage tests performed according to ASTM standards. This was done to explain repeated pipe breaks. This work led to improved standards for gasket designs and installation techniques in the province of subatmospheric transient pressures.[55–57]

Laboratory model dateline: The method of characteristics is based on a finite difference technique (1–3) where pressures are computed along the pipe for each time step,

$$V_p = \frac{1}{2}\left[\left(V_{L_e} + V_{R_i}\right) + \frac{g}{c}\left(H_{L_e} - H_{R_i}\right) - \left(\frac{f \cdot \Delta t}{2D}\right)\left(V_{L_e}\left|V_{L_e}\right| - V_{R_i}\left|V_{R_i}\right|\right)\right] \tag{8.1}$$

$$H_p = \frac{1}{2}\left[\frac{c}{g}\left(V_{L_e} + V_{R_i}\right) + \left(H_{L_e} - H_{R_i}\right) - \frac{c}{g}\left(\frac{f \cdot \Delta t}{2D}\right)\left(V_{L_e}\left|V_{L_e}\right| - V_{R_i}\left|V_{R_i}\right|\right)\right], \tag{8.2}$$

Surge pressure (ΔH) is a function of independent variables (X), such as

$$\Delta H \approx f, T, C, V, g, D, \tag{8.3}$$

For a model definition in this book (Tables 8.1–8.6), relation between surge pressure and pulsation (as a function) and several factors (as variables) have been investigated. Then CFD software has evaluated transient flow as a function of following parameters: ρ, K, d, $C1$, E_e, V, f, T, C, g, T_p. Regression software has fitted the function curve and provides regression analysis (Fig. 8.2).

TABLE 8.1 Model Description of Regression Software.

Model name		MOD_2
Dependent variable	1	Bar
Equation	1	Linear
	2	Logarithmic
	3	Inverse
	4	Quadratic
	5	Cubic
	6	Compound[a]
	7	Power[a]
	8	*S*[a]
	9	Growth[a]
	10	Exponential[a]
	11	Logistic[a]
Independent variable		m/s
Constant		Included
Variable whose values label observations in plots		Unspecified
Tolerance for entering terms in equations		0.0001

[a]The model requires all nonmissing values to be positive.

TABLE 8.2 Model Description of Regression Software.

	N
Total cases	24
Excluded cases[a]	0
Forecasted cases	0
Newly created cases	0

[a]Cases with a missing value in any variable are excluded from the analysis.

TABLE 8.3 Model Description of Regression Software.

		Variables	
		Dependent	**Independent**
		bar	**m/s**
Number of positive values		24	23
Number of zeros		0	1[a,b]
Number of negative values		0	0
Number of missing values	User-missing	0	0
	System-missing	0	0

[a]The inverse or *S* model cannot be calculated. [b]The logarithmic or power model cannot be calculated.

TABLE 8.4 Model Description of Regression Software.

Equation	**Model summary**					**Parameter estimates**			
	***R* square**	**F**	**df1**	**df2**	**Sig.**	a_0	a_1	a_2	a_3
Linear $y = a_0 + a_1x$	0.418	15.831	1	22	0.001	6.062	0.571		
Logarithmic[a] $\log y = \log(a) - (b)\log x$									
Inverse[b] $y = f^{-1}(y)$									
Quadratic $y = a_0 + a_1x + a_2x^2$	0.487	9.955	2	21	0.001	6.216	−0.365	0.468	
Cubic $y = a_0 + a_1x + a_2x^2 + a_3x^3$	0.493	10.193	2	21	0.001	6.239	0.000	−0.057	0.174
Compound $A = Ce^{kt}$	0.424	16.207	1	22	0.001	6.076	1.089		
Power *a* $y = cx^p$									

TABLE 8.4 *(Continued)*

Equation	Model summary					Parameter estimates			
	R square	F	df1	df2	Sig.	a_0	a_1	a_2	a_3
S^b $y = f_0(T, X, U)$									
Growth $(dA/dT) = KA$	0.424	16.207	1	22	0.001	1.804	0.085		
Exponential $e^x = \lim_{n\to\infty}\left(1+\frac{1}{n}\right)^n$	0.424	16.207	1	22	0.001	6.076	0.085		
Logistic $f(x) = \frac{L}{1+e^{-k(x-x_0)}}$	0.424	16.207	1	22	0.001	0.165	0.918		

Dependent variable: bar. The independent variable is m/s.

[a]The independent variable (m/s) contains nonpositive values. The minimum value is 0.00. The logarithmic and power models cannot be calculated.

[b]The independent variable (m/s) contains values of zero. The inverse and *S* models cannot be calculated.

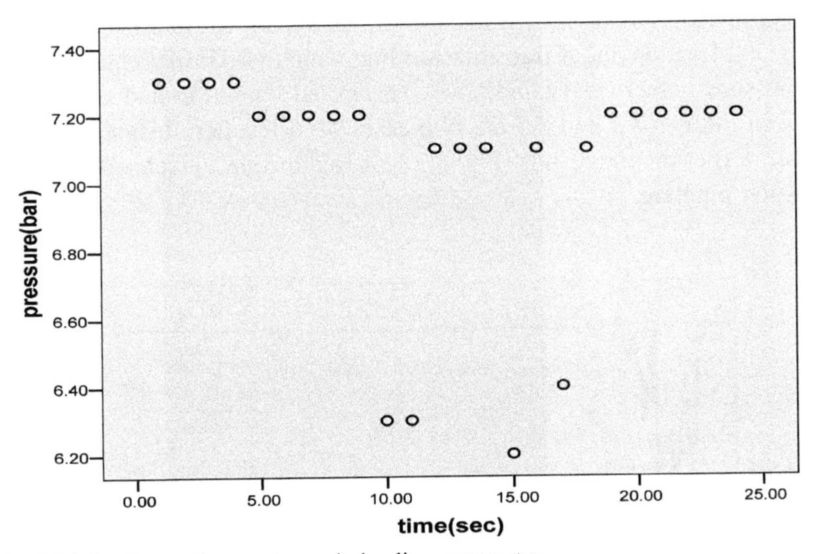

FIGURE 8.2 Regressions on transmission lines parameter.

Pulsation generally occurs when a liquid's motive force is generated by reciprocating or peristaltic positive displacement pumps. It is most commonly caused by the acceleration and deceleration of the pumped fluid. Installing a pulsation dampener can provide the most cost-efficient and effective choice. It can prevent the damaging effects of pulsation. The most current pulsation dampener design is the hydropneumatic dampener, consisting of a pressure vessel containing a compressed gas, generally air, or nitrogen separated from the process liquid by a bladder or diaphragm.[58–67] Valve closure can result in pressures well over the steady state values, while valve opening can cause seriously low pressures, possibly so low that the flowing liquid vaporizes inside the pipe. In this chapter, we have shown the maximum and minimum piezometric pressures, relative to atmospheric. This was observed in pipeline with respect to the time and location at which they were occurred. Results have been accounted helpful in design and determination of maximum (or minimum) expected pressures. Maximum or minimum points were due to the valve closure or opening and current pulsation. The models of the pipeline guided route selection, conceptual and detailed design of pipeline. Long-distance water transmission lines must be economical, reliable, and expandable. These results were retained to provide hydraulic input to a network-wide optimization and risk-reduction strategy for main pipeline. Data includes multibooster pressurized lines with surge protection ranging from check valves to gas vessels. Experimental results have been ensured reliable water transmission. The main assumption was based on the exploration about transmission line which was broken but equipped by pressure vessel (real condition). The initial condition and comparison between Figures 8.4 and 8.5 are two cases of "Elevation-distance transient curve." Experimental results (Fig. 8.3) ensured the gas vessels effective role for water pipeline.

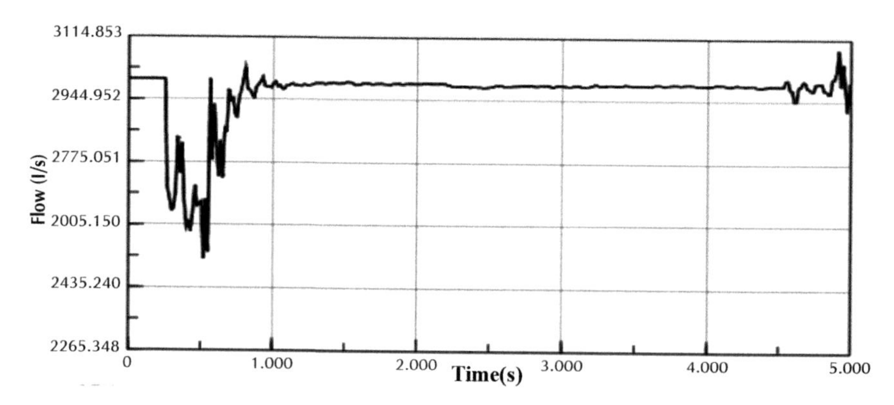

FIGURE 8.3 Field tests model: flow–time transient curve.

Results showed water-column separation and the entrance of air into pipeline. This pressure was too high for old piping, and it must be considered as hazard for piping.

Interpenetration was investigated by comparison between theoretical hydraulics analysis and bench-scale laboratory pilot[68–70] (Figs. 8.4 and 8.5).

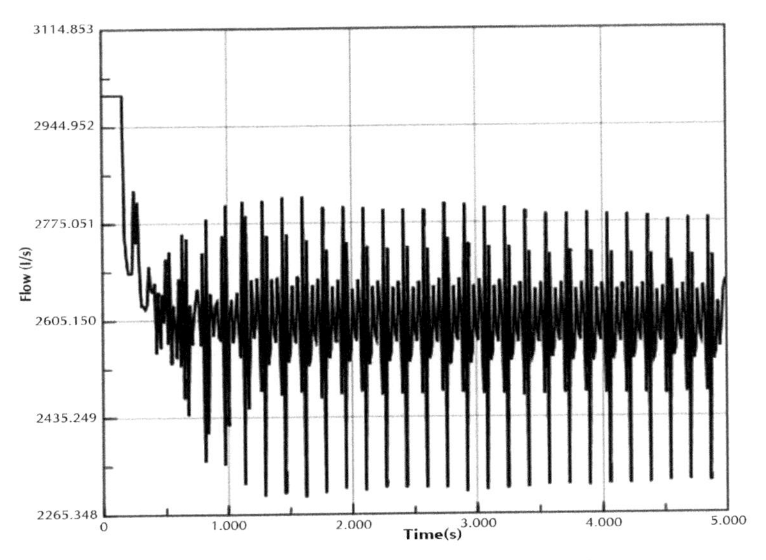

FIGURE 8.4 Field tests model: flow–time and volume–time transient curve.

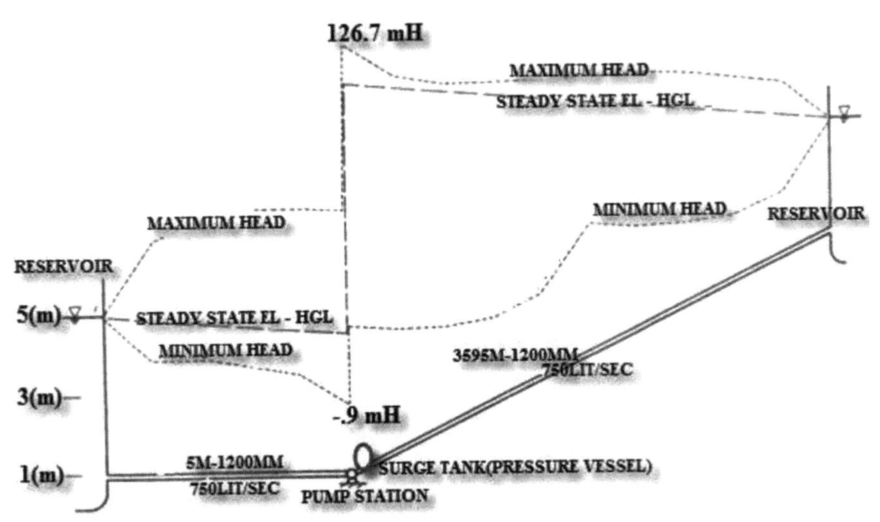

FIGURE 8.5 Max and min pressure due to pump turned off.

TABLE 8.5 Experimental Results.

No. of Pipe	From		To	Length
	Points	Point	Point	(m)
P2	5	P2:J6	P2:J7	60.7
P3	21	P3:J7	P3:J8	311.0
P4	2	P4:J3	P4:J4	1.0
P5	2	P5:J4	P5:J26	.5
P6	8	P6:J26	P6:J27	108.7
P7	3	P7:J27	P7:J6	21.5
P8	2	P8:J8	P8:J9	15.0
P9	23	P9:J9	P9:J10	340.7
P10	14	P10:J10	P10:J11	207.0
P11	22	P11:J11	P11:J12	339.0
P12	22	P12:J12	P12:J13	328.6
P13	4	P13:J13	P13:J14	47.0
P14	38	P14:J14	P14:J15	590.0
P15	4	P15:J15	P15:J16	49.0
P16	15	P16:J16	P16:J17	224.0
P17	2	P17:J17	P17:J18	18.4
P18	2	P18:J18	P18:J19	14.6
P19	2	P19:J19	P19:J20	12.0
P20	32	P20:J20	P20:J21	499.0
P21	17	P21:J21	P21:J22	243.4
P22	11	P22:J22	P22:J23	156.0
P23	3	P23:J23	P23:J24	22.0
P24	6	P24:J24	P24:J28	82.0
P25	4	P25:J28	P25:N1	35.6
P0	2	P0:J1	P0:J2	.5
P1	2	P1:J2	P1:J3	.5

TABLE 8.6 Experimental Results Ensured the Gas Vessels.

End Point	Max. Press. (mH)	Min. Press. (mH)	Max. Head (m)	Min. Head (m)
P2:J6	124.1	97.0	160.4	133.4
P2:J7	123.7	96.7	160.0	133.0
P3:J7	123.7	96.7	160.0	133.0
P3:J8	121.9	95.1	158.0	131.2
P4:J3	126.7	99.5	162.2	135.0
P4:J4	125.1	97.8	162.3	135.0
P5:J4	125.1	97.8	162.3	135.0
P5:J26	125.1	97.8	162.3	135.0
P6:J26	125.1	97.8	162.3	135.0
P6:J27	124.1	97.0	160.5	133.5
P7:J27	124.1	97.0	160.5	133.5
P7:J6	124.1	97.0	160.4	133.4
P8:J8	121.9	95.1	158.0	131.2
P8:J9	119.9	93.1	157.9	131.1
P9:J9	119.9	93.1	157.9	131.1
P9:J10	117.4	90.8	155.7	129.2
P10:J10	117.4	90.8	155.7	129.2
P10:J11	115.8	89.4	154.4	128.0
P11:J11	115.8	89.4	154.4	128.0
P11:J12	112.6	86.4	152.2	126.0
P12:J12	112.6	86.4	152.2	126.0
P12:J13	109.1	83.1	150.1	124.1
P13:J13	109.1	83.1	150.1	124.1
P13:J14	107.5	81.5	149.8	123.8
P14:J14	107.5	81.5	149.8	123.9
P14:J15	100.9	60.9	146.1	106.1
P15:J15	100.9	60.9	146.1	106.1
P15:J16	102.3	62.3	145.8	105.8

TABLE 8.6 *(Continued)*

End Point	Max. Press. (mH)	Min. Press. (mH)	Max. Head (m)	Min. Head (m)
P16:J16	102.3	62.3	145.8	105.8
P16:J17	99.3	59.2	144.3	104.3
P17:J17	99.3	59.2	144.3	104.3
P17:J18	101.2	61.1	144.2	104.1
P18:J18	101.2	61.1	144.2	104.1
P18:J19	101.8	61.7	144.1	104.0
P19:J19	101.8	61.7	144.1	104.0
P19:J20	99.8	59.8	144.0	104.0
P20:J20	99.8	59.8	144.0	104.0
P20:J21	98.3	58.1	140.9	100.6
P21:J21	98.3	58.1	140.9	100.6
P21:J22	94.7	54.4	139.3	99.0
P22:J22	94.7	54.4	139.3	99.0
P22:J23	68.4	28.1	138.3	98.0
P23:J23	68.4	28.1	138.3	98.0
P23:J24	56.3	15.9	138.1	97.7
P24:J24	56.3	15.9	138.1	97.7
P24:J28	42.4	0.0	137.6	95.2
P25:J28	42.4	0.0	137.6	95.2
P25:N1	16.7	16.7	112.6	112.6
P0:J1	0.0	0.0	40.6	40.6
P0:J2	27.5	2.1	63.0	37.6
P1:J2	27.5	2.1	63.0	37.6
P1:J3	27.5	2.1	63.0	37.6

Point	Vapor or air	Max. Vola (m^3)	Curr. Vola (m^3)	Curr. Flw. (cm)
P25:J28	Air	198.483	.000	2.666

[a]Flow variation.

8.3.1 COMPARISON OF PRESENT RESEARCH RESULTS WITH OTHER EXPERT'S RESEARCH

Comparison of present research results with other expert's research results shows similarity according to Figure 8.6.

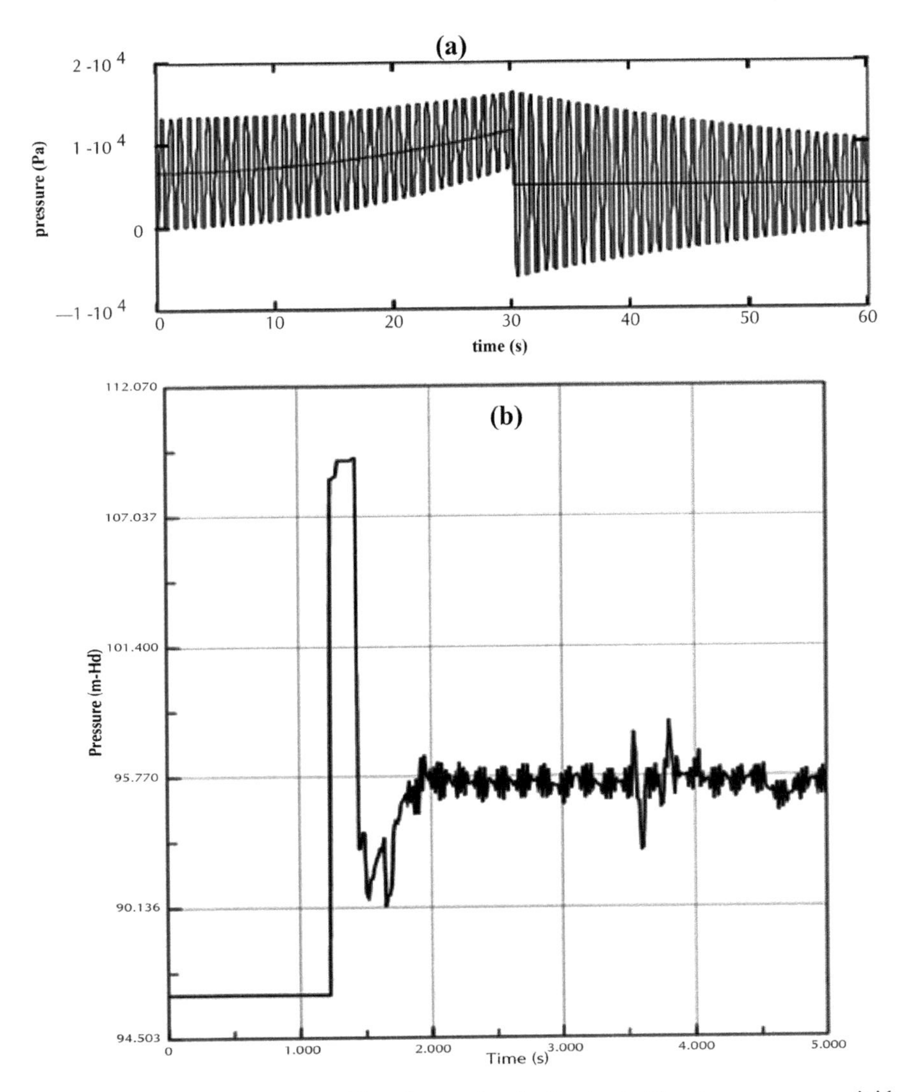

FIGURE 8.6 (a) Starting flow driven by 2 m/s velocity rise in 30 s time; average = rigid column; oscillation at midpoint,[60] (b) (present work).

8.4 CONCLUSION

According to lengths and pressure-wave speeds of pipes in present work (field tests), it seems that an advanced optimization algorithm is necessary. Algorithm considers lengths and pressure-wave speeds of pipes that must be installed at water transmission line. Research suggests that advanced flow and pressure sensors with high-speed data loggers must be installed in water transmission line. Data loggers must be linked to "PLC" in water pipeline. It will be recorded pulsation and fast transients, down to 5 ms for water flows interpenetration. Pressure transient recorder is a specialized data logger for monitoring rapid pressure changes in water pipe systems which is better to supplied in portable mode. Pressure Transient Loggers must be completely waterproof, submersible and battery powered. It must be maintenance free for at least 5 years. Pressure transient spikes are major cause of bursts and the associated expense of repair, water lost, and interruptions to supply. Conventional loggers do not log rapidly enough to identify these transients which often last only seconds. The changes of maximum flow rate and pressure drop in the pipes of hydraulic model compared with the flow rate and pressure drop problems related to the control valves operation in the water pipe systems. The results were compared with any future changes in the network calibration.

KEYWORDS

- **distribution network model**
- **model calibration**
- **hydraulic modeling**
- **computer mapping**
- **pressure lines**

REFERENCES

1. Streeter, V. L.; Wylie, E. B. *Fluid Mechanics*; McGraw-Hill Ltd.: New York, NY, 1979; pp 492–505.
2. Leon Arturo, S. An Efficient Second-Order Accurate Shock-Capturing Scheme for Modeling One and Two-Phase Water Hammer Flows. Ph.D. Thesis, March 29, 2007; pp 4–44.

3. Mala, G.; Li, D.; Dale, J. D. Heat Transfer and Fluid Flow in Microchannels, *J. Heat Transfer* **1997,** *40*, 3079–3088.
4. Xu, B.; Ooi, K. T.; Mavriplis, C.; Zaghloul, M. E. Viscous Dissipation Effects for Liquid Flow in Microchannels. *Micorsystems* **2002,** *113*, 53–57.
5. Pickford, J. *Analysis of Surge*; Macmillian: London, 1969; pp 153–156.
6. American Society of Civil Engineers. *Pipeline Design for Water and Wastewater*; American Society of Civil Engineers: New York, 1975, 54 p.
7. IWK. *IWK Sustainability Report 2012–2013*, 2012–2013.
8. Andreasen, P. Chemical Stabilization. *Sludge into Biosolids—Processing, Disposal and Utilization*. IWA Publishing: London, 2001.
9. CIWEM—The Chartered Institution of Water and Environmental Management. *Sewage Sludge: Stabilization and Disinfection—Handbooks of UK Wastewater Practice*. CIWEM Publishing: Lavenham, 1996.
10. Halalsheh, M.; Wendland, C. Integrated Anaerobic–Aerobic Treatment of Concentrated Sewage. *Efficient Management of Wastewater—Its Treatment and Reuse in Water Scarce Countries*. Springer Publishing Co.: New York, NY, 2008.
11. ISWA. *Handling, Treatment and Disposal of Sludge in Europe. Situation Report 1*. ISWA Working Group on Sewage Sludge and Water Works: Copenhagen, 1995.
12. Matthews, P. Agricultural and Other Land Uses. *Sludge into Biosolids—Processing, Disposal and Utilization*. IWA Publishing: London, 2001.
13. Holland, F. A.; Bragg, R. *Fluid Flow for Chemical Engineers*; Edward Arnold Publishers: London, 1995; pp 1–3.
14. Lee, T. S.; Pejovic, S. Air Influence on Similarity of Hydraulic Transients and Vibrations. *ASME J. Fluid Eng.* **1996,** *118* (4), 706–709.
15. Li, J.; McCorquodale, A. Modeling Mixed Flow in Storm Sewers. *J. Hydraul. Eng.* **1999,** *125* (11), 1170–1180.
16. Minnaert, M. On Musical Air Bubbles and the Sounds of Running Water. *Phil. Mag.* **1933,** *16* (7), 235–248.
17. Moeng, C. H.; McWilliams, J. C.; Rotunno, R.; Sullivan, P. P.; Weil, J.; Investigating 2D modeling of atmospheric convection in the PBL. *J. Atmos. Sci.* **2004,** *61*, 889–903.
18. Tuckerman, D. B.; Pease, R. F. W. High Performance Heat Sinking for VLSI. *IEEE Electron. Device Lett.* **1981,** *DEL-2*, 126–129.
19. Nagiyev, F. B.; Khabeev, N. S. Bubble Dynamics of Binary Solutions. *High Temp.* **1988,** *27* (3), 528–533.
20. Shvarts, D.; Oron, D.; Kartoon, D.; Rikanati, A.; Sadot, O. Scaling laws of Nonlinear Rayleigh–Taylor and Richtmyer–Meshkov Instabilities in Two and Three Dimensions. *C. R. Acad. Sci. Paris IV* **2000,** *719*, 312.
21. Cabot, W. H.; Cook, A. W.; Miller, P. L.; Laney, D. E.; Miller, M. C.; Childs, H. R. Large Eddy Simulation of Rayleigh–Taylor Instability. *Phys. Fluids* **2005,** *17*, 91–106.
22. Cabot, W. *Physics of Fluids*; University of California, Lawrence Livermore National Laboratory: Livermore, CA, 2006; pp 94–550.
23. Goncharov, V. N. Analytical Model of Nonlinear, Single-Mode, Classical Rayleigh–Taylor Instability at Arbitrary Atwood Numbers. *Phys. Rev. Lett.* **2002,** *88* (134502), 10–15.
24. Ramaprabhu, P.; Andrews, M. J. Experimental Investigation of Rayleigh–Taylor Mixing at Small Atwood Numbers. *J. Fluid Mech.* **2004,** *502*, 233 p.

25. Clark, T. T. A Numerical Study of the Statistics of a Two-Dimensional Rayleigh–Taylor Mixing Layer. *Phys. Fluids* **2003,** *15*, 2413.
26. Cook, A. W.; Cabot, W.; Miller, P. L. The Mixing Transition in Rayleigh–Taylor Instability. *J. Fluid Mech.* **2004,** *511*, 333.
27. Waddell, J. T.; Niederhaus, C. E.; Jacobs, J. W. Experimental Study of Rayleigh–Taylor Instability: Low Atwood Number Liquid Systems with Single-Mode Initial Perturbations. *Phys. Fluids* **2001,** *13*, 1263–1273.
28. Weber, S. V.; Dimonte, G.; Marinak, M. M. Arbitrary Lagrange–Eulerian Code Simulations of Turbulent Rayleigh–Taylor Instability in Two and Three Dimensions. *Laser Part. Beams* **2003,** *21*, 455 p.
29. Dimonte, G.; Youngs, D.; Dimits, A.; Weber, S.; Marinak, M. A Comparative Study of the Rayleigh–Taylor Instability Using High-Resolution Three-Dimensional Numerical Simulations: The Alpha Group Collaboration. *Phys. Fluids* **2004,** *16*, 1668.
30. Young, Y. N.; Tufo, H.; Dubey, A.; Rosner, R. On the Miscible Rayleigh–Taylor Instability: Two and Three Dimensions. *J. Fluid Mech.* **2001,** *447* (377), 2003–2500.
31. George, E.; Glimm, J. Self-similarity of Rayleigh–Taylor Mixing Rates. *Phys. Fluids* **2005,** *17* (054101), 1–3.
32. Oron, D.; Arazi, L.; Kartoon, D.; Rikanati, A.; Alon, U.; Shvarts, D. Dimensionality Dependence of the Rayleigh–Taylor and Richtmyer–Meshkov Instability Late-Time Scaling Laws. *Phys. Plasmas* **2001,** *8*, 2883 p.
33. Nagiyev, F. B.; Khabeev, N. S. Bubble Dynamics of Binary Solutions. *High Temp.* **1988,** *27* (3), 528–533.
34. Nagiyev, F. B. Damping of the Oscillations of Bubbles Boiling Binary Solutions. In *Mater. VIII Resp. Conf. Mathematics and Mechanics*, Baku, October 26–29 1988; pp 177–178.
35. Nagyiev, F. B.; Kadyrov, B. A. Small Oscillations of the Bubbles in a Binary Mixture in the Acoustic Field. *Math. AN Az. SSR Ser. Physicotech. Mate. Sci.* **1986,** *1*, 23–26.
36. Nagiyev, F. B. Dynamics, Heat and Mass Transfer of Vapor-Gas Bubbles in a Two-Component Liquid. In *Turkey-Azerbaijan petrol Semin.*, Ankara, Turkey, 1993; pp 32–40.
37. Nagiyev, F. B. The Method of Creation Effective Coolness Liquids. In *Third Baku international Congress*, Baku, Azerbaijan Republic, 1995; pp 19–22.
38. Nagiyev, F. B. The Linear Theory of Disturbances in Binary Liquids Bubble Solution. *Dep. VINITI* **1986,** *405* (86), 76–79.
39. Nagiyev, F. B. Structure of Stationary Shock Waves in Boiling Binary Solutions. *Math. USSR, Fluid Dyn.* **1989,** *1*, 81–87.
40. Rayleigh. On the Pressure Developed in a Liquid during the Collapse of a Spherical Cavity. *Philos. Mag. Ser. 6* **1917,** *34* (200), 94–98.
41. Perry, R. H.; Green, D. W.; Maloney, J. O. *Perry's Chemical Engineers Handbook*, seventh ed.; McGraw-Hill: New York, 1997; pp 1–61.
42. Nigmatulin, R. I. *Dynamics of Multiphase Media. Nauka* **1987,** *1* (2), 12–14.
43. Kodura, A.; Weinerowska, K. The Influence of the Local Pipeline Leak on Water Hammer Properties. In *Materials of the II Polish Congress of Environmental Engineering*, Lublin, 2005; pp 125–133.
44. Kane, J.; Arnett, D.; Remington, B. A.; Glendinning, S. G.; Baz'an, G. Two-Dimensional versus Three-Dimensional Supernova Hydrodynamic Instability Growth. *Astrophys. J.* **2000,** *528*, 989.
45. Quick, R. S. Comparison and Limitations of Various Water hammer Theories. *J. Hyd. Div., ASME,* **1933,** *May*, 43–45.

46. Jaeger, C. *Fluid Transients in Hydro-Electric Engineering Practice*; Blackie and Son Ltd.: London, 1977; pp 87–88.
47. Jaime Suárez, A. *Generalized Water Hammer Algorithm For Piping Systems With Unsteady Friction*; 2005, pp 72–77.
48. Fok, A.; Ashamalla, A.; Aldworth, G. Considerations in Optimizing Air Chamber for Pumping Plants. IN *Symposium on Fluid Transients and Acoustics in the Power Industry*, San Francisco, USA. December 1978; pp 112–114.
49. Fok, A. Design Charts for Surge Tanks on Pump Discharge Lines. In *BHRA 3rd Int. Conference on Pressure Surges*, Bedford, England, March 1980; pp 23–34.
50. Fok, A. Water Hammer and its Protection in Pumping Systems. In *Hydrotechnical Conference*, CSCE, Edmonton, May 1982; pp 45–55.
51. Fok, A. A Contribution to the Analysis of Energy Losses in Transient Pipe Flow. Ph.D. Thesis, University of Ottawa, 1987; pp 176–182.
52. Brunone, B.; Karney, B. W.; Mecarelli, M.; Ferrante, M. Velocity Profiles and Unsteady Pipe Friction in Transient Flow. *J. Water Resour. Plann. Manage.* **2000,** *126* (4), 236–244.
53. Koelle, E.; Luvizotto, Jr., E.; Andrade, J. P. G. Personality Investigation of Hydraulic Networks using MOC—Method of Characteristics. In *Proceedings of the 7th International Conference on Pressure Surges and Fluid Transients*. Harrogate Durham: United Kingdom, 1996, pp 1–8.
54. Filion, Y.; Karney, B. W. A Numerical Exploration of Transient Decay Mechanisms in Water Distribution Systems. In *Proceedings of the ASCE Environmental Water Resources Institute Conference*, American Society of Civil Engineers, Roanoke, Virginia, 2002; p 30.
55. Hamam, M. A.; McCorquodale, J. A. Transient Conditions in the Transition from Gravity to Surcharged Sewer Flow. *Can., J. of Civil Eng.* **1982,** 65–98.
56. Savic, D. A.; Walters, G. A. Genetic Algorithms Techniques for Calibrating Network Models. *Report No. 95/12*, Centre for Systems and Control Engineering, 1995; pp 137–146.
57. Walski, T. M.; Lutes, T. L. Hydraulic Transients Cause Low-Pressure Problems. *J. Am. Water Works Assoc.* **1994,** *75* (2), 58.
58. Lee, T. S.; Pejovic, S. Air Influence on Similarity of Hydraulic Transients and Vibrations. *ASME J. Fluid Eng.* **1996,** *118* (4), 706–709.
59. Chaudhry, M. H. *Applied Hydraulic Transients*; Van Nostrand Reinhold Co.: New York, NY, 1979; pp 1322–1324.
60. Parmakian, J. *Water Hammer Analysis*; Dover Publications, Inc.: New York, NY, 1963; pp 51–58.
61. Abbassi, B.; Al Baz, I. Integrated Wastewater Management: A Review. *Efficient Management of Wastewater—Its Treatment and Reuse in Water Scarce Countries*. Springer Publishing Co.: New York, NY, 2008.
62. Farooqui, T. A. Evaluation of Effects of Water Reuse, on Water and Energy Efficiency of an Urban Development Area, Using an Urban Metabolic Framework. *Master of Integrated Water Management Student Project Report*. International Water Centre, 2015.
63. Ferguson, B. C.; Frantzeskaki, N.; Brown, R. R. A Strategic Program for Transitioning to a Water Sensitive City. *Landsc. Urb. Plann.* **2013,** *117*, 32–45.
64. Kenway, S.; Gregory, A.; McMahon, J. Urban Water Mass Balance Analysis. *J. Ind. Ecol.* **2011,** *15* (5), 693–706.

65. Renouf, M. A.; Kenway, S. J.; Serrao-Neumann, S.; Low Choy, D. Urban Metabolism for Planning Water Sensitive Cities. Concept for an Urban Water Metabolism Evaluation Framework. *Milestone Report.* Cooperative Research Centre for Water Sensitive Cities, 2015.
66. Serrao-Neumann, S.; Schuch, G.; Kenway, S. J.; Low Choy, D. Comparative Assessment of the Statutory and Non-statutory Planning Systems: South East Queensland, Metropolitan Melbourne and Metropolitan Perth, 2013. Cooperative Research Centre for Water Sensitive Cities: Melbourne.
67. Andrews, S.; Traynor, P. *Guidelines for Assuring Quality of Food and Water Microbiological Culture Media.* August 2004.
68. Eaton, A. D.; Rice, E. W.; Clescceri, L. S. *Standard Methods for the Examination of Water and Waste Water*, Part 9000, 2012.
69. Razavi, K. Instructions *Procedure Rural Water and Wastewater Quality Assurance Test Results of Khorasan Razavi*, 2011.

CHAPTER 9

QUANTUM-CHEMICAL CALCULATIONS OF THE MOLECULES 1-METHYLBICYCLO[4,1,0]HEPTANE 2,4-SPIROHEPTANE BY PM3 METHOD

V. A. BABKIN[1], D. S. ANDREEV[1], V. V. PETROV[1], E. V. BELOZEROVA[1], O. V. STOYANOV[2], and G. E. ZAIKOV[3*]

[1]*Sebryakovsky Branch, Volgograd State University of Architecture and Engineering, Volgograd, Russia*

[2]*Department of Technology of Plastic Masses, Kazan State Technical University, Kazan, Russia*

[3]*Institute of Biochemical Physics RAN, Moscow, Russia*

[*]*Corresponding author. E-mail: GEZaikov@yahoo.com*

CONTENTS

ABSTRACT

This chapter presents the quantum-chemical calculations of the molecules 1-methylbicyclo[4,1,0]heptane, 2,4-spiroheptane which has been performed by the PM3 method with geometry optimization of all parameters by the gradient method. Optimized geometric and electronic structure of these compounds is obtained. Their acid strengths (p*K*a = 28 and 30) are theoretically evaluated. We have established that the researched molecules relate to the class of very weak acids (p*K*a > 14).

9.1 INTRODUCTION

Compounds with low cycles of 1-methyl bicyclo[4,1,0]heptane and 2,4-spiroheptane have hardly been studied so far or they have been poorly understood by physicochemical experimental methods. However, theoretical methods have been used, and, in particular, the method of quantum chemistry, modified neglect of diatomic overlap, made the first attempt to study the geometric and electronic structure of these compounds.[1,2] Such calculations are obviously needed to study the structure of the above-cationic polymerization of monomers in the electronic nanolevel, since it is a well-known fact (Academician Butlerov A.M.) that the structure of chemical compounds determines their properties. Test monomers 1-methylbicyclo[4,1,0]heptane and 2,4-spiroheptane are compounds with small classical cycles. The quantum-chemical method PM3A adequately simulates geometric and electronic structure of cyclic compounds; it is specifically parameterized to calculate the cyclic compounds and multivalent atoms.[3]

In this context, the aim of this work is the quantum-chemical calculation of the studied monomers 1-methylbicyclo[4,1,0]heptane and 2.4-PM3 spiroheptane method MP3 with optimization of geometry for all parameters by the gradient method, embedded in Firefly program,[4] which is partly based on the source code GAMESS (US),[5] in the isolated molecules in the gas-phase approximation and theoretical evaluation of their acid power. MacMolPlt program[6] was used to represent the molecular model visually.

9.2 THE CALCULATION RESULTS

Optimized geometric and electronic structure and quantum-chemical characteristics of the molecules 1-methylbicyclo[4,1,0]heptane,

2,4-spiroheptane obtained by PM3 are shown in Figures 9.1 and 9.2 and Tables 9.1–9.3. Using the formula p*K*a = 42.936–165.11 $q^{H^+}_{max}$, obtained by the authors according to the procedure for method PM3, but at technique for AM1 method[7] [$q^{H^+}_{max}$ = +0.092 (1-methylbicyclo[4,1,0]heptane) and +0.078 (2,4-spiroheptane)]—with maximum charges on the hydrogen atoms, p*K* is a universal indicator of acidity—see Tables 9.1–9.3), we find the value equal to the acid strength which is p*K*a = 28 and 30.

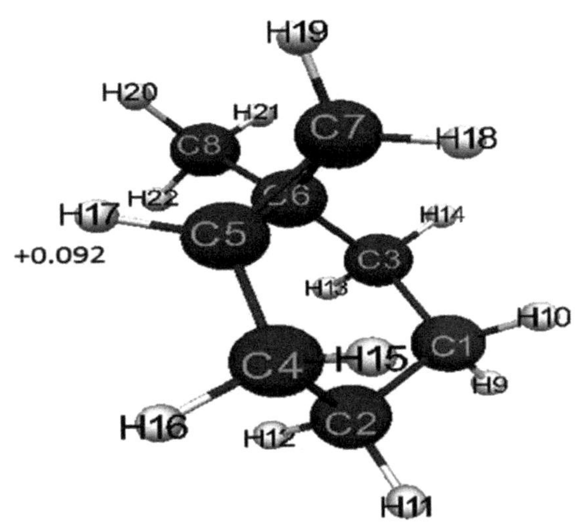

FIGURE 9.1 The geometric and electronic structure of molecules of 1-methylbicyclo[4,1,0] heptane (PM3 method) (E_0 = −112,291 kJ/mol).

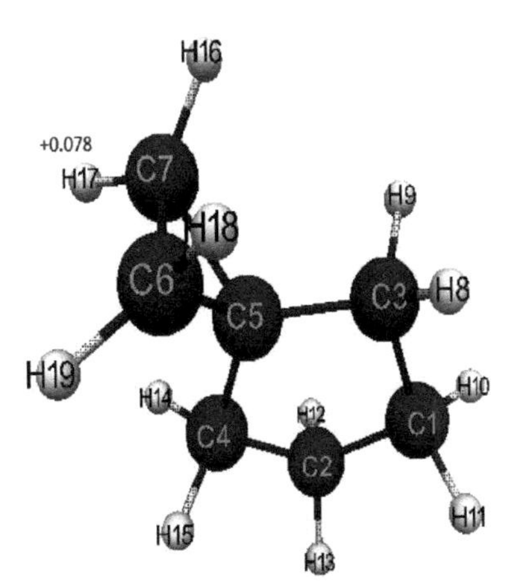

FIGURE 9.2 The geometric and electronic structure of the molecule 2,4-spiroheptane (PM3 method) (E_0 = −97,863 kJ/mol).

TABLE 9.1 The Optimized Bond Lengths, Bond Angles, and the Charges on the Atoms of the Molecule 1 Methylbicyclo[4,1,0]heptane (PM3 Method).

The bond length	*R* (A)	Valence angles	Grad	Atom	Charges on atoms of molecule
C(2)–C(1)	1.51	C(2)–C(4)–C(5)	112	C(1)	−0.098
C(1)–C(3)	1.52	C(4)–C(5)–C(6)	120	C(2)	−0.094
C(3)–C(6)	1.51	C(5)–C(6)–C(3)	119	C(3)	−0.058
C(6)–C(8)	1.50	C(6)–C(3)–C(1)	113	C(4)	−0.064
C(6)–C(7)	1.50	C(3)–C(1)–C(2)	110	C(5)	−0.127
C(6)–C(5)	1.50	C(5)–C(7)–C(6)	120	C(6)	−0.104
C(7)–C(5)	1.50	C(5)–C(6)–C(7)	60	C(7)	−0.155
C(5)–C(4)	1.50	C(5)–C(6)–C(8)	59	C(8)	−0.076
C(4)–C(2)	1.52	C(7)–C(6)–C(8)	117	H(9)	0.048
H(9)–C(1)	1.10	C(8)–C(6)–C(3)	112	H(10)	0.051
H(10)–C(1)	1.10	C(2)–C(1)–H(9)	110	H(11)	0.047
H(11)–C(2)	1.10	C(2)–C(1)–H(10)	109	H(12)	0.051
H(12)–C(2)	1.10	C(1)–C(2)–H(11)	110	H(13)	0.052
H(13)–C(3)	1.10	C(1)–C(2)–H(12)	109	H(14)	0.051
H(14)–C(3)	1.10	C(1)–C(3)–H(13)	109	H(15)	0.050
H(15)–C(4)	1.10	C(1)–C(3)–H(14)	110	H(16)	0.047
H(16)–C(4)	1.10	C(2)–C(4)–H(15)	109	H(17)	0.092
H(17)–C(5)	1.10	C(2)–C(4)–H(16)	109	H(18)	0.083
H(18)–C(7)	1.09	C(4)–C(5)–H(17)	112	H(19)	0.077
H(19)–C(7)	1.09	C(6)–C(7)–H(18)	119	H(20)	0.040
H(20)–C(8)	1.09	C(6)–C(7)–H(19)	119	H(21)	0.041
H(21)–C(8)	1.09	C(6)–C(8)–H(20)	111	H(22)	0.042
H(22)–C(8)	1.09	C(6)–C(8)–H(21)	110		
		C(6)–C(8)–H(22)	111		

TABLE 9.2 The Optimized Bond Length, Bond Angles, and the Charges on the Atoms of the Molecule 2,4-spiroheptane (PM3 Method).

The bond length	R (A)	Valence angles	Grad	Atom	Charges on atoms of molecule
C(1)–C(2)	1.52	C(5)–C(3)–C(1)	106	C(1)	−0.094
C(1)–C(3)	1.52	C(5)–C(4)–C(2)	106	C(2)	−0.094
C(3)–C(5)	1.51	C(3)–C(1)–C(2)	106	C(3)	−0.059
C(2)–C(4)	1.52	C(1)–C(2)–C(4)	106	C(4)	−0.059

TABLE 9.2 *(Continued)*

The bond length	R (A)	Valence angles	Grad	Atom	Charges on atoms of molecule
C(4)–C(5)	1.51	C(2)–C(4)–C(5)	106	C(5)	−0.111
C(5)–C(7)	1.50	C(4)–C(5)–C(3)	108	C(6)	−0.148
C(5)–C(6)	1.50	C(5)–C(7)–C(6)	60	C(7)	−0.148
C(6)–C(7)	1.49	C(5)–C(6)–C(7)	60	C(8)	0.052
H(8)–C(3)	1.10	C(1)–C(2)H(13)	111	H(9)	0.051
H(9)–C(3)	1.10	C(1)–C(2)H(12)	110	H(10)	0.049
H(10)–C(1)	1.10	C(2)–C(4)H(14)	111	H(11)	0.047
H(11)–C(1)	1.10	C(2)–C(4)H(15)	111	H(12)	0.047
H(12)–C(2)	1.10	C(5)–C(3)–H(8)	111	H(13)	0.049
H(13)–C(2)	1.10	C(5)–C(3)–H(9)	110	H(14)	0.052
H(14)–C(4)	1.10	C(3)–C(1)H(10)	111	H(15)	0.051
H(15)–C(4)	1.10	C(3)–C(1)H(11)	110	H(16)	0.077
H(16)–C(7)	1.09	C(5)–C(7)H(17)	119	H(17)	0.078
H(17)–C(7)	1.09	C(5)–C(7)H(16)	119	H(18)	0.078
H(18)–C(6)	1.09	C(5)–C(6)H(18)	119	H(19)	0.077
H(19)–C(6)	1.09	C(5)–C(6)H(19)	119		

TABLE 9.3 Total Energy (E_0, kJ/mol), the Maximum Charge on the Hydrogen Atom ($q_{max}^{H^+}$) and Universal Indicator of Acidity (p*K*) Monomers (PM3 Method).

No.	Monomer	E_0	$q_{max}^{H^+}$	p*K*a
1	1-Methylbicyclo[4,1,0]heptanes	−97,863	+0.092	28
2	2,4-Spiroheptane	−112,291	+0.078	30

9.3 CONCLUSION

Due to the work done, we have managed to perform quantum-chemical calculations of molecules 1-methylbicyclo[4,1,0]heptane using 2,4-spiroheptane method PM3. Besides, optimized geometric and electronic structure of these compounds has been determined. Also, their acid strength (p*K*a = 28 and 30) has been theoretically estimated. These data qualitatively correlate with the results of the experiments.[1,2] Furthermore, it has been determined that the studied monomers 1-methylbicyclo[4,1,0]heptane and 2,4-spiroheptane belong to the class of very weak acids H (p*K*a > 14).

KEYWORDS

- **quantum chemical calculation**
- **method PM3**
- **1-methylbicyclo[4,1,0]heptane**
- **2,4-spiroheptane**
- **acid strength**

REFERENCES

1. Babkin, V. A.; Shamin, S. M. *Quantum Chemical Calculations of Molecules 1 Methylbicyclo[4,1,0]heptane. The Collection of Articles: Quantum-Chemical Calculation of Unique Molecular System*; CRC Press: Boca Raton, FL, 2014; vol. 3, pp 138–140.
2. Babkin, V. A.; Shamin, S. M. *Quantum Chemical Calculations of the Molecule 2,4-Spiroheptane. The Collection of Articles: Quantum-Chemical Calculation of Unique Molecular System*; Apple Academic Press: Waretown, NJ, 2014; vol. 3, pp 121–124.
3. Tsirelson, V. G. *Quantum Chemistry. Molecules Molecular Systems and Solids*; BKL Publisher: Moscow, 2010; 422 p.
4. Granovsky, A. A. *Firefly Version 8*. http://classic.chem.msu.su/gran/firefly/index.html.
5. Shmidt, M. W.; Baldrosge, K. K.; Elbert, J. A.; Gordon, M. S.; Enseh, J. H.; Koseki, S.; Matsvnaga, N.; Nguyen, K. A.; Su, S. J.; et al. Advances in Electronic Structure Theory: GAMESS a Decade Later. *J. Comput. Chem.* **1993,** *14*, 1347–1363.
6. Bode, B. M.; Gordon, M. S. MacMolPlt: A Graphical User Interface for GAMESS. *J. Mol. Graphics* **1998,** *16*, 133–138.
7. Babkin, V. A.; Andreyev, D. S.; Fomichev, V. T.; Stammering, G. E.; Muhamedzyanova, E. R. About Correlation Universal Indicator of Acidity with a Maximum Charge on the Hydrogen Atom, H-Acid. AM Method. *Kazan Bull. Kazan Technol. Univ.* **2012,** *10*, 15–18.

CHAPTER 10

MODEL-BASED INVESTIGATION OF TRANSPORT PHENOMENA IN WDNs

KAVEH HARIRI ASLI[1] and HOSSEIN HARIRI ASLI[2]

[1]*Department of Mathematics and Mechanics, National Academy of Science of Azerbaijan "AMEA," Baku, Azerbaijan*

[2]*Civil Engineering Department, Faculty of Engineering, University of Guilan, Rasht, Iran*

Corresponding author. E-mail: hariri_k@yahoo.com

CONTENTS

ABSTRACT

The increase in flow of velocities in the pipes will increase the risk of hydraulic shock. The emergency power shutdown of engine pumps is the main reason which causes the pressure wave in pipelines. A hydraulic shock arises from the changes in the degree of opening the valves. This may lead to the destruction of the pipeline and violations of the normal operation of dewatering systems. The establishment of controllable, manageable areas [district meter areas (DMAs)] within a distribution system, whose demands are easily monitored, has been found to be extremely helpful for effective leakage control and supply management. On the other hand, this work showed proper analysis to provide a dynamic response to the shortcomings of the system. It also performed the design protection equipment to manage the transition energy and determine the operational procedures to avoid transients. Consequently, the results will help to reduce the risk of system damage or failure at the water pipeline.

10.1 INTRODUCTION

The problems of protection against hydraulic shock solution are formed by increasing the safety margin of pipes. This can be attributed to the imperfect design of the device proposed for protecting pipelines from hydraulic shock. It is a complex process, nonlinearly depending on factors such as the moment of inertia of the rotor pump unit. It is related to the length of pressure pipe, and so on.

The plants with lower height probably have high relative increase in pressure. Therefore, it is necessary the protection of dewatering equipment necessary to apply the device for protection against hydraulic shock. The suggestion of the suitable device against hydraulic shock must be recognized in the design of a new pumping installation. The value of excess pressure (working pressure of the system) in the hydraulic impact must be revealed. This problem can be solved in several ways. It usually requires a lot of time consuming and complex calculation that a person do not have. For this purpose, an abroad program has been developed for calculating water hammer in the pipeline. But as a rule, the cost of the software is sufficiently large and, therefore, design organizations and research institutions have limited ability to use software products of foreign firms.[1–5]

The sudden opening of the valve (i.e., the rapid exchange of fluid velocity) causes a decrease in pressure occurs in oscillatory process and pressure change. In cases with the impact phenomenon, the system must be equipped with devices that are not allowed to make an instant decrease in velocity (valve-type shut-off devices). This device shall be restricting the spread of pressure wave attack.

Water loss may be defined as that water which having been obtained from a source and put into a supply and distribution system is lost via leaks or is allowed to escape or is taken there from for no useful purpose. "Water loss" is usually considered as "leakage" and "water-loss reduction" referred to as "leakage control." Water loss is usually quantified on the following basis:

Water loss = (Quantity of water put in to supply)
– (Nondomestic usage + Domestic consumption)

Nonrevenue Water

To allow for leakage and quantities termed as "other water uses," the term nonrevenue water (called NRW) or "unaccounted for water" is used (called UFW). This is a good way of distinguishing it from the useful water supplied to both domestic and nondomestic consumers, which is grouped together as "accounted for" water. The classic leakage control formula is

$$U = S - \left(M + \left(A \times P\right)\right), \tag{10.1}$$

where U is the unknown or unaccounted for quantities of water including leakage; S is the sum of all water inputs into a system; M is the sum of all water accounted for by measure (metered supplies, domestic, and nondomestic); A is the average domestic usage per capita of population; and P is the population supplied (nonmetered).

Total integrated flow formula:

$$U = S - \left(M + A \times P\right), \tag{10.2}$$

where U is the unaccounted for water; S is the total volume supplied; M is the metered use; A is the per capita use; P is the population supplied unmetered.

Old iron mains still form the majority of mains and they are the worst culprits for leakage. They suffer from both external and internal corrosion attack which progressively weakens them. Iron mains can then crack and leak or holes form due to the corrosion process. Once the leakage occurs, which may be finally precipitated by an increase in pressure, flow, or temporary change, it will worsen. This may occur steadily, or rapidly degenerate into a large burst. Cases of subterranean caverns beneath metallic roadways are known where the escaping water hollows out a void by its pressure jet. Concrete lining of iron pipes mains virtually stops internal corrosion but have no effect on external corrosion. Asbestos cement pipes mains normally fail by cracking, acting as a beam under load, and the subsequent collar repair can be a source of future trouble.

The unplasticized polyvinyl chloride (UPVC) pipes are not thought to contribute largely to the total water lost. Failures in the early plastic pipes have been frequent in large diameter sizes, and the pipes usually fail by shattering. Joint ring failure is sometimes a problem.

Medium-density polycthylene pipes are still a relatively recent introduction but their performance to date is excellent, provided they are jointed properly. Furthermore, polyethylene pipes are still being improved, which can only be good for the future.

Steel pipes mains only form a small proportion of mains and these are usually in aqueducts with cathodic protection. Steel fails usually with pin holing, necessitating welded patch repairs.

Clearly soils influence corrosion and leakage rates. Some light soils scarcely affect the pipes while others such as alluvium are very aggressive. Trench back-fill of sulfate-rich ashes is especially corrosive. Water fed into supply should be carefully controlled for quality. It should be checked to ensure that it is not plumbo-solvent. Certain natural waters have a higher rate of attack on iron pipe than others. Seasonal variations in climate have a marked effect upon leakage levels. For instance, a hard winter induces ground movement in the "freeze/thaw" cycle, and this causes a high number of bursts. In a similar way, a long drought causes ground movement, and again often results in an increase in the number of bursts. This may distort leakage estimates for particular years. Sudden saturation of dried out subsoil can also cause problems through local "heave." Mining subsidence can create successive tension and compression of the pipe work causing joint movement or failure of the pipe. Removal of support from thrust blocks can lead to excessive joint movement. It should be noted that this can also be created by excavation adjacent to the thrust block destroying passive ground pressure at the supporting face. Dissimilar metals between pipes and fittings (e.g., just a position of copper and stainless steel) can cause galvanic corrosion. This must be avoided by reference to guidance in bylaws/regulations/standards, etc. Electrical earthing of buildings to the water fittings has been prohibited in the United Kingdom since 1961. It was common before that time, and faulty electrical fittings can create a "to earth" potential onto water pipes which, in turn, will create corrosion and eventual leakage. It should also be noted that this now obsolete practice can make service pipes (and mains to some extent) electrically live and dangerous. Temporary earth loops must be used.

A properly designed distribution system should prevent some vulnerability to leakage at the outset. Such design would assess the need for cathodic protection of steel and ductile iron mains. It would ensure that all mains with unrestrained flexible joints had appropriately sized and positioned concrete thrust blocks at all changes of direction and blank ends. All mains and services should be laid with the correct amount of cover to the surface, and appropriately distanced from other underground services. The use of marker tape sited 300 mm above the main will alert excavation to the presence of the main, thus preventing incidental damage and ensuing leakage. Where plastic pipe is used, such tape should have a metallic strip incorporated to assist with location equipment. Correct sizing of mains at the outset, considering such factors as peak flow, fire-fighting requirements and future development, will prevent the temptation to "force" more water through by increasing pressures at a later date. Oversized mains also need to be avoided, particularly from a water quality point of view. There is no substitute for good workmanship of the initial installation in preventing future leakage. Pipe handling, bedding, laying, jointing, and backfilling must be to a high standard. Extra care should be given to repair work, as a repair does represent a potential weakness to the integrity of the system. It is obvious that all materials used in the distribution system must comply with relevant standards for long-term usefulness be of a high quality, be appropriate to the surrounding conditions, and be of the correct operational capabilities. It should also be ensured that the same standards apply to repair materials, and that poor substitutes are not used for permanent repairs. High pressure equals high leakage. This factor is very important in leakage control and will be discussed in more detail in a subsequent section. This may seem obvious but it is very important to remember that a small increase in the size of the leak has a big effect in terms of volume leaked. Figure 10.1 illustrates this for a constant pressure of water. The longer a leak is left to run, the bigger the hole will get.

A speedy location and repair of leaks is essential to reduce waste levels. A leak running for a long time can waste just as much water as catastrophic trunk main burst which is repaired quickly. Time before discovery, time to detect, and time to repair are the major components. Leakage will only be reduced by sustained, determined detection, and rapid repair. "Find and Fix Fast" is an appropriate axiom. Severe pressures can be generated by the rapid operation of isolating valves, thus precipitating bursts and leakage.

Ironically, rapid recharging of a system following leakage repair work can cause further damage and leakage. Valve closures and mains recharging work should therefore be carried out in a steady, controlled manner. This is particularly relevant when mains scraping and relining is taking place. Similar care must be taken during mains flushing, swabbing, and air scouring. The aging process cannot be stopped and increasing leakage is indicative of deteriorating structural condition. It should therefore be recognized that a realistic and consistent level of renewal of the infrastructure is an essential part of leakage strategy development. This may be achieved by targeted mains relining (where iron pipes are in use and corrosion is mostly internal) or by targeted mains replacement. The former has little, if any, impact on leakage rates from those mains, whereas the latter should eradicate it for a substantial period of time if done well. The modern techniques of mains replacement have substantially cut excavation and backfilling costs. It is essential that the renewal of service pipes is included in such work for the greatest benefit. Leakage grows with time, and without action to curb it would grow to a point where supplies would be unsustainable. Passive control, that is, the repair of bursts and leakage showing on the surface, and the elimination of poor pressure and flow complaints, is the minimum possible response. This is required to prevent damage to persons and property and to maintain supplies to customers. The actual leakage level reached will depend on how quickly low pressure and flow will be experienced and other factors, such as how quickly leakage appears on the surface and is reported. For any given area in the distribution system, there will be a characteristic growth rate. This characteristic growth rate will be affected by changes in the physical elements of the system, such as rehabilitation of mains, renewal of service pipes, and changes in pressure. Sooner or later, leakage control must be associated with a program of mains renewal to maintain the supply/demand balance. However, improvement of mains and services is expensive and clearly, for the system as whole, is very much a long-term strategy. Reduction in pressure is also effective in reducing both the volume of leakage and its rate of growth, although there is some doubt whether the latter effect persists in the long term. The scope for pressure reduction is, of course, limited, given that adequate supplies to customers must be maintained. To reduce the natural level of leakage at any pressure, a program of leakage detection must be planned, coordinated and implemented. The effects of the introduction of various levels or frequencies of leakage detection are again illustrated in Figure 10.1. This shows clearly the need to maintain a consistent level of effort if the required leakage level is to be maintained. It is not sufficient to put in a high level of resource for a short period, as any slackening of effort will lead to an increase in leakage over a period of time. Given that no two water distribution systems are identical in terms of physical or economic characteristics, it is not possible to determine the most appropriate leakage control policy in a general manner. The best policy for any given system will depend on its particular characteristics.

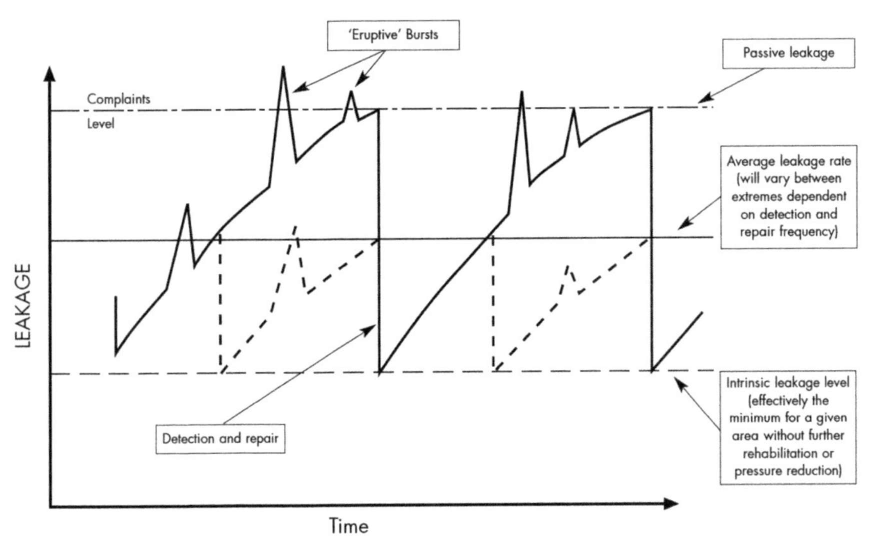

FIGURE 10.1 Graph to show growth of leakage with time.

The economic balance of searching for and repairing leakage, and of controlling it to an acceptable level, is a complex issue. Typically, a leakage percentage of below 10% or even 15% may not be economic to pursue. In other words, the effect of hunting down, identifying, and repairing the leakage costs more than the value of the water saved. These remarks need to be heavily qualified however. For instance, a modern housing estate could have a serious problem with 10% leakage, whereas an old area with a stubborn leakage of 30% may require a mains renewal scheme. Each area will have its own intrinsic economic leakage level. In the United Kingdom, historically, a quantity equivalent to 55 l/prop/day was deemed too expensive to find and repair and was termed "acceptable leakage." It was further suggested that of the "acceptable leakage," the quantity of leakage which was undetectable approximately 30 1/prop/day. In addition to the volume of water lost, its scarcity and marginal cost per megaliter are vital factors. In an area of rising demand, needing to promote, build and commission a new source, intensive leakage control activity would be essential. In an area which relied upon pumped supplies with high electricity costs, a high degree of leakage control would make sense and have priority over an area with plentiful supplies fed by gravity. It is increasingly accepted that an active approach of searching for leakage is preferential in cost/benefit terms to a passive approach of only reacting when the situation has deteriorated.

This relates not only to the water supplier's distribution system but also to private pipe work where customers are encouraged to carry out repairs on any leakage detected. Active control would usually involve the monitoring of flows in a distribution network by using a system of permanently installed distribution meters. If unexpectedly high flows of water are observed, these are immediately investigated; leakage detection teams being carefully directed to ensure that leakage is maintained within defined criteria (such criteria being prepared using an acceptable cost/benefit basis). It is obvious that monitoring which does not initiate further action is unproductive. It will also be unproductive if, when further action is worthwhile, resources are not available to proceed with location of the leakage. An active policy requires expenditure on meter installations, etc. and the day-to-day operating costs of leakage detection teams. The following benefits should be achieved:

- It minimizes leakage and, hence, reduces the loss of water in monetary terms.
- It results in an overall reduction of water demand.
- Limited water resources are conserved for legitimate use and rationing, etc. is avoided.
- It reduces operating costs (savings on electrical power and chemical treatment costs).
- Work is planned (rather than acting in response to emergency).
- Dangerous leakage is minimized (e.g., freezing water on highway).
- Customer perception is improved.
- Capital expenditure requirements on treatment works, reservoirs, and mains are reduced.

Because of their potential, it is worth noting that active leakage detection in the future is likely to increasingly employ acoustic loggers, some permanently installed. This could result in larger meter areas and, hence, fewer district meters. A well-managed active leakage detection policy ensures that the cost of the leakage detection teams and the repayments of the capital necessary to establish the system is exceeded by the value of the water saved.

It is applicable if:

- The cost of water production is high.
- The sources of water have limited capacity and cannot meet normal and/or foreseen demand.
- Bursts are "invisible" due to the strata.
- The quantity of water being put into supply is increasing at an unacceptable rate.

Leakage reduction and control is a long-term activity and should be regarded as a part of good distribution management. Occasional short bursts of effort are unlikely to produce lasting results because distribution systems continue to deteriorate for one reason or another. If only the obvious leaks are repaired, leakage levels will still increase, as will consumer problems. The development of a long-term leakage control strategy is therefore essential if water supply and distribution systems are to be effectively managed. Such development needs to be flexible, with occasional reviews to ensure that the strategy adopted is the most appropriate one for the situation. Cost/benefit analysis is important in this regard. "Active" leakage control (i.e., finding and repairing leaks before their presence becomes obvious or generates problems) has been found to be a cost-effective method of reducing water supply deficiencies. A planned approach should result in lower complaint costs, and lower repair and maintenance costs. The establishment of controllable, manageable areas [district meter areas (DMAs)] within a distribution system, whose demands are easily monitored, has been found to be extremely helpful for effective leakage control and supply management. It forces plans to be updated, locating mains and buried fittings. It introduces new valves to give better operational control. It locates illegal connections and identifies malfunctioning meters and public supplies. In the very process of this setting up work, leakages and wastages are found and repaired. It enforces good housekeeping. Regard has to be given, however, to the minimization of dead ends and their associated quality problems. Leakage reduction requires a dedicated core of highly trained, specialist personnel using appropriate "state-of-the-art" equipment and techniques. Local knowledge is essential together with an understanding of the day-to-day operation of the distribution system and demand patterns. Support can be obtained from specialist agencies/contractors, given precise briefs and targets. Personnel motivation, good communication, and synchronization of activities and continuous feedback of decisions/results cannot be over-emphasized. This is vital for understanding, efficiency, and success. Everyone should be included, from planners to repair teams. The organization of leakage control personnel can vary widely. Distribution personnel can either be organized as a specialist team, or be integrated into general distribution system operational duties, and spend only part of their time on leakage control. It is generally accepted that to properly pursue active leakage control and to meet agreed monitoring/detection frequencies, it is necessary to set up specialist teams. However, general operational duties cannot and should not be entirely divorced from leakage control. Technical support is required for design and modification of district metering, computer systems support, compilation of base data for DMAs, production of reports, overall-performance monitoring, production of drawings, system records updating, and for problem solving. Clerical support is required for computer input and administrative duties such as serving notices relating to private pipe repairs. Skilled and knowledgeable technical support is crucial if the mass of data now regularly available is to be handled and analyzed to the best advantage for the leakage reduction effort. Good leakage control depends upon good and progressively improving data. To achieve this, it is necessary to establish and keep an audit trail of data, building from individual DMAs and their component data up to the regional total figures for the water supplier. These can be collected in the two data streams of

- Aggregated night-lines/"bottom-up" calculation
- Total integrated flow/"top-down" calculation.

Network analysis is the term used to describe the "analysis of water flows and head losses in a pressurized distribution system under a given set of demand conditions on the system."

A network is the collection of pipes, valves, booster pumps, and service reservoirs forming the water distribution system. Due to the complexity of most distribution systems, it was normal to simplify the system by considering only the key mains. With the development in recent years of computer hardware and software, it is now possible to include all reservoirs and mains in a distribution system, and all the various control features, with their operating constraints and regimes. The demands and demand patterns on a network are also vital ingredients and are made up of a number of components:

1. domestic demand;
2. metered industrial/commercial demand; and
3. unaccounted for water including leakage.

This is the process of calculating the flows and head losses in a network for a given set of demand conditions. Two types of analysis are normally used:

- In a snapshot analysis, the flows and head losses are considered at only a single given set of demand conditions. This is frequently expressed as a single time interval.

Dynamic or extended time

- In each dynamic analysis, the flows and head losses are considered for a series of varying demand conditions. This is frequently a 24-h time period and is the sort of analysis that is now most commonly used. The power and speed of computing for network analysis continues to improve.

A network model is basically an intelligent mains record drawing—allowing one to access hydraulic data as well as the position of the mains in the ground. A model represents everything we know about a particular distribution system. It will have been calibrated by the model builders to ensure that within reason, the model gives the same flows and pressures as the real system. This is done by comparing the results from the model with huge amounts of data from field tests. It is essential to know the system of configuration on the calibration day—that is, the day chosen as the most "typical" from the field test. The calibration process will find any significant problems with the model's representation of the distribution system, but not all of them. To calibrate a model, it is necessary to get the pressures right within one meter at virtually all points in the system at all times of day. Once the model is created, it has to be converted to what is known as an *average day model*. To do this, the model builder converts the demands on the model to average demands by comparing the demands for that area with the test day.

If there is a disagreement between the computed flows and the measured flows, a number of factors can be involved. The more common are listed below:

- Incorrect estimates for model demands
- Incorrect assumptions for hydraulic resistances
- Wrong pipe lengths or diameters
- Unsuspected network cross-connection
- Closed valves/opened valves

- Bypasses around pressure reduction valve (PRV) or meters
- Restrictions in mains
- Pressure measurement on "rider" main.

The process of model building can thus uncover many problems which may go unnoticed until a burst occurs, often wasting time and money.

Network analysis is a powerful tool for the effective management of distribution systems. Once a model exists, it allows any user to experiment with system changes before they are tried out on the ground. These could be such things as checking what reinforcements are needed to supply a new development, so that levels of service are not affected somewhere else, perhaps miles away. The model could help maximize the utilization of low cost supplies, and in pumped distribution systems, minimize the cost of pumping. It could also ensure that levels of service are achieved at customer taps by identifying areas of inadequate or excessive pressures, and areas of high leakage; corrective measures could then be simulated. It might be used for planning a trunk main shut-off, with effects over wide areas, perhaps to see how long the reservoir storage will last. It can be used to check on rehabilitation problems—reline, renew, or upsize. It can also be used to design pressure reduction or to alter distribution areas. As the techniques improve, it will also be used to investigate water quality problems. Network models can already tell us how old the water is throughout a system and how that changes during the day. They can also be used to tell us how different source waters blend in the system at different times of the day. All these might point to problem areas and show the results on water quality of system changes. Network models give us a better picture of the system operation, and help improve levels of service. A lot of money can be saved on capital schemes by using models to find out what size mains are really needed, or to sometimes find ways of not laying new mains at all. Network models may be useful in locating large leaks by comparing modeled pressures against actual. Large leaks cause a lowering of pressures. Network models are not perfect, but they are the only tool available to provide such detailed hydraulic information. In the past we often had to guess about the behavior of complex systems. Distribution management is an important activity which has considerable impact on customers. The costs of distribution operations are high. It is therefore vital that management decisions are taken in a framework of knowledge and understanding of how the system operates. The development of DMAs as part of a structured operation of the distribution system allows the network to be operated in a planned way. This planned approach inevitably leads to better understanding and control of the distribution system, updated and more comprehensive records, fewer consumer complaints, and closer control of labor. Such an approach helps to ensure that distribution managers can meet the primary objectives to the maximum benefit of the customer and the water supplier. DMAs are the basic building blocks of a zoned distribution system. They provide a manageable unit by which the distribution customer and performance information can be linked to other activities and data systems.

Their fundamental characteristic is that their boundaries are closed except for defined, measured inputs and outputs. Ideally, this should be a single metered input, but this is not always achievable in practice. DMAs in the United Kingdom are generally between 1000 and 5000 properties in size, and they have similar topography with limited head loss within their area even and allow pressure and leakage to be managed most effectively. Larger areas are possible from a detection point of view if acoustic logging is part of the detection policy employed. The principles of DMA design and structure are very simple. Nevertheless, where possible, the design should be checked using network analysis to ensure that pressures are sustained at all likely demands, that no unnecessarily long water retention periods are created and that water quality variations are within an acceptable range—larger areas usually means less "dead ends." System record plans are required, preferably at a scale of 1:2500, together with property count data. This information is used, together with the local system operator's knowledge, to define the boundaries of each DMA. Other important considerations in this process are as

1. to cross the fewest number of distribution mains (helped by using natural boundaries such as railway lines, canals and major roads), thereby reducing the number of meters used and the number of closed valves (which can lead to water quality problems);
2. to avoid districts with high outflows (this leads to inaccuracy in calculation of district demand as any changes in demand will be a small proportion of the total flow measured).

Having defined the limits of a DMA, it will normally be necessary to trial the area in practice. It will be necessary to ensure that all stop (stand shut, boundary) valves perform correctly, and that satisfactory flows and pressures are maintained throughout the DMA.[6–16]

10.2 MATERIALS AND METHODS

In practice, DMAs often have to be checked very carefully during establishment. Unforeseen difficulties may be found, such as buried, or closed valves, or even unknown pipes. These problems are often discovered when the DMA is first modeled and anomalies in the model are investigated. Once satisfactorily piloted, the DMA can be fully established. This will require

1. the installation of flow meters at all inlets and outlets;
2. the closing and marking of all boundary valves;
3. the installation of flushing, or "OXO," points; and
4. the updating of plans, records, and related information systems.

The simple checklist below can be used to ensure that all of these activities are performed before a DMA is commissioned.

For a DMA, both rigid-column flow and transient analyses will treat in pipe networks. Hydraulic transients can lead to the following physical phenomena High or low transient pressures that may arise in the piping and connections in the share of second. They often alternate from highest to lowest levels. High pressures are a consequence of the collapse of steam bubbles or cavities are similar to steam pump cavitations. It can yield the tensile strength of the pipes. It can also penetrate the groundwater into the pipeline.[17–58]

High-speed flows are also very fast pulse pressure. It leads to temporary but very significant transient forces in the bends and other devices that can make a connection to deform. Even strain buried pipes under the influence of cyclical pressures may lead to deterioration of joints and lead to leakage. In the low-pressure pumping stations at downstream, a very rapid closing of the valve, known as shut-off valve, may lead to high-pressure transient flows.

Water column, usually, are separated with sharp changes in the profile or the local high points. It is because of the excess of atmospheric pressure. The spaces between the columns are filled with water or the formation of steam (e.g., steam at ambient temperature) or air, if allowed admission into the pipe through the valve. Collapse of cavitation bubbles or steam can cause the dramatic impact of rising pressure on the transition process. If the water column is divided very quickly, it could in turn lead to rupture of the pipeline. Vapors cavitation may also lead to curvature of the pipe. High-pressure wave can also be caused by the rapid removal of air from the pipeline. Steam bubbles or cavities are generated during the hydraulic transition. The level of hydraulic pressure or energy gradient or pressure in some areas could fall low enough to reach the top of the pipe. It leads to subatmospheric pressure or even full-vacuum pressures. Part of the water may undergo a phase transition, changing from liquid to steam, while maintaining the vacuum pressure. This leads to a temporary separation of the water column. When the system pressure increases, the columns of water rapidly approach to each other. The pair reverts to the liquid until vapor cavity completely dissolved. This is the most powerful and destructive power phenomenon. If system pressure drops to vapor pressure of the liquid, the fluid passes into the vapor, leading to the separation of liquid columns. Consequently, the vapor pressure is a fundamental parameter for hydraulic transient modeling. The vapor pressure varies considerably at high temperature or altitude. Fortunately, for typical water pipelines and networks, the pressure does not reach such values. If the system is at high altitude or if it is the industrial system, operating at high temperatures or pressures, it should be guided by a table or a state of vapor pressure curve vapor liquid. The single-phase (pure liquid) hydraulic transient equations are the method of characteristics finite differences (FDs), wave-characteristic method, finite elements (FE), and finite volume. One difficulty that commonly arises relates to the selection of an appropriate level of time step to use for the analysis. The obvious trade-off is between computational speed and accuracy. In general, for the smaller the time step, there is the longer the run time, but the greater is the numerical accuracy. An evaluation of surge or pressure wave in elastic is the case with the free water bubble. It started with the solving of approximate equations by numerical solutions of the nonlinear Navier–Stokes equations based on the MOC. Then, it derived the Zhukovsky formula and velocity of surge or pressure wave in an elastic case with the high value of free water bubble. So the numerical modeling and simulation which was defined by "MOC" provided a set of results. Basically, the "MOC" approach transforms the water hammer partial differential equations into the ordinary differential equations along the characteristic lines.

It is defined as the combination of momentum equation and continuity equation for determining the velocity and pressure in a one-dimensional flow system. The solving of these equations produces a theoretical result that usually corresponds quite closely to actual system measurements (10.3–10.37):

$$P\Delta A-\left(P+\frac{\partial P}{\partial S}\cdot\Delta S\right)\Delta A-W\cdot\sin\theta-\tau\cdot\Delta S\cdot\pi\cdot d=\frac{W}{g}\cdot\frac{\mathrm{d}V}{\mathrm{d}t} \qquad (10.3)$$

Both sides are divided by m and with assumption:

$$\frac{\partial Z}{\partial S} = +\sin\theta$$

$$-\frac{1}{\partial}\cdot\frac{\partial P}{\partial S} - \frac{\partial Z}{\partial S} - \frac{4\tau}{\gamma D} = \frac{1}{g}\cdot\frac{dV}{dt}, \tag{10.4}$$

$$\Delta A = \frac{\Pi \cdot D^2}{4}.$$

If fluid diameter assumed equal to pipe diameter, then:

$$\frac{-1}{\gamma}\cdot\frac{\partial P}{\partial S} - \frac{\partial Z}{\partial S} - \frac{4\tau_0}{\gamma \cdot D}, \tag{10.5}$$

$$\tau_0 = \frac{1}{8}\rho\cdot f\cdot V^2,$$

$$-\frac{1}{\gamma}\cdot\frac{\partial P}{\partial S} - \frac{\partial Z}{\partial S} - \frac{f}{D}\cdot\frac{V^2}{2g} = \frac{1}{g}\cdot\frac{dV}{dt}, \tag{10.6}$$

$$V^2 = V\,|\,V\,|, \quad \frac{dV}{dt} + \frac{1}{\rho}.\frac{\partial P}{\partial S} + g\frac{dZ}{dS} + \frac{f}{2D}V|V| = 0. \tag{10.7}$$

(Euler equation)

For finding (V) and (P), we need "conservation of mass law":

$$\rho AV - \left[\rho AV - \frac{\partial}{\partial S}(\rho AV)dS\right] = \frac{\partial}{\partial t}(\rho A dS) - \frac{\partial}{\partial S}(\rho AV)dS = \frac{\partial}{\partial t}(\rho A dS), \tag{10.8}$$

$$-\left(\rho A\frac{\partial V}{\partial S}dS + \rho V\frac{\partial A}{\partial S}dS + AV\frac{\partial \rho}{\partial S}AS\right) = \rho A\frac{\partial}{\partial t}(dS) + \rho dS\frac{\partial A}{\partial t} + AdS\frac{\partial p}{\partial t}, \tag{10.9}$$

$$\frac{1}{\rho}\left(\frac{\partial \rho}{\partial t} + V\frac{\partial \rho}{\partial S}\right) + \frac{1}{A}\left(\frac{\partial A}{\partial t} + V\frac{\partial A}{\partial S}\right) + \frac{1}{dS}\cdot\frac{\partial}{\partial t}(dS) + \frac{\partial V}{\partial S} = \circ$$

with

$$\frac{\partial \rho}{\partial t} + V\frac{\partial \rho}{\partial S} = \frac{d\rho}{dt} \quad \text{and} \quad \frac{\partial A}{\partial t} + V\frac{\partial A}{\partial S} = \frac{dA}{dt}$$

$$\frac{1}{\rho}\cdot\frac{d\rho}{dt} + \frac{1}{A}\cdot\frac{dA}{dt} + \frac{\partial V}{\partial S} + \frac{1}{dS}\cdot\frac{1}{dt}(dS) = \circ. \tag{10.10}$$

$K = \left(\mathrm{d}\rho / (\mathrm{d}\rho / \rho)\right)$ (Fluid module of elasticity), then

$$\frac{1}{\rho} \cdot \frac{\mathrm{d}\rho}{\mathrm{d}t} = \frac{1}{k} \cdot \frac{\mathrm{d}\rho}{\mathrm{d}t}. \tag{10.11}$$

Put (10.7) into (10.8), then

$$\frac{\partial V}{\partial S} + \frac{1}{k} \cdot \frac{\mathrm{d}\rho}{\mathrm{d}t} + \frac{1}{A} \cdot \frac{\mathrm{d}A}{\mathrm{d}t} + \frac{1}{\mathrm{d}S} \cdot \frac{\mathrm{d}}{\mathrm{d}t}(\mathrm{d}S) = \circ$$

$$\rho \frac{\partial V}{\partial S} + \frac{\mathrm{d}\rho}{\mathrm{d}t} \rho \begin{bmatrix} \frac{1}{k} + \frac{1}{A} \cdot \frac{\mathrm{d}A}{\mathrm{d}\rho} \\ + \frac{1}{\mathrm{d}S} \cdot \frac{\mathrm{d}}{\mathrm{d}\rho}(\mathrm{d}S) \end{bmatrix} = \circ, \tag{10.12}$$

$$\rho \left[\frac{1}{k} + \frac{1}{A} \cdot \frac{\mathrm{d}A}{\mathrm{d}t} + \frac{1}{\mathrm{d}S} \cdot \frac{\mathrm{d}}{\mathrm{d}\rho}(\mathrm{d}S) \right] = \frac{1}{C^2}.$$

Then

$$C^2 \frac{\partial V}{\partial S} + \frac{1}{\rho} \cdot \frac{\mathrm{d}\rho}{\mathrm{d}t} = \circ. \tag{10.13}$$

(Continuity equation)

Partial differential equation is solved by method of characteristics (MOC):

$$\frac{\mathrm{d}p}{\mathrm{d}t} = \frac{\partial p}{\partial t} + \frac{\partial p}{\partial S} \cdot \frac{\mathrm{d}S}{\mathrm{d}t}, \tag{10.14}$$

$$\frac{\mathrm{d}V}{\mathrm{d}t} = \frac{\partial V}{\mathrm{d}t} + \frac{\partial V}{\partial S} \cdot \frac{\mathrm{d}S}{\mathrm{d}t}, \tag{10.15}$$

then

$$\left| \frac{\partial V}{\partial t} + \frac{1}{\rho} \frac{\partial p}{\partial S} + g \frac{\mathrm{d}z}{\mathrm{d}S} + \frac{f}{2D} V|V| = \circ, \right. \tag{10.16}$$

$$\left| C^2 \frac{\partial V}{\partial S} + \frac{1}{\rho} \frac{\partial P}{\partial t} = \circ. \right. \tag{10.17}$$

By linear combination of

$$\lambda \left(\frac{\partial V}{\partial t} + \frac{1}{\rho} \frac{\partial p}{\partial S} + g \cdot \frac{\mathrm{d}z}{\mathrm{d}S} + \frac{f}{2D} V|V| \right) + C^2 \frac{\partial V}{\partial S} + \frac{1}{\rho} \frac{\partial p}{\partial t} = \circ, \tag{10.18}$$

$$\left(\lambda\frac{\partial V}{\partial t}+C^2\frac{\partial V}{\partial S}\right)+\left(\frac{1}{\rho}\cdot\frac{\partial \rho}{\partial t}+\frac{\lambda}{\rho}\cdot\frac{\partial P}{\partial S}\right)+\lambda\cdot g\cdot\frac{\mathrm{d}z}{\mathrm{d}S}+\frac{\lambda\cdot f}{2D}V|V|=\circ, \quad (10.19)$$

$$\lambda\frac{\partial V}{\partial t}+C^2\frac{\partial V}{\partial S}=\lambda\frac{\mathrm{d}V}{\mathrm{d}t}\Rightarrow\lambda\frac{\mathrm{d}S}{\mathrm{d}t}=C^2, \quad (10.20)$$

$$\frac{1}{\rho}\cdot\frac{\partial p}{\partial t}+\frac{\lambda}{\rho}\cdot\frac{\partial \rho}{\partial S}=\frac{1}{\rho}\cdot\frac{\mathrm{d}\rho}{\mathrm{d}t}\Rightarrow\frac{\lambda}{\rho}=\frac{1}{\rho}\cdot\frac{\mathrm{d}S}{\mathrm{d}t}, \quad (10.21)$$

$|C^2/\lambda=\lambda$ (by removing $\mathrm{d}S/\mathrm{d}t$), $\lambda=\pm C$.

For $\lambda=\pm C$ from eq 10.18, we have

$$C\frac{\mathrm{d}V}{\mathrm{d}t}+\frac{1}{\rho}\cdot\frac{\mathrm{d}p}{\mathrm{d}t}+C\cdot g\cdot\frac{\mathrm{d}z}{\mathrm{d}S}+C\cdot\frac{f}{2D}V|V|=\circ$$

With dividing both sides by "C":

$$\frac{\mathrm{d}V}{\mathrm{d}t}+\frac{1}{c\cdot\rho}\frac{\mathrm{d}P}{\mathrm{d}t}+g\cdot\frac{\mathrm{d}z}{\mathrm{d}S}+\frac{f}{2D}V|V|=\circ. \quad (10.22)$$

For $\lambda=-C$, by (10.16):

$$\frac{\mathrm{d}V}{\mathrm{d}t}-\frac{1}{c\cdot\rho}\frac{\mathrm{d}p}{\mathrm{d}t}+g\frac{\mathrm{d}Z}{\mathrm{d}S}+\frac{f}{2D}V|V|=\circ. \quad (10.23)$$

If $\rho=\rho\cdot g\,(H-Z)$, then

$$\left|\begin{array}{l}\dfrac{\mathrm{d}V}{\mathrm{d}t}+\dfrac{g}{c}\cdot\dfrac{\mathrm{d}H}{\mathrm{d}t}+\dfrac{f}{2D}V|V|=\circ\\ \text{if: }\dfrac{\mathrm{d}S}{\mathrm{d}t}=C\end{array}\right., \quad (10.24), (10.25)$$

$$\left|\begin{array}{l}\dfrac{\mathrm{d}V}{\mathrm{d}t}+\dfrac{g}{c}\cdot\dfrac{\mathrm{d}H}{\mathrm{d}t}+\dfrac{f}{2D}V|V|=\circ\\ \text{if: }\dfrac{\mathrm{d}S}{\mathrm{d}t}=-C\end{array}\right.. \quad (10.26), (10.27)$$

The method of characteristics is a finite-difference technique where pressures were computed along the pipe for each time step.

Calculation automatically subdivided the pipe into sections (intervals) and selected a time interval for computations. Equations are the characteristic equation.

If: $f = 0$, then equation will be

$$\frac{dV}{dt} - \frac{g}{c} \cdot \frac{dH}{dt} = \circ$$

or

$$dH = \left(\frac{C}{g}\right) dV \text{ (Zhukousky)}, \tag{10.28}$$

If the pressure at the inlet of the pipe and along its length is equal to p_0, then slugging pressure undergoes a sharp increase:

$$\Delta p\text{: } p = p_0 + \Delta p.$$

The Zhukousky formula is as following:

$$\Delta p = \left(\frac{C \cdot \Delta V}{g}\right). \tag{10.29}$$

The speed of the shock wave is calculated by the formula:

$$C = \sqrt{\frac{g \cdot (E_W / \rho)}{1 + (d/t_W) \cdot (E_W / E)}}. \tag{10.30}$$

By finite-difference method of water

Hammer:

$$c^+ : \frac{(V_p - V_{L_e})}{(T_p - \circ)} + \left(\frac{g}{c}\right)\frac{(H_p - H_{L_e})}{(T_p - \circ)} + \frac{fV_{L_e}|V_{L_e}|}{2D} = \circ, \tag{10.31}$$

$$c^- : \frac{(V_p - V_{R_i})}{(T_p - \circ)} + \left(\frac{g}{c}\right)\frac{(H_p - H_{R_i})}{(T_p - \circ)} + \frac{fV_{R_i}|V_{R_i}|}{2D} = \circ, \tag{10.32}$$

$$c^+ : (V_p - V_{L_e}) + \left(\frac{g}{c}\right)(H_p - H_{L_e}) + (f \cdot \Delta t)\left(\frac{f \cdot V_{L_e}|V_{L_e}|}{2D}\right) = \circ, \tag{10.33}$$

$$c^- : (V_p - V_{R_i}) + \left(\frac{g}{c}\right)(H_p - H_{R_i}) + (f \cdot \Delta t)\left(\frac{fV_{R_i}|V_{R_i}|}{2D}\right) = \circ, \tag{10.34}$$

$$V_p = \frac{1}{2}\begin{bmatrix} \left(V_{L_e} + V_{R_i}\right) \\ + \frac{g}{c}\left(H_{L_e} - H_{R_i}\right) \\ \left(-\left(\left(f \cdot \ddot{A}t\right)/2D\right)\right)\left(V_{L_e}\left|V_{L_e}\right| - V_{R_i}\left|V_{R_i}\right|\right) \end{bmatrix}, \tag{10.35}$$

$$H_p = \frac{1}{2}\begin{bmatrix} \frac{c}{g}\left(V_{L_e} + V_{R_i}\right) \\ + \left(H_{L_e} - H_{R_i}\right) \\ -\frac{c}{g}\left(\left(f \cdot \ddot{A}t\right)/2D\right)\left(V_{L_e}\left|V_{L_e}\right| - V_{R_i}\left|V_{R_i}\right|\right) \end{bmatrix}, \tag{10.36}$$

$$\Delta S = \frac{L}{N} \quad \text{and} \quad \Delta t = \frac{\Delta s}{C} \tag{10.37}$$

where $V_{L_e}, V_{R_i}, H_{L_e}, H_{R_i}, f, D$ are initial conditions parameters.

They are applied for solution at steady state condition. Computational fluid dynamics software equations calculation starts with pipe length L divided by N parts:

For internal points P_N through $P_{N'}$, H and P_1 are found.

Hence:

At P_1, there is only one characteristic line (c^-)

At (c^+), there is only one characteristic line (c^+).

10.3 RESULTS AND DISCUSSION

For finding V and V at P_1 and P_{N+1}, the boundary conditions are used. The challenge of selecting a time step is made difficult in pipeline systems by two conflicting constraints. They are defined by dynamic model for calculating many boundary conditions, such as obtaining the head and discharge at the junction of two or more pipes. It is necessary that the time step be common to all pipes. The second constraint arises from the nature of the "MOC." If the adjective terms in the governing equations are neglected (as is almost always justified), the "MOC" requires that ratio of the distance x to the time step t be equal to the wave speed in each pipe. In other words, the Courant number should ideally be equal to one and must not exceed one by stability reasons and simulation of max pressure variation:

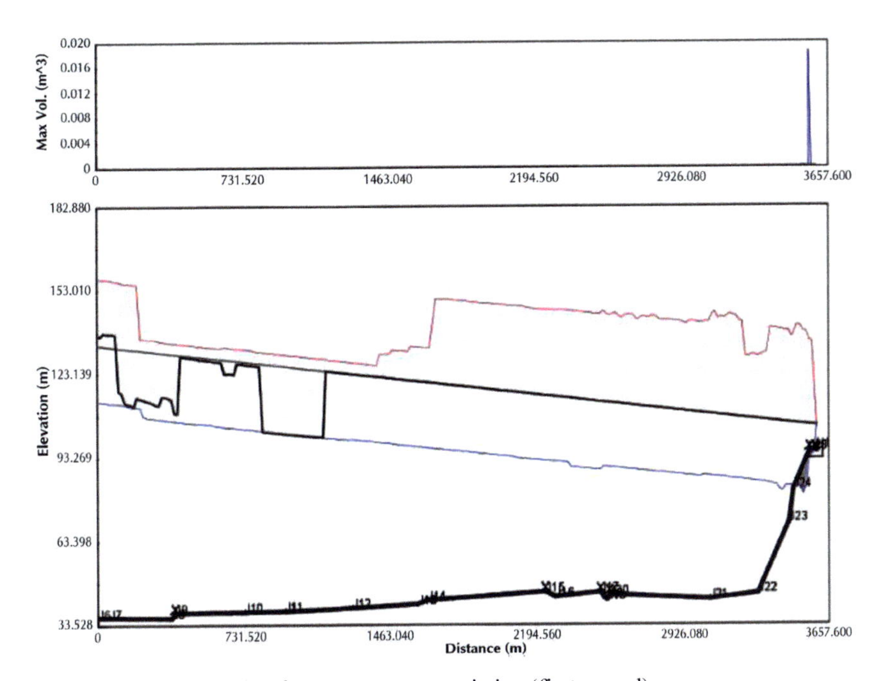

FIGURE 10.2 Simulation for max pressure variation (first record).

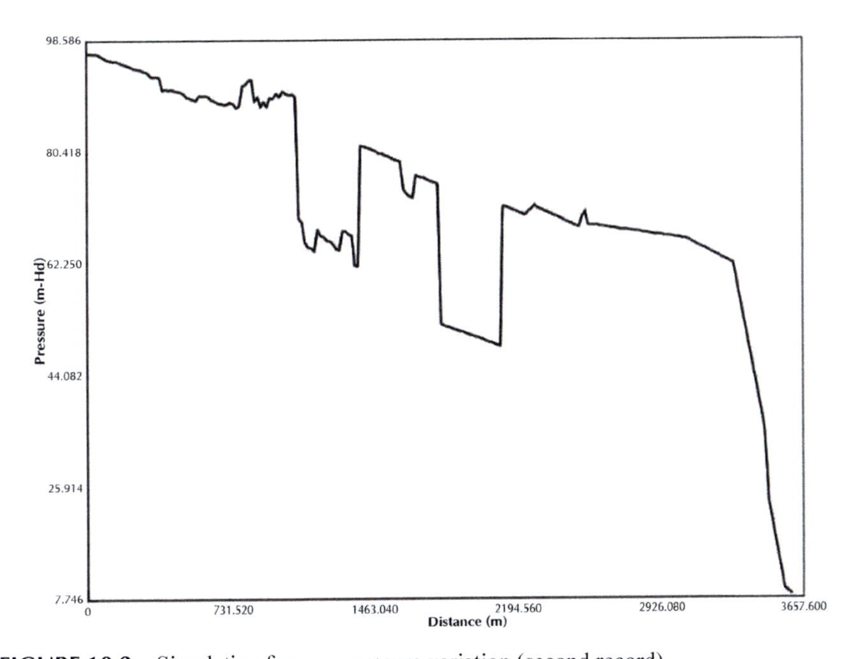

FIGURE 10.3 Simulation for max pressure variation (second record).

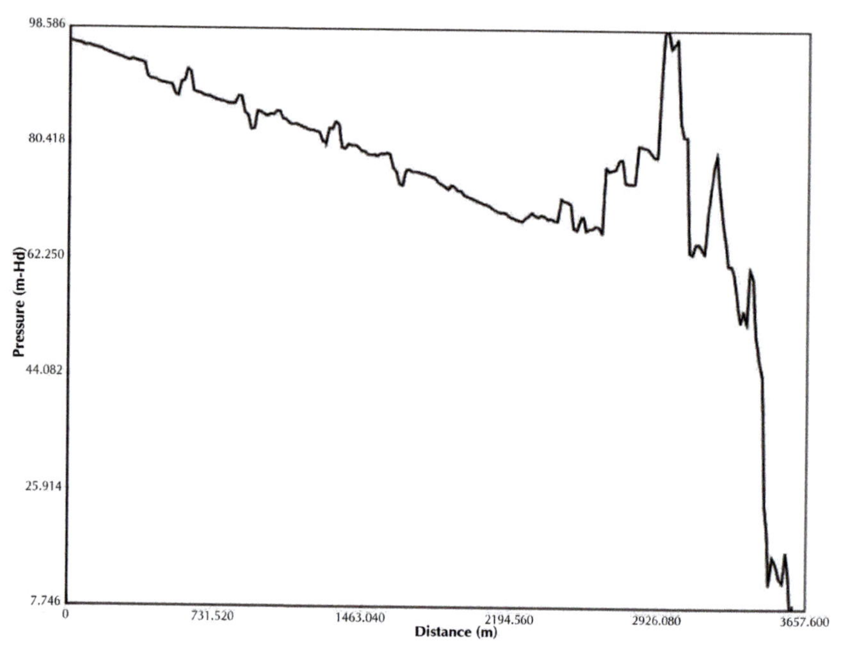

FIGURE 10.4 Simulation for max pressure variation (third record).

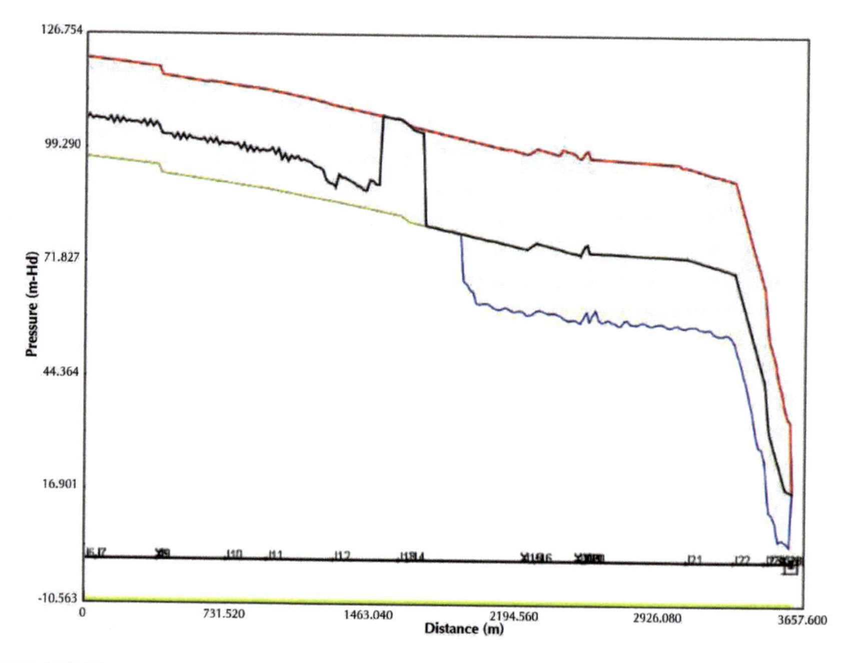

FIGURE 10.5 Simulation for max pressure variation (fourth record).

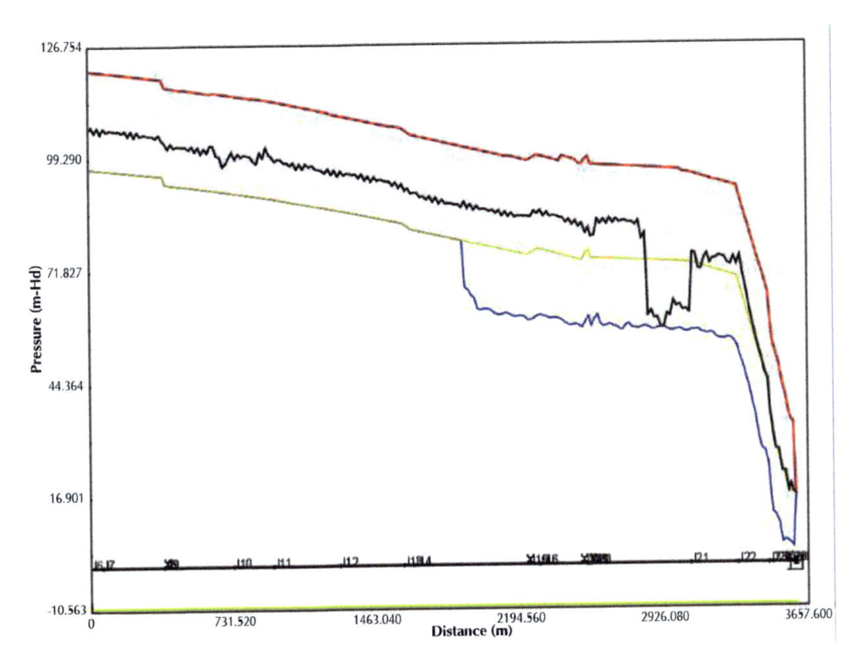

FIGURE 10.6 Simulation for max pressure variation (fifth record).

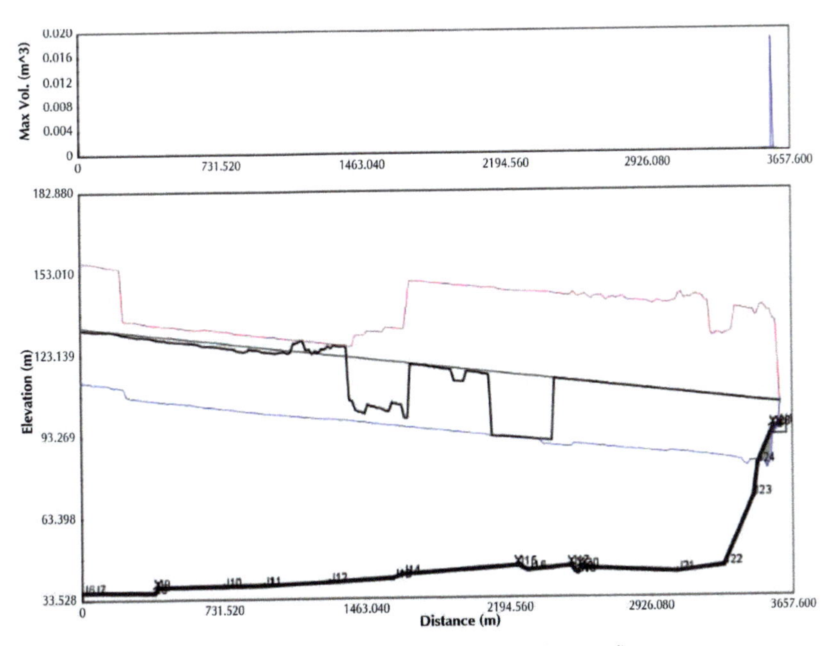

FIGURE 10.7 Simulation for max pressure variation (sixth record).

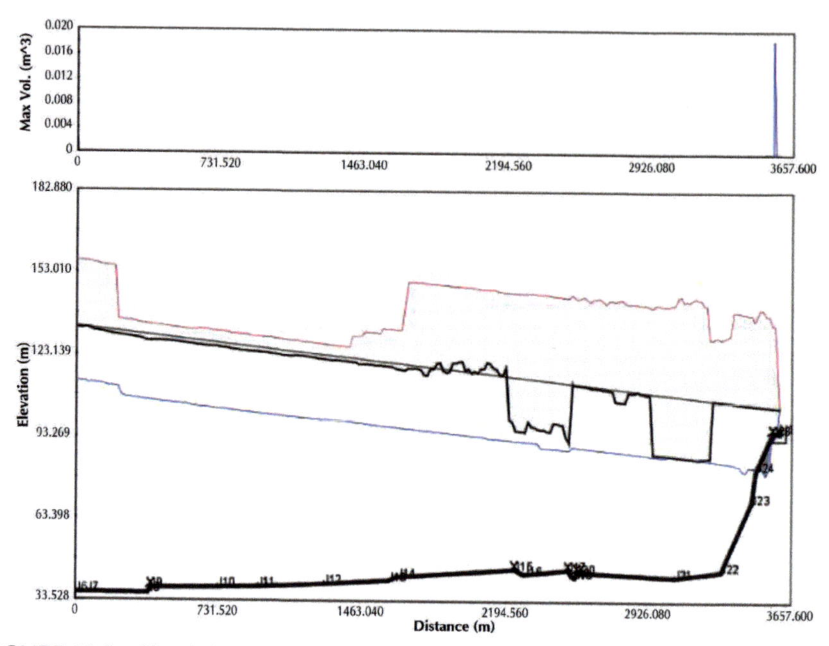

FIGURE 10.8 Simulation for max pressure variation (seventh record).

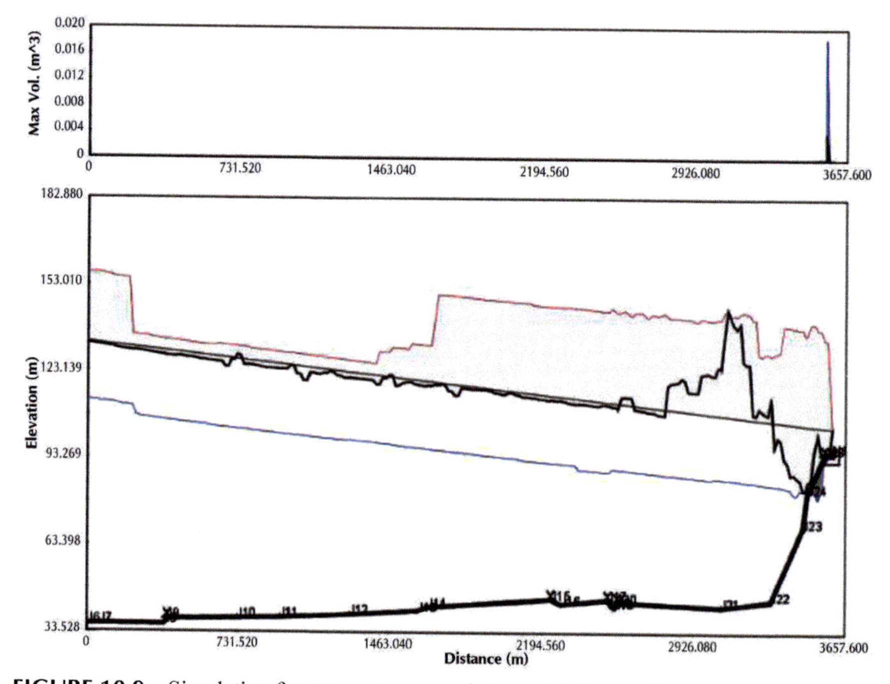

FIGURE 10.9 Simulation for max pressure variation (eighth record).

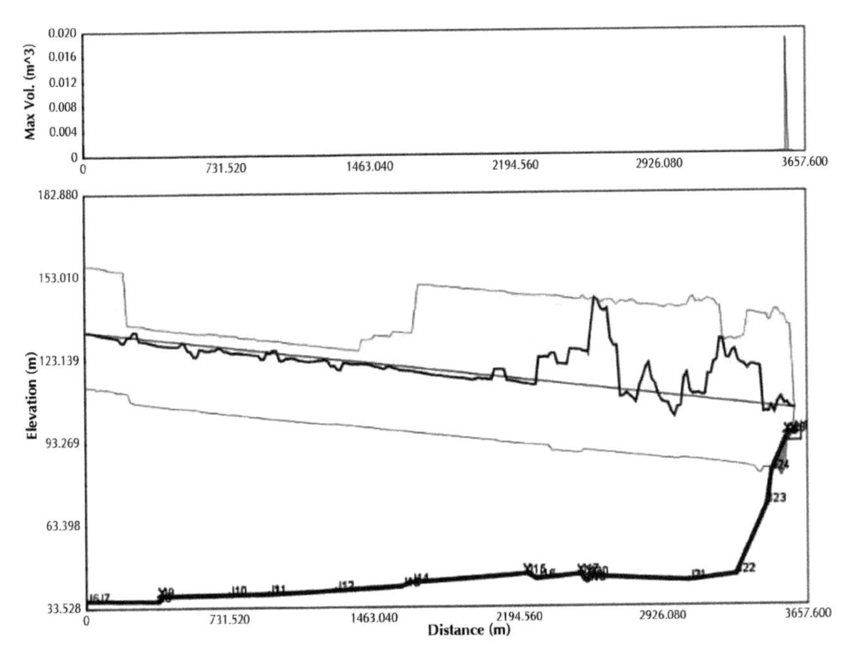

FIGURE 10.10 Simulation for max pressure variation (ninth record).

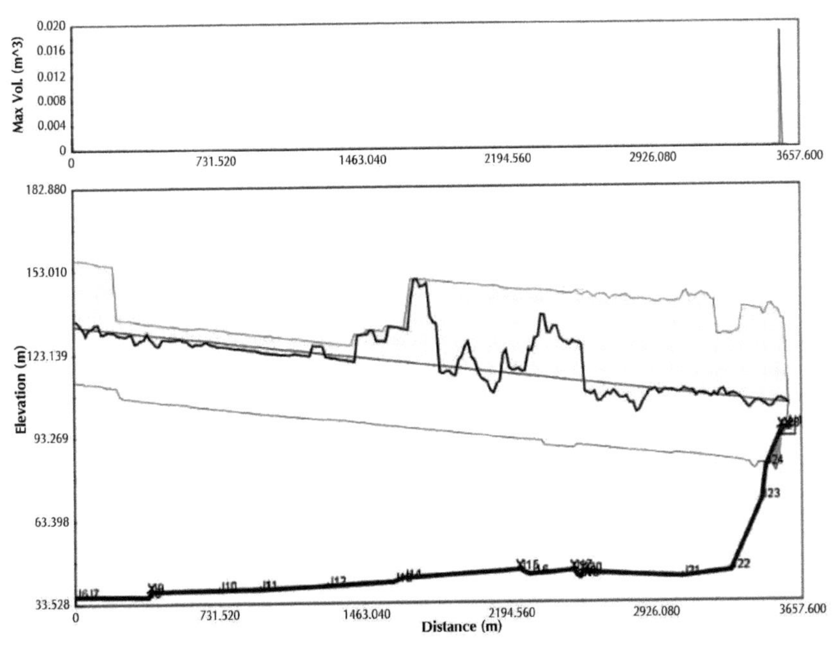

FIGURE 10.11 Simulation for max pressure variation (10th record).

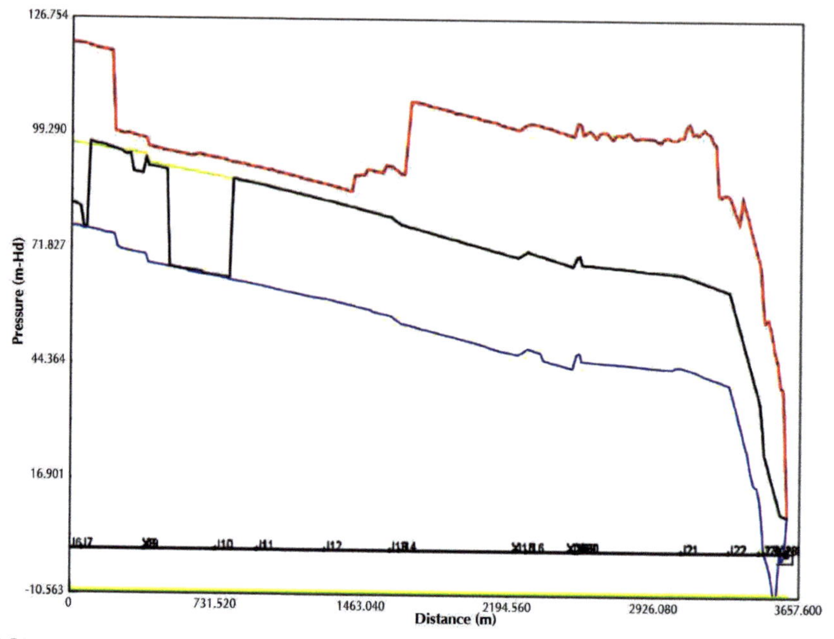

FIGURE 10.12 Simulation for max pressure variation (11th record).

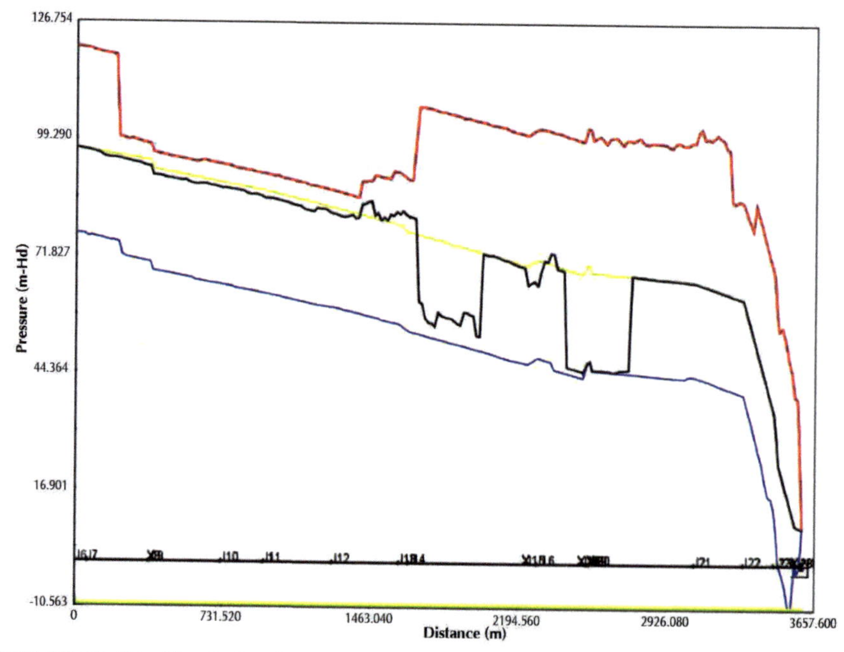

FIGURE 10.13 Simulation for max pressure variation (12th record).

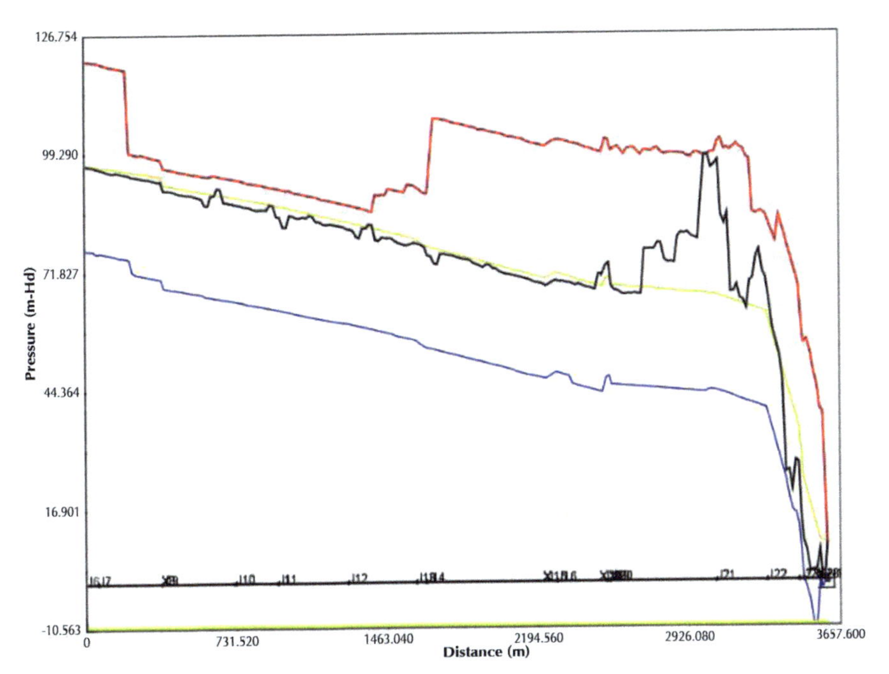

FIGURE 10.14 Simulation for max pressure variation (13th record).

A methodology to estimate wave speed in each pipe and its relation with background minimum night flows (MNFs) for water networks in individual DMAs, given all relevant local characteristics (mains length, number of households and nonhouseholds, pressure), are potentially of significant value. It could indicate the night flow at which it is no longer appropriate to allocate resources to try to locate significant unreported bursts in that DMA. Such a methodology also provides an independent check on the MNF achieved when a DMA is initially set up, after the "best practice" of thoroughly checking the DMA for leaks by step-testing and sounding has been carried out. The background night flow losses (when no bursts exist in a DMA) can be calculated for any DMA (given L (length of mains in km), N (number of properties), AZNP (average zonal night pressure in m)) from the following equation:

$$\text{NFLB (l/h)} = (\text{C1} \times L + (\text{C2} + \text{C3}) \times N) \times \text{PCF}, \tag{10.38}$$

Using the following values of C1, C2, and C3 from Table 10.1 and the pressure correction factor (PCF) from Table 10.2, based on the UK research of the 1980s, are calculated.

TABLE 10.1 Night Flow Losses.

Background losses component	Units	Low	Average	High
C1: Dist mains	l/km/hr	20	40	60
C2: Common pipes	l/prop/hr	1.5	3	4.5
C3: Supply pipes	l/prop/hr	0.5	1	1.5

TABLE 10.2 Pressure Correction Factors.

AZNP (m)	20	30	40	50	60	70	80	90	100
PCF	0.329	0.529	0.753	1.00	1.271	1.565	1.884	2.226	2.592

Once established, DMAs need to be maintained. For two adjacent DMAs, the opening of a single boundary stop valve is sufficient to destroy the accuracy of DMA-demand monitoring. A regular regime of meter readings, boundary valve checks, and pressure monitoring must therefore be established for each DMA. For leakage control purposes, it is necessary to establish the number of domestic properties, and the demand of major industrial users within each DMA. This requires regular, usually weekly, reading of DMA meters and loggers, preferably with the input of the information into a computer analysis program. Careful inspection of the meter and logger readings can quickly spot any unusual results. This can be used to trigger leak detection follow-up work. Simple management procedures must be introduced to ensure that the integrity of the DMA is maintained, otherwise the cost and effort of establishment and monitoring will be wasted. The following details are worth noting for effective management:

- All boundary valves should be kept tight closed and a regular checking program should be followed
- All boundary valves should be clearly marked and identified
- Valves within the DMA should be fully open
- Status quo should be re-established after bursts, rehabilitation, or other operational necessity.
- High-pressure DMAs should be examined for pressure reduction.
- Logger readings of low pressure should be investigated to determine whether leakage is indicated.
- Leakage within the DMA, whether visible or not showing, should be repaired promptly.
- DMA meters should not be valued out.
- DMA meters and loggers should be operating normally.

- PRV areas should be properly isolated and operating.
- Plans should be up to date and show new property.

The principal benefit of DMAs is that the key characteristics (e.g., demand, quality, and cost) of a well-defined area of the distribution system can be closely monitored. The results of this monitoring allow management action to be prioritized and targeted on where it is most cost effective. Specifically DMAs impact on

1. leakage control;
2. pressure management and levels of service;
4. asset maintenance and renewal;
5. the monitoring and maintenance of water quality; and
6. the planning and programming of repair and maintenance work.

Perhaps, the most important benefit of DMAs is a little less tangible. Together with a zoned approach to distribution management, they provide a better knowledge of how the system works and how water gets to the customers in an appropriate condition. This allows the water supplier to focus attention on those activities which produce most benefits to customers—a pro-active rather than a reactive approach. For example, flow reversals and retention times can be minimized and more consistent pressures established. This results in a better knowledge of the system, improved demand management, better and more consistent service to customers, all at a lower, long-term cost to the water supplier. All inflows and outflows of zones are measured continuously, including the effect of any increase or decrease in storage. Zones are too large to identify small leakages, as again these will be swamped by normal daily variations. However, they could possibly identify major leakage, especially if daily readings are collected. Zone metering may also be useful for comparing the performance of different leakage control teams, or for collecting together data for parts of this system with similar characteristics such as unit cost, age, urban/rural character. Within each zone, there will be several DMAs. District metering may be considered as the first level of metering which can be used for leakage detection, the previous two levels being used for performance assessment and monitoring rather than detection.

The original concept of district metering was to measure the total volume entering the DMA between the reading intervals and hence to calculate the average daily demand. This would then be compared to previous readings, and also to the readings for all other DMAs for that period to assess climatic effects. A significant increase in demand, not generally reflected across the system, would signify a likely increase in leakage. Normally, a second cycle of readings would be taken to confirm the result before further action was taken.

This procedure suffered from a number of disadvantages:

1. It was insensitive as leakage would not be identified until it exceeded a significant proportion of the daily demand, normally at least 10%.
2. The time taken to identify the leakage and initiate further action would be two reading intervals.
3. It was not possible to differentiate between increases in leakage and increases in metered consumption, except for very large consumers whose meters may have been read as district meters.
4. Elimination of climatic factors and holiday effects was difficult, and very much a matter of judgment and experience.

Due to the large numbers of meters likely to be involved, it may not be economic for all these meters to be on telemetry, in which case data must be collected by site visit. The frequency of data collection and analysis may itself be limited by the amount of resources which can be economically justified to undertake this activity. This can be varied with the leakage growth characteristics of the area. However, district meters are now usually fitted with data loggers which will record, in addition to the total flow, the night flow over a specified period for a number of nights. This immediately achieves a better than fivefold improvement in the sensitivity of the method in the original concept, as night flows will normally be less than 20% of the average daily flow and will suffer less variation due to demand. The time taken to identify leakage is reduced to one reading interval as the night flow readings will confirm the leakage, unless it occurred at the end of the period. The effect of climatic variation is significantly reduced, although care may be needed on occasions when garden sprinklers may be left on overnight. Differentiation between leakage and metered use is easier, as any increase in metered use is less likely to take place at night, particularly at weekends. Logger manufacturers usually provide powerful software to analyze and manipulate recorded data. The equipment and economics associated with data collection are changing. Some Water Suppliers are beginning to move in favor of automated, remote, and centralized interrogation of intelligent data loggers at meters, monitoring pressure as well as flow. The primary use of net night flow data is to provide operational data on which to decide on the need for further action. The MNF can be readily measured with reasonable accuracy for both district and waste meter areas, allowing small changes in flow volumes to be observed. Determination of the night metered consumption is more difficult. In many areas, it will be negligible and can be ignored. Where it is not deemed negligible, the alternative methods available to determine it are as follows:

1. Use a percentage of average daily consumption. This is satisfactory where the total nondomestic consumption is relatively small.
2. Measure MNF immediately prior to and during a "bank holiday" period. The difference will give the night consumption of industrial users who shut down for the holiday. Some allowance will still be required for commercial users with an element of domestic type consumption and for industrial users with continuous processes.
3. Do a telephone survey of major consumers to determine whether there is significant night usage, for example, replenishment of factory storage tanks. Some users may be able to supply night consumption data. In addition, on large complex sites there is a possibility of misuse of water (e.g., unauthorized use of fire mains), and it may be prudent to check such connections before embarking on leak location work.
4. Take night meter readings of the major nondomestic users. Use data loggers where the meters are logger compatible—consider changing/converting old meters on major users where this is not the case.
5. Trade effluent data may provide useful information. It must also be remembered, however, that while domestic consumption is reduced to a minimum by measuring flows at night, it is not eliminated entirely. Research in the United Kingdom suggests an allowance of about 21/prop/h, which includes minor undetectable leakage such as dripping taps and passing ball cocks. This consumption is included in the net night flow figure. The increasing use of domestic appliances overnight using economy electricity tariffs is also a factor which may need consideration.

Pressure is one of the most frequently measured parameters in the water industry, often being measured alongside flow. Many methods of measurement are in usage, but pressure transducers have become the most common means in distribution systems. They operate by converting fluid pressures into electrical signals.

Pressure measurement typically takes place for:

- General monitoring of the distribution system.
- Specific monitoring at critical points (levels of service).
- Particular consumer problems of inadequate pressure.
- Co-ordination with particular flow tests, for example, new housing estates, high-rise flats, industrial consumers, fire-fighting installations, and fire hydrants.
- Network-analysis calibration.

Pressure management is a major element in a leakage management strategy. Pressure reduction is probably the simplest and most immediate way of reducing leakage within the distribution system. Its benefits are immediate. Even where already practiced, it is likely to be worthwhile to re-examine and reset equipment and schemes to take advantage of progressive technical developments, and local-system alterations. Pressure management can be accomplished in a number of ways and not just via the installation of a new PRV. In fact, the generation of pressure almost always costs money, so reducing pressure by means of a PRV is intrinsically inefficient.

The following options should be considered first:

- Re-zoning the area supplied to match input head to topography and minimize system losses. This may include boosting to a smaller, critical area, reinforcing or reconditioning mains to allow low pressure zones to be extended, or transferring demand zones to an alternative source with a lower overall head. Network analysis could greatly facilitate this investigation.
- Matching pump output curves to closely match distribution demands. This could include resizing pumps to match known demands, or staged or variable speed pumping, or closed loop control using flow or pressure signals.
- Installation of break pressure tanks. These generally have a higher capital cost and are a potential contamination risk. On account of this, they are no longer used in the United Kingdom. Having considered these three options, mechanical pressure control devices, typically PRVs, provide the next stage in a pressure control strategy. Pressure control can:
    - reduce leakage;
    - reduce pressure-related consumption such as hand washing, car washing, etc.;

- reduce the frequency of bursts, at least in the immediate future—subsequent savings in repair costs can exceed those due to reduced leakage;
- stabilize pressure, decreasing the possibility of pipe work movement and fatigue type failures, and possibly eliminating certain household plumbing problems;
- provide a more constant service to customers—large diurnal pressure variations may give customers an impression of a poorly managed service, and unnecessarily high-pressures raise customers' expectations and perceptions of what is adequate;
- enable a company to standardize on pipes and fittings which have a lower pressure rating and are therefore cheaper;
- assist demand management when flow restriction is necessary, that is, during drought.

Some examples of the problems that can potentially arise, and their consequences, are listed below. Some of these can be designed out of the system. In correctly configured systems, this is typically a result of restrictions and blockages of individual supplies. Flow and pressure tests at the property affected will reveal the location of the problem which can then be dealt with in the normal way. Partly closed stop taps and valves are a typical problem. Poor pressures may also be the result of pipe work simply being undersized, perhaps through corrosion. They may also occur by the setting up of the PRV area severing the normal interlinking of the system. This should be assessed beforehand at the area design stage.[59–68]

Noise can be a problem close to PRV installations. Noise is usually associated with small valve openings and may be associated with cavitation problems. Attention to pipe work detail and valve settings can reduce noise levels but it is best avoided by correct selection and sitting. Noise through a PRV does create difficulties for leak detection work in the vicinity because of its interference. Blockages can occur as a result of mains material becoming trapped in the PRV. This may result in failure of the control and actuating mechanism and loss of pressure control, leading to excessively high or low pressures. Attention to the maintenance of filters and correct flushing are necessary to avoid blockages in distribution systems which are prone to solids contamination. It is generally recommended that planned preventative maintenance be carried out on a 6-month basis. Valves without close mechanical tolerances are less susceptible to this type of failure. Closing of valves between the PRV and a remote pressure monitoring point will result in the PRV attempting to rectify the apparent loss of pressure at the remote point. Typically, this occurs when valves are shut in the course of a routine repair. The results of exposing the system to maximum pressures at moderate flows will usually be a series of burst mains. This situation should be avoided by ensuring that inspectors, in particular, are aware of pressure control systems and follow appropriate procedures before closing critical valves. Network models can also be used to simulate valve closures prior to operation on site to help understand how the system will react.[69–76]

Under certain circumstances, surges of pressure and flow can cause unpredictable PRV behavior. This can result, with certain valves, in the piston exceeding its travel and jamming in the fully open or closed position. Usually, the surges which cause this type of failure result from valve or pump operations which should be examined to minimize the risk. The provision of "stops" to limit the travel in mechanical systems can be helpful.

Ordnance survey data alone is insufficient in planning an area from a topographical point of view a tall building survey should be undertaken. In areas where existing flats rely upon a high-pressure mains supply, pressure reduction may only be possible if the supplier is willing to bear costs of pumping and plumbing modifications. Where small boosters are already feeding multistory buildings, the lowering of pressures may cause the boosters to operate more regularly. The rapid payback of investment and reduction in leakage reduced incidence of burst mains and reduced customer complaints. A PRV can be defined as a mechanical device which will give a reduced outlet (downstream) pressure for a range of flow rates and upstream pressures. All PRVs have certain features in common. These are a means of controlling the flow (the valve), a means of sensing the pressure differential between the inlet and the outlet, and a means of actuating the valve. A variety of more or less sophisticated means of providing these features have been developed by manufacturers.

The two principal categories of PRV are fixed outlet and flow-modulated, each with several variations. Generally, fixed outlet characteristics maintain approximately the same value of downstream pressure over a range of flow rates. The pressure has to be set so that level of service pressure is maintained at the target point for the maximum design flow rate. The resultant AZNP will be at a higher value than a flow modulated pressure in a similar system, since in the latter case, pressures can be optimized for minimum demand. In reality, some fixed outlet valves are not always capable of maintaining a constant outlet pressure, particularly at low flow when some rise in outlet pressure can be experienced. A "pilot" can assist in providing the necessary variable throttling affect to keep a constant outlet pressure as inlet pressures and flows vary. Two pilots with a timed changeover can give a "day" and "night" setting of outlet pressure. Flow-modulated PRVs vary the outlet pressure in such a manner that a constant head can be maintained at a target point in the distribution system for a range of flow rates and inlet pressures. The activating mechanism is responsible for regulating the outlet pressure and may be mechanical or electronic or a combination of both. "Look-up" tables or telemetry may be involved in the outlet pressure control. Generally speaking, where head losses across the target area exceed 10 m (night time/no-flow pressure minus day-time peak flow pressure) flow-modulated devices will provide greater net benefit (in spite of the extra cost) and are to be preferred.

Because of advances in control practice and communications, control systems for PRVs are becoming more complex and more effective. The valves are now fitted primarily to reduce leakage and to some extent pressure dependent consumption, rather than the traditional reason of protecting the downstream infrastructure. Water suppliers must be seen to be operating efficiently and effectively and must be able to cost justifies their level of leakage and works designed to manage leakage, particularly to their customers who want to see their costs minimized. Leakage is often seen as synonymous with waste, and reducing leakage is seen as a means of saving money. Water lost through leakage has a value and so reducing the level of leakage offers benefits. However, eliminating leakage completely is impracticable and the cost of reducing it to low levels may exceed the cost of producing the water saved. Conversely, when little effort is expended on active leakage control, leakage levels will rise to levels where the cost of the water lost predominates. Water suppliers must therefore strike a balance between the cost of reducing leakage and the value of the water saved. The level of leakage at which it would cost more to make further reductions than to produce the water from another source is what is known as the economic level of leakage (ELL). Operating at ELL means that the total cost to the customer of supplying water is minimized, and suppliers are operating efficiently. This means that leakage reduction should be pursued to the point where the long-run marginal cost of leakage control is equal to the long-run marginal benefit of the water saved. The latter depends on the long-run marginal costs of augmenting supplies by alternative means, including an assessment of the environmental benefits. The ELL is not fixed for all time. It depends on a wide range of factors, which will vary over time. For example, the cost of detecting and repairing leaks will fall as new technology is introduced. This will cause the ELL to fall. Conversely, if total demand falls to a point where there is a large surplus of water, it may not be economic to reduce leakage, unless the water can be sold to other suppliers.

The marginal cost of leakage control is, therefore, the additional cost required to reduce leakage levels in an area by one unit. The difficulty in calculating the actual or marginal cost of leakage control is that only one point is known, namely the current operating conditions. Here, the costs increase in some exponential form as the level of leakage is reduced. Before any economic optimum can be derived, a method to estimate costs away from the current level must be established. A possible approach is to assume that in any water supply zone, levels of leakage could range between two extremes:

- A base level of leakage where all bursts are repaired, and the only leaks running are those which cannot be detected by the current method of active leakage control. This base, or intrinsic level, can be approximated to by measuring the level attained following an intensive program of detection and repair in a specific area.
- At the other extreme, if no money were spent on active leakage control (ALC), the level of leakage would be that controlled by customer-reported bursts.

Between these two extremes is the actual level of leakage and the cost of ALC in the water supply area. An equation can be produced based on the two extremes and the actual data point to give a form of ALC-cost curve as illustrated, which is also based upon UK data.

Measurement of the current leakage level and cost will give sufficient information to use the following equation, which typifies this curve and is taken from the UKWIR "managing leakage" from the following equation:

$$\text{Total cost of leakage control} = C = (-1/d)\ 1n\ ((L - L_b)/(L_p - L_b))$$

$$\text{where } d \text{ is a constant and} = (-1/C_a)\ 1n\ ((L_a - L_b)/(L_p - L_b)), \qquad (10.39)$$

The other terms are as follows:

L is the level of leakage (m^3/prop/year)
C is the cost of leakage control (£/prop/year)
L_a is the actual level of leakage for the area (m^3/prop/year)
C_a is the actual cost of leakage control for the area (£/prop/year)
L_b is the base level of leakage (m^3/prop/year)
L_p is the passive level of leakage (m^3/prop/year).

Once a graph has been drawn, it is possible to estimate the level of leakage, corresponding to any given level of resource input, and hence calculate the optimum level of control activity. This process will establish the optimum ELL for the chosen control method. To move to the optimum level of leakage will require a one-off additional expenditure on burst repairs. Costs associated with burst repairs will remain constant for any area in steady state conditions, but it is likely that by intensifying the method of active leakage control, an increasing number of background leaks will be found and include in the repairs. This "one-off" cost of repairs should be included in any project appraisal study. Assuming that burst occurrences remain constant over time, then once this "backlog" of leak repairs has been made, overall repair costs will return to the level that existed before. The possible exception to this is that pressure reduction could produce a new, lower level of burst occurrence. The calculation of the above optimum level assumes that the method of active leakage control and system pressure is held constant. It is equally possible to conduct similar studies to investigate the effect of pressure control or changing to a new method of active leakage control if there is a need to reduce leakage further. In each case, there will be a series of one-off costs to establish the new approach, including installation of PRVs or district meters, staff training, district reconfiguration, and backlog of repairs. The assessment of the optimum level of leakage then will reflect the cost of water and the new shape of the relationship between the cost of active leakage control and the leakage level. The marginal cost of active leakage control, at any level of leakage, can only be confirmed when the new policy is implemented from the following equation:

$$\text{Unit capital cost} = (\text{TDCC} \times r^2)/[(1 + r) \times 365 \times d], \tag{10.40}$$

(in £/m^3)

where TDCC is the total discounted capital cost (£); r is the discount rate; and d is the yearly growth in demand (m^3/day)

In addition to the capital cost, it is also necessary to include fixed operating costs when calculating the TDCC, as these costs will also be deferred if schemes are put back in the program.

10.4 CONCLUSION

Generally, the economic optimum level can be compared to the present level of leakage, and the supplier can then set targets for leakage control in conjunction with other corporate policies on customer metering, mains rehabilitation, resource development, and pressure control. To do so, they will need to appraise the investment required for these various different supply and demand management solutions, and the benefits which are expected to accrue. Due to the complexity of the issues, it is not possible to generalize to provide standardized formulae for setting leakage targets. Thus, there is a need to examine each system to determine the most appropriate method of leakage control and to plan the required capital investment, manpower and revenue resource. However, any supplier who is prepared to commit resources to collecting the required data and to carry out the analysis and appraisals, will develop a greater understanding of the factors which are important to target setting. They will also be less likely to have unrealistic or uneconomic targets imposed on them from outside, or fall into the trap of setting leakage targets themselves without full consideration of the practicalities of achieving them, or the economic consequences. There are many possible ways of setting a leakage target. These can include targets based on minimum night flows, areas with excess pressure, areas with expensive water or the most urbanized areas. The setting of economic targets, that is, a level of leakage which provides the most economic mix of leakage-related costs is independent of variations in physical factors such as property density, pressure, etc. and can provide clear information upon which sound management decisions may be based. However, it is recognized that there may be social, environmental, and political factors which dictate the target leakage level, as well as economic ones related to the suppliers' own operating environment. This has given rise to a broader concept of "the most appropriate leakage target," being described as "that level of leakage which, over a long-term planning horizon, provides the least cost combination of demand management and resource development, while adequately providing a low risk of security of supply to customers, and not unduly over-abstracting water from the environment."

KEYWORDS

- **WDNs (water distribution networks)**
- **hydraulic shock**
- **pressure wave**
- **program logic control**
- **minimum night flow**

REFERENCES

1. Streeter, V. L.; Wylie, E. B. *Fluid Mechanics*; McGraw-Hill Ltd., New York, NY, 1979; pp 492–505.
2. Leon Arturo, S. An Efficient Second-Order Accurate Shock-Capturing Scheme for Modeling One and Two-Phase Water Hammer Flows. Ph.D. Thesis, 29 March 2007; pp 4–44.
3. Adams, T. M.; Abdel-Khalik, S. I.; Jeter, S. M.; Qureshi, Z. H. An Experimental Investigation of Single-Phase Forced Convection in Microchannels. *Int. J. Heat Mass Transfer* **1998,** *41*, 851–857.
4. Peng, X. F.; Peterson, G. P. Convective Heat Transfer and Flow Friction for Water Flow in Microchannel Structure. *Int. J. Heat Mass Transfer* **1996,** *36*, 2599–2608.
5. Mala, G.; Li, D.; Dale, J. D. Heat Transfer and Fluid Flow in Microchannels. *J. Heat Transfer* **1997,** *40*, 3079–3088.
6. Xu, B.; Ooi, K. T.; Mavriplis, C.; Zaghloul, M. E. Viscous Dissipation Effects for Liquid Flow in Microchannels. *Microsystems* **2002,** 53–57.
7. Pickford, J. *Analysis of Surge*; Macmillan: London, 1969; pp 153–156.
8. American Society of Civil Engineers. *Pipeline Design for Water and Wastewater*. American Society of Civil Engineers: New York, 1975, 54 p.
9. Xu, B.; Ooi, K. T.; Mavriplis, C.; Zaghloul, M. E.; Viscous Dissipation Effects for Liquid Flow in Microchannels. *Microsystems* **2002,** 53–57.
10. Fedorov, A. G.; Viskanta, R. Three-Dimensional Conjugate Heat Transfer into Microchannel Heat Sink for Electronic Packaging. *Int. J. Heat Mass Transfer* **2000,** *43*, 399–415.
11. Tuckerman, D. B. Heat Transfer Microstructures for Integrated Circuits. Ph.D. Thesis, Stanford University, 1984; pp 10–120.
12. Harms, T. M.; Kazmierczak, M. J.; Cerner, F. M.; Holke, A.; Henderson, H. T.; Pilchowski, H. T.; Baker, K. Experimental Investigation of Heat Transfer and Pressure Drop through Deep Micro channels in a (1 0 0) Silicon Substrate. *Proc. ASME. Heat Transf. Div.* **1997,** *HTD 351*, 347–357.
13. Holland, F. A.; Bragg, R. *Fluid Flow for Chemical Engineers*; Edward Arnold Publishers: London, 1995; pp 1–3.
14. Lee TS, Pejovic, S. Air Influence on Similarity of Hydraulic Transients and Vibrations. *ASME J. Fluid Eng.* **1996,** *118* (4), 706–709.
15. Li, J.; McCorquodale, A. Modeling Mixed Flow in Storm Sewers. *J. Hydraul. Eng., ASCE* **1999,** 125 (11), 1170–1180.
16. Minnaert, M. On Musical Air Bubbles and the Sounds of Running Water. *Phil. Mag.* **1933,** 16 (7), 235–248.
17. Moeng, C. H.; McWilliams, J. C.; Rotunno, R.; Sullivan, P. P.; Weil, J. Investigating 2D Modeling of Atmospheric Convection in the PBL. *J. Atmos. Sci.* **2004,** *61*, 889–903.
18. Tuckerman, D. B.; Pease R. F. W. High Performance Heat Sinking for VLSI. *IEEE Electron. Device Lett.* **1981,** *DEL-2*, 126–129.
19. Nagiyev, F. B.; Khabeev, N. S, Bubble Dynamics of Binary Solutions. *High Temp.* **1988,** 27 (3), 528–533.
20. Shvarts, D.; Oron, D.; Kartoon, D.; Rikanati, A.; Sadot, O. Scaling Laws of Nonlinear Rayleigh–Taylor and Richtmyer–Meshkov Instabilities in Two and Three Dimensions. *C. R. Acad. Sci. Paris IV* **2000,** *719*, 312 p.

21. Cabot, W. H.; Cook, A. W.; Miller, P. L.; Laney, D. E.; Miller, M. C.; Childs, H. R. Large Eddy Simulation of Rayleigh–Taylor Instability. *Phys. Fluids* **2005,** *17*, 91–106.
22. Cabot, W. *Physics of Fluids*. University of California, Lawrence Livermore National Laboratory: Livermore, CA, 2006; pp 94–550.
23. Goncharov, V. N. Analytical Model of Nonlinear, Single-Mode, Classical Rayleigh–Taylor Instability at Arbitrary Atwood Numbers. *Phys. Rev. Lett.* **2002,** 88 (134502), 10–15.
24. Ramaprabhu, P.; Andrews, M. J. Experimental Investigation of Rayleigh–Taylor Mixing at Small Atwood Numbers. *J. Fluid Mech.* **2004,** *502*, 233 p.
25. Clark, T. T. A Numerical Study of the Statistics of a Two-Dimensional Rayleigh–Taylor Mixing Layer. *Phys. Fluids* **2003,** *15*, 2413.
26. Cook, A. W.; Cabot, W.; Miller, P. L. The Mixing Transition in Rayleigh–Taylor Instability. *J. Fluid Mech.* **2004,** *511*, 333.
27. Waddell, J. T.; Niederhaus, C. E.; Jacobs, J. W. Experimental Study of Rayleigh–Taylor Instability: Low Atwood Number Liquid Systems with Single-Mode Initial Perturbations. *Phys. Fluids* **2001,** *13*, 1263–1273.
28. Weber, S. V.; Dimonte, G.; Marinak, M. M. Arbitrary Lagrange–Eulerian Code Simulations of Turbulent Rayleigh–Taylor Instability in Two and Three Dimensions. *Laser Part. Beams* **2003,** *21*, 455 p.
29. Dimonte, G.; Youngs, D.; Dimits, A.; Weber, S.; Marinak, M. A Comparative Study of the Rayleigh–Taylor Instability Using High-Resolution Three-Dimensional Numerical Simulations: The Alpha Group Collaboration. *Phys. Fluids* **2004,** 16, 1668.
30. Young, Y. N.; Tufo, H.; Dubey, A.; Rosner, R. On the Miscible Rayleigh–Taylor Instability: Two and Three Dimensions. *J. Fluid Mech.* **2001,** *447* (377), 2003–2500.
31. George, E.; Glimm, J. Self-similarity of Rayleigh–Taylor Mixing Rates. *Phys. Fluids* **2005,** *17* (054101), 1–3.
32. Oron. D.; Arazi, L.; Kartoon, D.; Rikanati, A.; Alon, U.; Shvarts, D. Dimensionality Dependence of the Rayleigh–Taylor and Richtmyer–Meshkov Instability Late-Time Scaling Laws. *Phys. Plasmas* **2001,** *8*, 2883 p.
33. Nigmatulin, R. I.; Nagiyev, F. B.; Khabeev, N. S. Effective Heat Transfer Coefficients of the Bubbles in the Liquid Radial Pulse. In Mater. Second-Union. Conf. Heat Mass Transfer, "Heat Massoob-Men in the Biphasic With." *Minsk* **1980,** *5*, 111–115.
34. Nagiyev, F. B.; Khabeev, N. S. Bubble Dynamics of Binary Solutions. *High Temp.* **1988,** *27* (3), 528–533.
35. Nagiyev, F. B.; Damping of the Oscillations of Bubbles Boiling Binary Solutions. In *Mater. VIII Resp. Conf. Mathematics and Mechanics*, Baku, 26–29 October 1988; pp 177–178.
36. Nagyiev, F. B.; Kadyrov, B. A.; Small Oscillations of the Bubbles in a Binary Mixture in the Acoustic Field. *Math. AN Az. SSR Ser. Physicotech. Mat. Sci.* **1986,** *1*, 23–26.
37. Nagiyev, F. B.; Dynamics, Heat and Mass Transfer of Vapor-Gas Bubbles in a Two-Component Liquid. In *Turkey-Azerbaijan Petrol Semin.*, Ankara, Turkey, 1993; pp 32–40.
38. Nagiyev, F. B. The Method of Creation Effective Coolness Liquids. In *Third Baku International Congress*, Baku, Azerbaijan Republic, 1995; pp 19–22.
39. Nagiyev, F. B. The Linear Theory of Disturbances in Binary Liquids Bubble Solution. *Dep. VINITI* **1986,** *405* (86), 76–79.
40. Nagiyev, F. B. Structure of Stationary Shock Waves in Boiling Binary Solutions. *Math. USSR, Fluid Dyn.* **1989,** *1*, 81–87.

41. Rayleigh. On the Pressure Developed in a Liquid During the Collapse of a Spherical Cavity. *Philos. Mag. Ser. 6* **1917,** *34* (200), 94–98.
42. Perry. R. H.; Green, D. W.; Maloney, J. O. *Perry's Chemical Engineers Handbook*, seventh ed. McGraw-Hill: New York, 1997, pp 1–61.
43. Nigmatulin, R. I.; Dynamics of Multiphase Media, Moscow, *Nauka* **1987,** *1* (2), 12–14.
44. Kodura, A.; Weinerowska, K. The Influence of the Local Pipeline Leak on Water Hammer Properties. In *Materials of the II Polish Congress of Environmental Engineering*, Lublin, 2005; pp 125–133.
45. Kane, J.; Arnett, D.; Remington, B. A.; Glendinning, S. G.; Baz'an, G. Two-Dimensional versus Three-Dimensional Supernova Hydrodynamic Instability Growth. *Astrophys. J.* **2000,** 528–989.
46. Quick, R. S. Comparison and Limitations of Various Water Hammer Theories. *J. Hyd. Div., ASME* **1933,** *May*, 43–45.
47. Jaeger, C. *Fluid Transients in Hydro-Electric Engineering Practice*; Blackie and Son Ltd.: Glasgow, 1977; pp 87–88.
48. Jaime Suárez, A. *Generalized Water Hammer Algorithm for Piping Systems with Unsteady Friction*; 2005, pp 72–77.
49. Fok, A.; Ashamalla, A.; Aldworth, G. Considerations in Optimizing Air Chamber for Pumping Plants. In *Symposium on Fluid Transients and Acoustics in the Power Industry*, San Francisco, USA. December 1978, 112–114.
50. Fok, A. Design Charts for Surge Tanks on Pump Discharge Lines. In *BHRA 3rd Int. Conference on Pressure Surges*, Bedford, England, March 1980; pp 23–34.
51. Fok, A. Water hammer and Its Protection in Pumping Systems. In *Hydrotechnical Conference*, CSCE, Edmonton, May, 1982; pp 45–55.
52. Fok, A. A Contribution to the Analysis of Energy Losses in Transient Pipe Flow. PhD. Thesis, University of Ottawa, 1987, 176–182.
53. Brunone, B.; Karney, B. W.; Mecarelli, M.; Ferrante, M. Velocity Profiles and Unsteady Pipe Friction in Transient Flow. *J. Water Resour. Plann. Manage., ASCE* **2000,** *126* (4), 236–244.
54. Koelle, E.; Luvizotto, Jr. E.; Andrade, J. P. G. Personality Investigation of Hydraulic Networks using MOC—Method of Characteristics. *Proceedings of the 7th International Conference on Pressure Surges and Fluid Transients*, Harrogate Durham: United Kingdom, 1996, pp 1–8.
55. Filion, Y.; Karney, B. W. A Numerical Exploration of Transient Decay Mechanisms in Water Distribution Systems. In *Proceedings of the ASCE Environmental Water Resources Institute Conference*, American Society of Civil Engineers, Roanoke, VA, 2002, pp 30.
56. Hamam, M. A.; Mc Corquodale, J. A. Transient Conditions in the Transition from Gravity to Surcharged Sewer Flow. *Can. J. Civil Eng.* **1982,** 65–98.
57. Savic, D. A.; Walters, G. A. Genetic Algorithms Techniques for Calibrating Network Models. *Report No. 95/12*; Centre for Systems and Control Engineering, 1995; pp 137–146.
58. Walski, T. M.; Lutes, T. L. Hydraulic Transients Cause Low-Pressure Problems. *J. Am. Water Works Assoc.* **1994,** *75* (2), 58.
59. Lee, T. S.; Pejovic, S. Air Influence on Similarity of Hydraulic Transients and Vibrations. *ASME J. Fluid Eng.* **1996,** *118* (4), 706–709.
60. Chaudhry, M. H. *Applied Hydraulic Transients*; Van Nostrand Reinhold Co.: New York, 1979; pp 1322–1324.

61. Parmakian, J. *Water Hammer Analysis*; Dover Publications, Inc.: New York, NY, 1963; pp 51–58.
62. Tuckerman, D. B.; Pease, R. F. W. High Performance Heat Sinking for VLSI. *IEEE Electron. Device Lett.* **1981,** *DEL-2*, 126–129.
63. Farooqui, T. A. Evaluation of Effects of Water Reuse, on Water and Energy Efficiency of an Urban Development Area, Using an Urban Metabolic Framework. *Master of Integrated Water Management Student Project Report*, International Water Centre, 2015.
64. Ferguson, B. C.; Frantzeskaki, N.; Brown, R. R. A Strategic Program for Transitioning to a Water Sensitive City. *Landsc. Urban Plann.* **2013,** *117*, 32–45.
65. Kenway, S.; Gregory, A.; McMahon, J. Urban Water Mass Balance Analysis. *J. Ind. Ecol.* **2011,** *15* (5), 693–706.
66. Renouf, M. A.; Kenway, S. J.; Serrao-Neumann, S.; Low Choy, D. Urban Metabolism for Planning Water Sensitive Cities. Concept for an Urban Water Metabolism Evaluation Framework. *Milestone Report.* Cooperative Research Centre for Water Sensitive Cities: Melbourne, 2015.
67. Serrao-Neumann, S.; Schuch, G.; Kenway, S. J.; Low Choy, D. *Comparative Assessment of the Statutory and Non-statutory Planning Systems: South East Queensland,* Metropolitan Melbourne and Metropolitan Perth. Cooperative Research Centre for Water Sensitive Cities: Melbourne, 2013.
68. Andrews, S.; Traynor, P. *Guidelines for Assuring Quality of Food and Water Microbiological Culture Media*, August 2004.
69. Eaton, A. D.; Rice, E. W.; Clescceri, L. S. *Standard Methods for the Examination of Water and Waste Water*, Part 9000, 2012.
70. Razavi, K. *Instructions Procedure Rural Water and Wastewater Quality Assurance Test Results of Khorasan Razavi*, 2011.
71. Abbassi, B.; Al Baz, I. *Integrated Wastewater Management: A Review. Efficient Management of Wastewater—Its Treatment and Reuse in Water Scarce Countries.* Springer Publishing Co.: New York, NY, 2008.
72. Andreasen, P. Chemical Stabilization. *Sludge into Biosolids—Processing, Disposal and Utilization.* IWA Publishing: London, 2001.
73. CIWEM—The Chartered Institution of Water and Environmental Management. *Sewage Sludge: Stabilization and Disinfection–Handbooks of UK Wastewater Practice.* CIWEM Publishing: Lavenham, 1996.
74. Halalsheh, M.; Wendland, C. Integrated Anaerobic–Aerobic Treatment of Concentrated Sewage. *Efficient Management of Wastewater—Its Treatment and Reuse in Water Scarce Countries.* Springer Publishing Co.: New York, NY, 2008.
75. ISWA. Handling, Treatment and Disposal of Sludge in Europe. *Situation Report 1.* ISWA Working Group on Sewage Sludge and Water Works: Copenhagen, 1995.
76. Matthews, P. Agricultural and Other Land Uses. *Sludge into Biosolids, Processing, Disposal and Utilization.* IWA Publishing: London, 2001.

PART II
Experimental Designs

CHAPTER 11

METAL CONTROL ON STRUCTURE AND FUNCTION OF Ni(Fe) DIOXYGENASES INCLUDED IN METHIONINE SALVAGE PATHWAY: ROLE OF TYR-FRAGMENT AND MACROSTRUCTURES IN MECHANISM OF CATALYSIS ON MODEL SYSTEMS

L. I. MATIENKO*, L. A. MOSOLOVA, V. I. BINYUKOV, E. M. MIL, and G. E. ZAIKOV

The Federal State Budget Institution of Science, N. M. Emanuel Institute of Biochemical Physics, Russian Academy of Sciences, 4 Kosygin Str., Moscow 119334, Russia

Corresponding author. E-mail: matienko@sky.chph.ras.ru

CONTENTS

ABSTRACT

The atomic force microscopy (AFM) method has been used for research possibility of the stable supramolecular nanostructures formation based on Ni(Fe) ARD dioxygenase models: iron complexes $Fe^{III}_x(acac)_y 18C6_m(H_2O)_n$, and nickel complexes $Ni_xL^1_y(L^1_{ox})_z(L^2)_n(H_2O)_m$, $\{Ni^{II}(acac)_2 \cdot \mathbf{L^2} \cdot PhOH\}$ [L^2 = MP, HMPA, MSt (M = Na, Li)], triple system $\{Ni^{II}(acac)_2 + \mathbf{Tyr} + PhOH\}$ (Tyr = L-tyrosine)—with the assistance of intermolecular H bonds. Role of Ni(Fe) macrostructures in mechanisms of catalysis is discussed.

11.1 INTRODUCTION

In recent years, the studies in the field of homogeneous catalytic oxidation of hydrocarbons with molecular oxygen were developed in two directions, namely, the free-radical chain oxidation catalyzed by transition metal complexes and the catalysis by metal complexes that mimic enzymes.[1,2] The findings on the mechanism of action of enzymes, and, in particular, dioxygenases and their models are very useful in the treatment of the mechanism of catalysis by metal complexes in the processes of oxidation of hydrocarbons with molecular oxygen. Moreover, as one will see below, the investigation of the mechanism of catalysis by metal complexes can give the necessary material for the study of the mechanism of action of enzymes.

The method of modifying the Ni^{II} and $Fe^{II,III}$ complexes used in the selective oxidation of alkylarens (ethylbenzene and cumene) with molecular oxygen to afford the corresponding hydroperoxides aimed at increasing their selectivity's has been first proposed by Matienko and new efficient catalysts of selective oxidation of ethylbenzene to α-phenyl ethyl hydroperoxide (PEH), as intermediates in the large-scale production of important monomers, were developed.[1,2]

The preservation of high activity of catalysts during oxidations seems to be due to formation of the stable supramolecular structures, based on catalytic active complexes, with assistance of intermolecular H-bonds.[3] This hypothesis is evidenced by us with AFM (atomic force microscopy) technique. Thus, we have offered the new approach to research of mechanism of homogenous catalysis, and the mechanism of action of enzymes also, with use of AFM method.[4,5] In this chapter, we discuss the possible role of macrostructures and Tyr-fragment in mechanism of Ni(ARD) dioxygenase based on experience data that we received at the first time on model systems.

11.2 EXPERIMENTAL

AFM SOLVER P47/SMENA/ with silicon cantilevers NSG11S (NT MDT) with curvature radius of 10 nm, tip height: 10–15 μm, and cone angle ≤ 22° in taping mode on resonant frequency 150 kHz was used.[4,5]

As substrate, the polished silicone surface special chemically modified was used.

Waterproof-modified silicone surface was exploit for the self-assembly driven growth due to H-bonding of complexes $Fe^{III}{}_x(acac)_y 18C6_m(H_2O)_n$, $Ni_xL^1{}_y(L^1{}_{ox})_z(L^2)_n(H_2O)_m$, $\{Ni^{II}(acac)_2 \cdot \mathbf{L}^2 \cdot PhOH\}$ (L^2 = MP, HMPA, MSt), systems $\{Ni^{II}(acac)_2$ + MP + Tyr$\}$ and $\{Ni^{II}(acac)_2\}$ + Tyr$\}$ with silicone surface. The saturated chloroform ($CHCl_3$) or water solutions of complexes was put on a surface, maintained some time, and then solvent was deleted from a surface by means of special method—spin-coating process.

In the course of scanning of investigated samples, it has been found that the structures are fixed on a surface strongly enough due to H-bonding. The self-assembly driven growth of the supramolecular structures on modified silicone surface on the basis of researched complexes, due to H-bonds and perhaps the other noncovalent interactions was observed.

11.3 DISCUSSION

11.3.1 THE ROLE OF HYDROGEN-BONDS IN MECHANISMS OF CATALYSIS AND HYDROCARBON OXIDATIONS, CATALYZED WITH NI(OR FE) COMPLEXES

As a rule, in the quest for axial modifying ligands L^2 that control the activity and selectivity of homogeneous metal complex catalysts, attention of scientists is focused on their steric and electronic properties. The interactions of ligands L^2 with L^1 taking place in the outer coordination sphere are less studied; the same applies to the role of hydrogen bonds, which are usually difficult to control.[6,7]

Secondary interactions (hydrogen bonding, proton transfer) play an important role in the dioxygen activation and its binding to the active sites of metalloenzymes.[8]

In designing catalytic systems that mimic the enzymatic activity, special attention should be paid to the formation of H-bonds in the second coordination sphere of a metal ion.

Transition metal β-diketonates are involved in various substitution reactions. Methine protons of chelate rings in β-diketonate complexes can be substituted by different electrophiles (E) (formally, these reactions are analogous to the Michael addition reactions).[9–11] This is a metal-controlled process of the C−C bond formation.[11] The complex $Ni^{II}(acac)_2$ is the most efficient catalyst of such reactions.

In our works, we have modeled efficient catalytic systems $\{ML^1_n + L^2\}$ (M = Ni, Fe, $L^1 = acac^-$, L^2 are crown ethers or quaternary ammonium salts, different electron-donating modifying extra-ligands) for ethylbenzene oxidation to α-phenyl ethyl hydroperoxide that was based on the established (for Ni complexes) and hypothetical (for Fe complexes) mechanisms of formation of catalytically active species and their operation.[1,2] The high activity of systems $\{ML^1_n + L^2\}$ is associated with the fact that during the ethylbenzene oxidation, the active primary $(M^{II}L^1_2)_x(L^2)_y$ complexes and heteroligand $M^{II}_xL^1_y(L^1_{ox})_z(L^2)_n(H_2O)_m$ complexes are formed to be involved in the oxidation process.

We established mechanism of formation of high-effective catalysts, heteroligand complexes $M^{II}_xL^1_y(L^1_{ox})_z(L^2)_n(H_2O)_m$. The axially coordinated electron-donating ligand L^2 controls the formation of primary active complexes $ML^1_2{\cdot}L^2$ and the subsequent reactions of β-diketonate ligands in the outer coordination sphere of these complexes. The coordination of an electron-donating extra-ligand L^2 with an $M^{II}L^1_2$ complex favorable for stabilization of the transient zwitter-ion $L^2[L^1M(L^1)^+O_2^-]$ enhances the probability of regioselective O_2 addition to the methine C−H bond of an acetylacetonate ligand activated by its coordination with metal ions. The outer-sphere reaction of O_2 incorporation into the chelate ring depends on the nature of the metal and the modifying ligand L^2 (see Refs. [1,2]). Thus for nickel complexes $Ni^{II}_xL^1_y(L^1_{ox})_z(L^2)_n$, the reaction of oxygenation of ligand $L^1 = acac^-$ follows a mechanism analogous to those of Ni^{II}-containing acireductone dioxygenase (ARD)[12] or Cu- and Fe-containing quercetin 2,3-dioxygenases.[13,14] Namely, incorporation of O_2 into the chelate acac-ring was accompanied by the proton transfer and the redistribution of bonds in the transition complex leading to the scission of the cyclic system to form a chelate ligand $L^1_{ox} = OAc^-$, acetaldehyde, and CO (in the Criegee rearrangement, Scheme 11.1).

In the effect of iron(II) acetylacetonate complexes $Fe^{II}_xL^1_y(L^1_{ox})_z(L^2)_n$, we have found (2) an analogy with the action of Fe^{II}-ARD[9] or Fe^{II}-acetyl acetone Dioxygenase (Dke1) (Scheme 11.2).[15] For iron complexes, oxygen adds to C−C bond (rather than inserts into the C=C bond as in the case of catalysis with nickel(II) complexes) to afford intermediate, that is, a

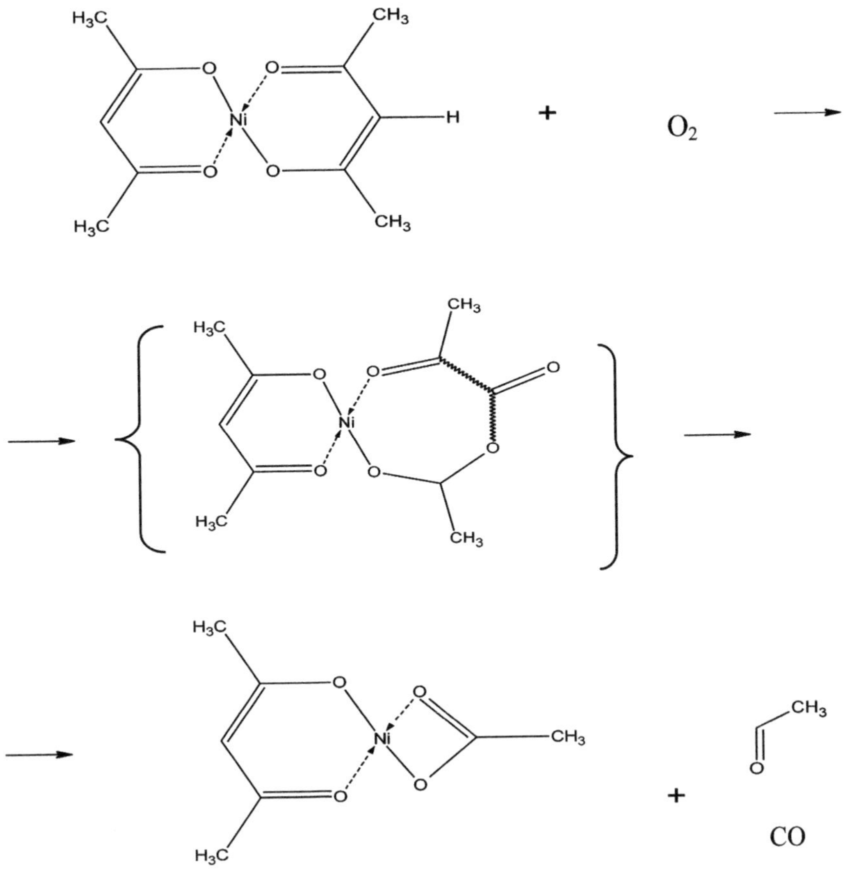

SCHEME 11.1 The reaction of oxygenation of acac$^-$ ligand in Ni(acac)$_2$ follows a mechanism analogous to those of Ni^{II}-containing ARD.

Fe complex with a chelate ligand containing 1,2-dioxetane fragment. The process is completed with the formation of the (OAc)$^-$ chelate ligand and methylglyoxal as the second decomposition product of a modified acac-ring (as it has been shown in Ref. [15], Scheme 11.2).

One of the most effective catalytic systems of the ethylbenzene oxidation to the PEH is the triple systems.[1,2] Namely, the phenomenon of a substantial increase in the selectivity (*S*) and conversion (*C*) of the ethylbenzene oxidation to the α-phenyl ethyl hydroperoxide upon addition of PhOH together with ligands *N*-methylpyrrolidone-2 (MP), hexamethylphosphorotriamide (HMPA) or alkali metal stearate MSt (M = Li, Na) to metal

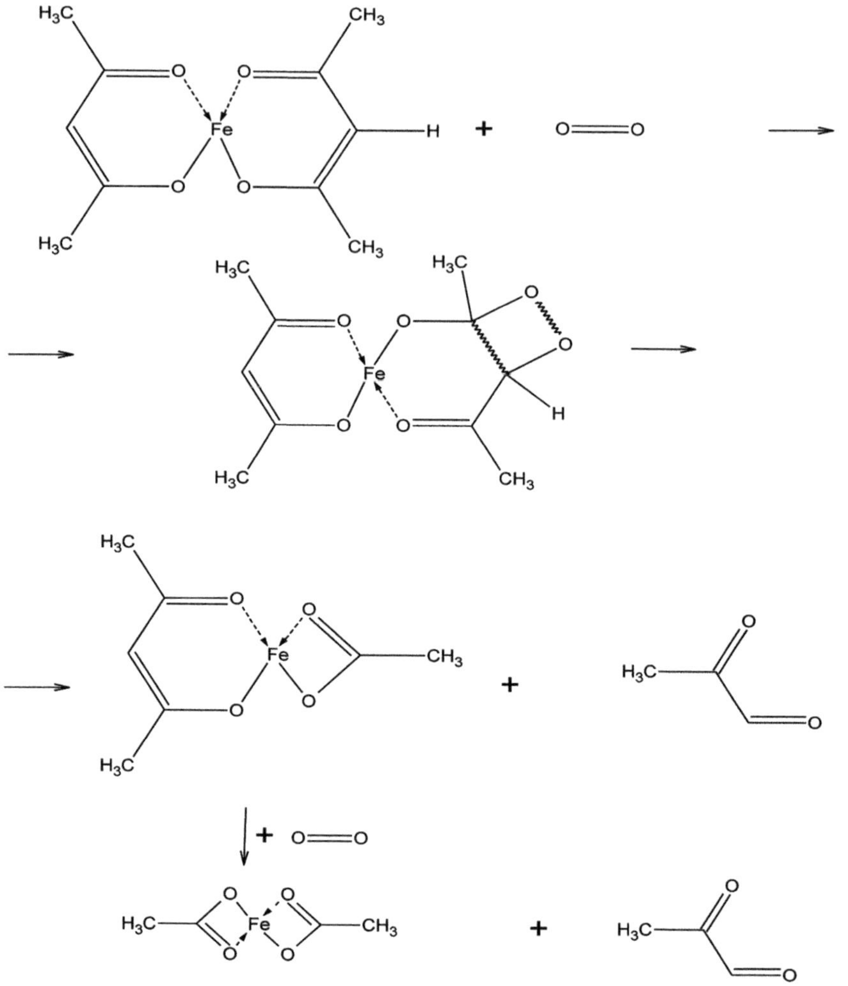

SCHEME 11.2 The reaction of acac⁻ ligand oxygenation in $Fe(acac)_2$ complex follows a mechanism analogous to the action of Fe^{II}-ARD or Fe^{II}-acetyl acetone dioxygenase (Dke1).

complex $Ni^{II}(acac)_2$ was discovered in works Matienko and Mosolova.[1,2] The role of intramolecular H-bonds was established by us in mechanism of formation of triple catalytic complexes $\{Ni(II)(acac)_2{\cdot}L^2{\cdot}PhOH\}$ (L^2 = MP) in the process of ethylbenzene oxidation with molecular oxygen.[2,3] The formation of triple complexes $Ni^{II}(acac)_2$ L^2 PhOH from the earliest stages of oxidation was established with kinetic methods.[1–3] In the course of the oxidation, the rates of products accumulation unchanged during the long period $t \leq 30–40$ h namely, the reaction rate remains practically the same during the oxidation process.[1–3] We assumed that the stability of complexes $Ni(acac)_2 \cdot L^2 \cdot PhOH$ in the process of ethyl benzene oxidation can be associated as one of reasons, with the supramolecular structures formation due to intermolecular H-bonds (phenol–carboxylate) (see below) and, possible, the other noncovalent interactions:

$$\{Ni^{II}(acac)_2 + L^2 + PhOH\} \rightarrow Ni(acac)_2 \cdot L^2 \cdot PhOH \rightarrow \{Ni(acac)_2 \cdot L^2 \cdot PhOH\}_n$$

In favor of formation of supramolecular macrostructures based on the triple complexes, $\{Ni(acac)_2 \cdot L^2 \cdot PhOH\}$ in the real catalytic ethyl benzene oxidation shows data of AFM-microscopy (see below).

11.3.2 ROLE OF SUPRAMOLECULAR NANOSTRUCTURES FORMED DUE TO H-BONDING, IN MECHANISMS OF CATALYSIS: MODELS OF NI(FE)ARD DIOXYGENASES

As mentioned before, the high stability of effective catalytic complexes, which formed in the process of selective oxidation of ethylbenzene to PEH at catalysis with $M^{II}{}_xL^1{}_y(L^1{}_{ox})_z(L^2)_n$, ($M = Ni, Fe$, $L^1 = acac^-$, $L^1{}_{ox} = OAc^-$, $L^2 =$ crown ethers or quaternary ammonium salts) complexes and triple systems $\{Ni^{II}(acac)_2 + L^2 + PhOH\}$ [L^2 = MP, HMPA, or alkali metal stearate MSt (M = Li, Na)] seems to be associated with the formation of supramolecular structures due to intermolecular H-bonds.

Hydrogen bonds play an important role in the structures of proteins and DNA, as well as in drug-receptor binding and catalysis.[16] Proton-coupled bicarboxylates tops the list as the earliest and still the best-studied systems suspected of forming low-barrier hydrogen bonds (LBHBs) in the vicinity of the active sites of enzymes [17 (20)]. These hydrogen-bonded couples can be depicted as $R{-}C({=}O){-}O^-{\cdots}H{-}O{-}C({=}O){-}R'$ and they can be abbreviated by the general formula $X^- \cdots HX$. Proton-coupled bicarboxylates appear in 16% of all protein X-ray structures. There are at least five X-ray structures showing short (and therefore strong) hydrogen bonds between an enzyme carboxylate and a reaction intermediate or transition state analogue bound at the enzyme active site. The authors[17] consider these structures to be the best de facto evidence of the existence of low-barrier hydrogen bonds stabilizing high-energy reaction intermediates at enzyme-active sites. Carboxylates figure prominently in the LBHB enzymatic story in part because all negative charges on proteins are carboxylates.

H-bonds are commonly used for the fabrication of supramolecular assemblies because they are directional and have a wide range of interaction energies that are tunable by adjusting the number of H-bonds, their relative orientation, and their position in the overall structure. H-bonds in the center of protein helices can be 20 kcal/mol due to cooperative dipolar interactions.[18,19]

The porphyrin linkage through H-bonds is the binding type generally observed in nature. One of the simplest artificial self-assembling supramolecular porphyrin systems is the formation of a dimer based on carboxylic acid functionality.[20]

11.3.2.1 THE POSSIBLE ROLE OF THE SELF-ASSEMBLING SUPRAMOLECULAR MACROSTRUCTURES IN MECHANISM OF ACTION OF ACIREDUCTONE DIOXYGENASES (ARDS) Ni(Fe)-ARD INVOLVED IN THE METHIONINE RECYCLE PATHWAY

The methionine salvage pathway (MSP) (Scheme 11.3) plays a critical role in regulating a number of important metabolites in prokaryotes and eukaryotes. ARDs Ni(Fe)-ARD are enzymes involved in the methionine recycle pathway, which regulates aspects of the cell cycle. The relatively subtle differences between the two metalloproteins complexes are amplified by the surrounding protein structure, giving two enzymes of different structures and activities from a single polypeptide (Scheme 11.3).[21] Both enzymes $Ni^{II}(Fe^{II})$-ARD are members of the structural super family, known as cupins, which also include Fe–acetyl acetone dioxygenase (Dke1) and cysteine dioxygenase. These enzymes that form structure super family of cupins use a triad of histidine ligands (His), and also one or two oxygen from water and a carboxylate oxygen (Glu), for binding with Fe (or Ni) center.[22]

11.3.2.1.1 Structural and Functional Differences between the Two ARDs Enzymes Are Determined by the Type of Metal Ion Bound in the Active Site of the Enzyme

The two acireductone dioxygenase enzymes (ARD and ARD′ share the same amino acid sequence, and only differ in the metal ions that they bind, which results in distinct catalytic activities. ARD has a bound Ni^{2+} atom while ARD′ has a bound Fe^{2+} atom. The apo-protein, resulting from removal of the bound

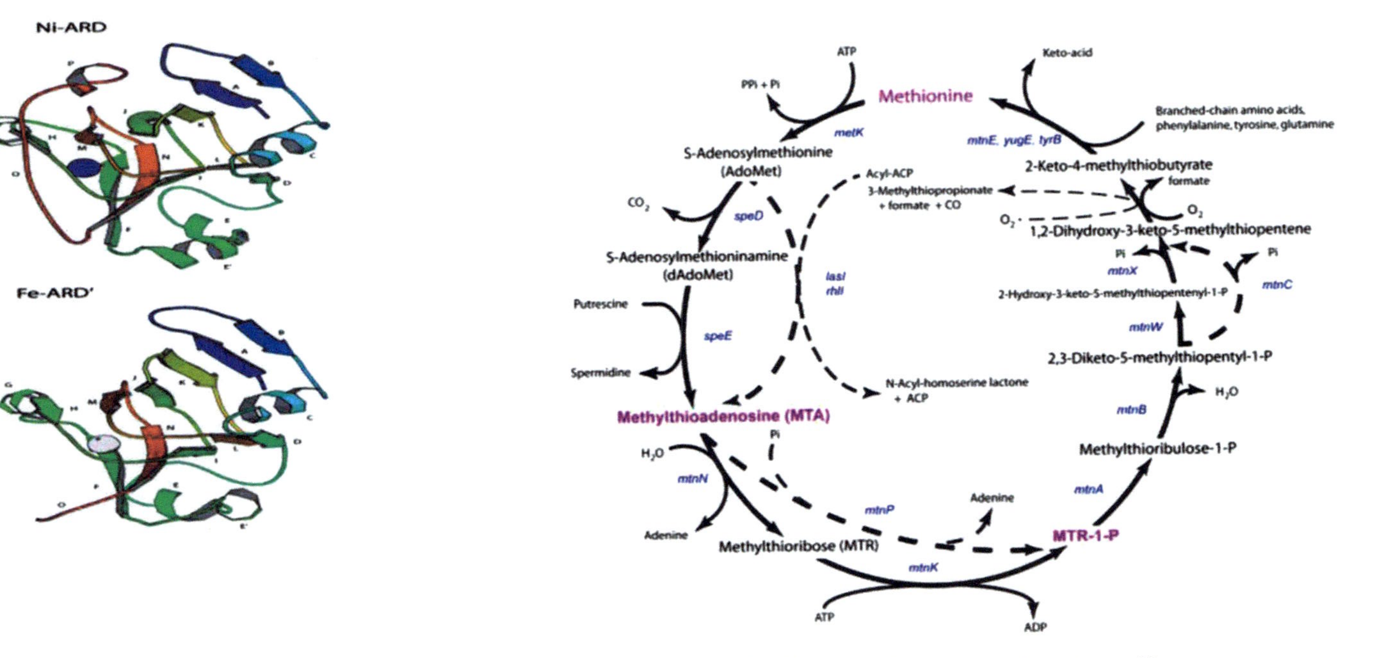

SCHEME 11.3 Acireductone dioxygenases Ni-ARD and Fe-ARD′ (a)[21] are involved in the methionine recycle pathway (b).

metal is identical and is catalytically inactive. ARD and ARD′ can be interconverted by removing the bound metal and reconstituting the enzyme with the alternative metal. ARD and ARD′ act on the same substrate, the acireductone, 1,2-dihydroxy-3-keto-5-methylthiopentene anion, but they yield different products. ARD′ catalyzes a 1,2-oxygenolytic reaction, yielding formate and 2-keto-4-methylthiobutyrate, a precursor of methionine, and thereby part of the MSP, while Ni-ARD catalyzes a 1,3-oxygenolytic reaction, yielding formate, carbon monoxide (CO), and 3-methylthiopropionate, an off-pathway transformation of the acireductone. The role of a reaction catalyzed by the enzyme ARD, still not clear.

We assumed that one of the reasons for the different activity of $Ni^{II}(Fe^{II})$-ARD in the functioning of enzymes in relation to the common substrates (acireductone and O_2) can be the association of catalyst in various macrostructure due to intermolecular H-bonds.

The Fe^{II}ARD operation seems to comprise the step of oxygen activation ($Fe^{II} + O_2 \rightarrow Fe^{III} - O_2^{-}\cdot$) (by analogy with Dke1 action[15]). Specific structural organization of iron complexes may facilitate the following regioselective addition of activated oxygen to Aci-reductone ligand and the reactions leading to formation of methionine. Association of the catalyst in macrostructures with the assistance of the intermolecular H-bonds may be one of reasons of reducing Ni^{II}ARD activity in mechanisms of $Ni^{II}(Fe^{II})$ ARD operation.[21,22] Here, we demonstrate for the first time a specific structural organization of functional models of iron (nickel) enzymes.

The possibility of the formation of stable supramolecular nanostructures based on iron (nickel) heteroligand complexes due to intermolecular H-bonds we researched with the AFM method.[4,5]

In Figures 11.1 and 11.2, three-dimensional and two-dimensional AFM image of the structures on the basis of iron complex with 18C6 $Fe^{III}_x(acac)_y 18C6_m(H_2O)_n$, formed at putting a uterine solution on a hydrophobic surface of modified silicone are presented. It is visible that the generated structures are organized in certain way forming structures resembling the shape of tubule microfiber cavity (Fig. 11.2c). The heights of particles are about 3–4 nm. In control experiments, it was shown that for similar complexes of nickel $Ni^{II}(acac)_2 \cdot 18C6 \cdot (H_2O)_n$ (as well as complexes $Ni_2(OAc)_3(acac) \cdot MP \cdot 2H_2O$), this structure organization is not observed. It was established that these iron constructions are not formed in the absence of the aqueous environment. Earlier, we showed the participation of H_2O molecules in mechanism of $Fe^{III,II}_x(acac)_y 18C6_m(H_2O)_n$ transformation by analogy with Dke1 action and also the increase in catalytic activity of iron complexes

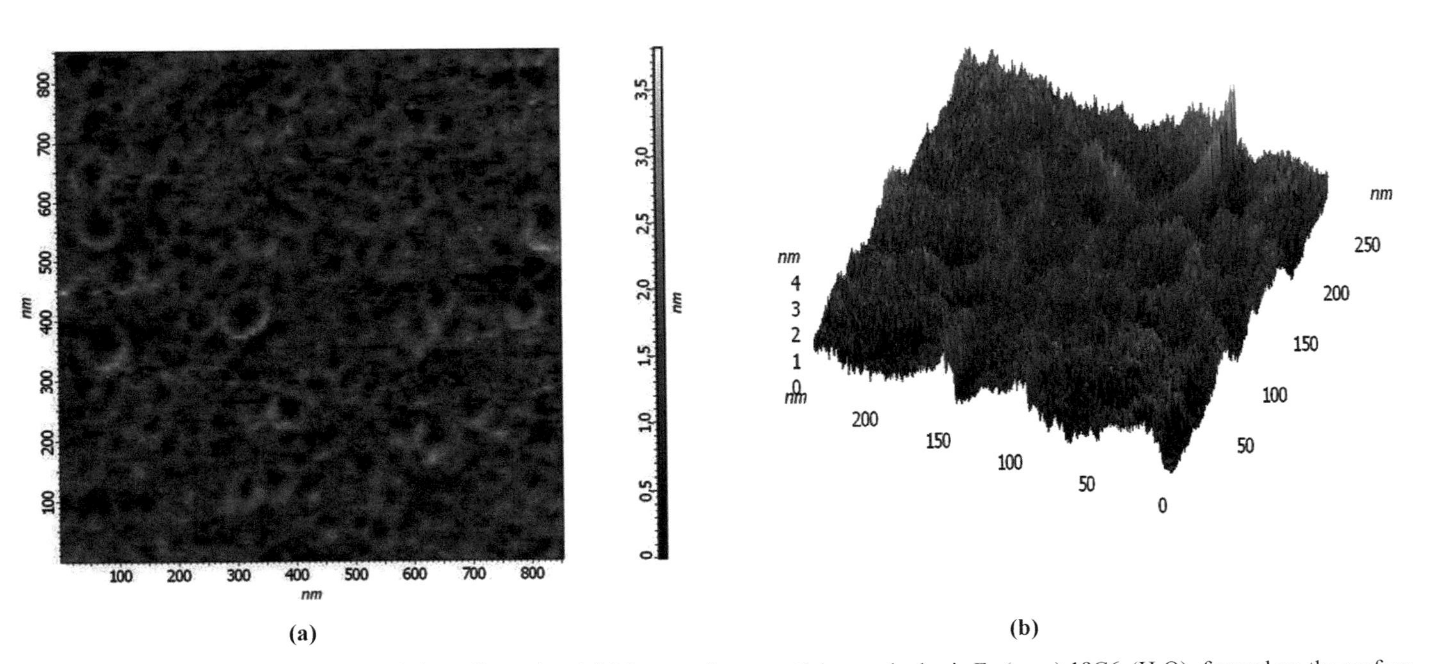

FIGURE 11.1 The AFM two- (a) and three-dimensional (b) image of nanoparticles on the basis $Fe_x(acac)_y18C6_m(H_2O)_n$ formed on the surface of modified silicone.

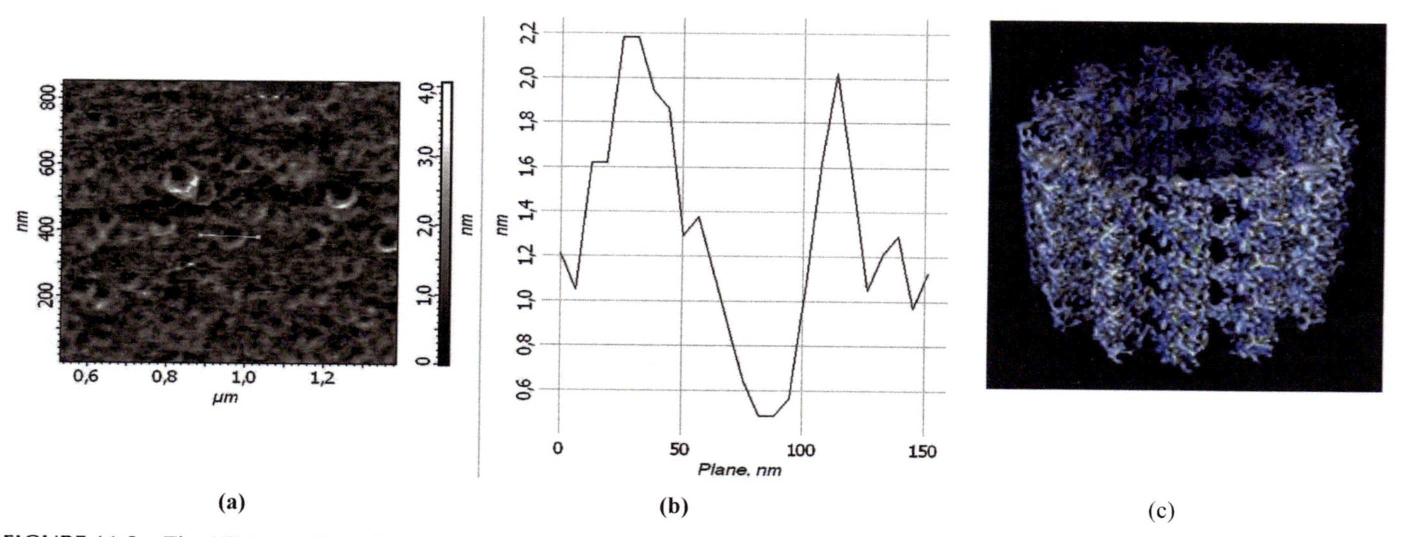

FIGURE 11.2 The AFM two-dimensional image (a) of nanoparticles on the basis $Fe_x(acac)_y18C6_m(H_2O)_n$ formed on the hydrophobic surface of modified silicone. The section of a circular shape with fixed length and orientation is about 50–80 nm (b). (c) The structure of the cell microtubules.

($Fe^{III}_x(acac)_y18C6_m(H_2O)_n$, $Fe^{II}_x(acac)_y18C6_m(H_2O)_n$ and $Fe^{II}_xL^1_y(L^1_{ox})_z(18C6)_n(H_2O)_m$) in the ethyl benzene oxidation in the presence of small amounts of water.[23] After our works in the chapter,[24] it was found that the possibility of decomposition of the β-diketone in iron complex by analogy with Fe-ARD′ action increases in aquatic environment. That apparently is consistent with data, published by us earlier in our initial work.[23]

Unlike catalysis with iron dioxygenase, mechanism of catalysis by the Ni^{II}ARD does not include activation of O_2, and oxygenation of acireductone leads to the formation of products not being precursors of methionine.[21] In our previous works, we have shown that formation of multidimensional forms based on nickel complexes can be one of the ways of regulating the activity of two enzymes.[4] The association of complexes $Ni_2(AcO)_3(acac)\cdot MP\cdot 2H_2O$, which is functional and structure model of Ni-ARD, to supramolecular nanostructure due to intermolecular H-bonds (H_2O–MP, H_2O–(OAc^-)(or ($acac^-$)), is demonstrated in Figure 11.3. All structures (Fig. 11.3) are various on heights from the minimal 3–4 nm to ~20–25 nm for maximal values (in the form reminding three almost merged spheres).[4]

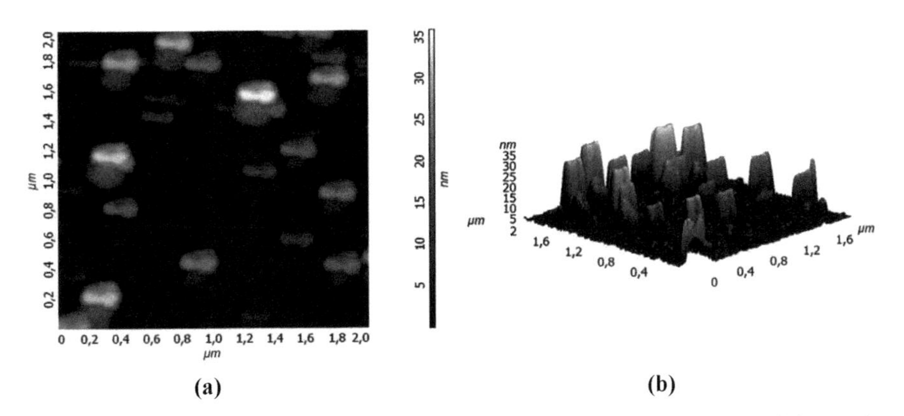

FIGURE 11.3 The AFM two- (a) and three-dimensional (b) image of nanoparticles on the basis $Ni_2(AcO)_3(acac)\cdot L^2\cdot 2H_2O$ formed on the hydrophobic surface of modified silicone.

As one can see on Figure 11.4 in case of binary complexes {$Ni(acac)_2\cdot MP$} we also observed formation of nanostructures due to H-bonds. But these nanoparticles differ on forms and are characterized with less height: h ~ 8 nm (Fig. 11.4) as compared with nanostructures on the basis of complexes $Ni_2(AcO)_3(acac)\cdot L^2\cdot 2H_2O$ (Fig. 11.3).

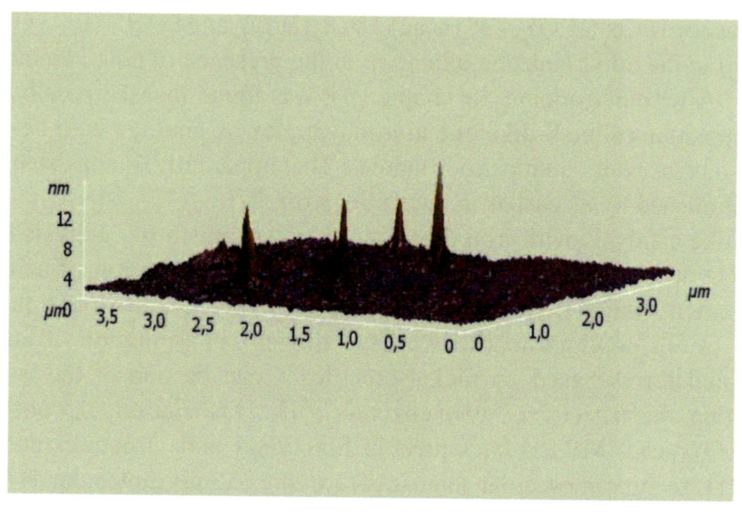

FIGURE 11.4 The AFM of three-dimensional image (5.0 × 5.0 (µm)) of nanoparticles on the basis {$Ni(acac)_2$·MP} formed on the surface of modified silicone. Data presented in the figure will be published in an article which is in print.

11.3.2.2 POSSIBLE EFFECT OF TYR-FRAGMENT, BEING IN THE SECOND COORDINATION SPHERE OF METAL COMPLEX

Here, we assume that it may be necessary to take into account the role of the second coordination sphere, including Tyr-fragment (see Fig. 11.5[21]). We are for the first time suggesting the participation of Tyrosine moiety in mechanisms of action of $Ni^{II}(Fe^{II})$ARD enzymes.

It is known that Tyrosine residues are located in different regions of protein by virtue of the relatively large phenol amphiphatic side chain capable of (a) interacting with water and participating in hydrogen bond formation and (b) undergoing cation–π and nonpolar interactions.[25] The versatile physicochemical properties of tyrosine allow it to play a central role in conformation and molecular recognition.[26] Moreover, tyrosine has special role by virtue of the phenol functionality: for example, it can receive phosphate groups in target proteins by way of protein tyrosine kinases, and it participates in electron transfer processes with intermediate formation of tyrosyl radical.

Tyrosine can take part in different enzymatic reactions. Recently, it has been researched role of tyrosine residues in mechanism of heme oxygenase (HO) action. HO is responsible for the degradation of a histidine-ligated ferric protoporphyrin IX (Por) to biliverdin, CO, and the free ferrous ion.

The role of reactions of tyrosyl radical formation which occurs after oxidation Fe(III)(Por) to Fe(IV) = O(Por(+)) in mechanism of human heme oxygenase isoform-1 (hHO-1) and the structurally homologous protein from *Corynebacterium diphtheriae* (cdHO) are described.[27]

It is assumed that Tyr-fragment may be involved in substrate H-binding in step of O_2-activation by iron catalyst, and this can decrease the oxygenation rate of the substrate in the case of Homoprotocatechuate 2,3-dioxygenase action.[28]

Tyr-fragment is discussed as important in methyl group transfer from *S*-adenosylmethionine (AdoMet) to dopamine.[29] The experimental findings with the model of Methyltransferase and structural survey imply that methyl CH···O hydrogen bonding (with participation of Tyr-fragment) represents a convergent evolutionary feature of AdoMet-dependent methyltransferases, mediating a universal mechanism for methyl transfer.[30]

Tyrosine residue Tyr149 is found in the Met-turn for astacin endpeptidases and serralizines. Tyr149 giving a proton, forms a hydrogen bond with zinc and becomes the fifth ligand. This switch plays a specific role, participating in the stabilization of the transition state during the binding of the substrate to the enzyme.[31]

In the case of Ni-dioxygenase ARD, Tyr-fragment, involved in the mechanism, can reduce the Ni^{II}ARD-activity. The structure of the active center of Ni^{II}ARD with Tyr residue in II coordination sphere is shown in Figure 11.5.

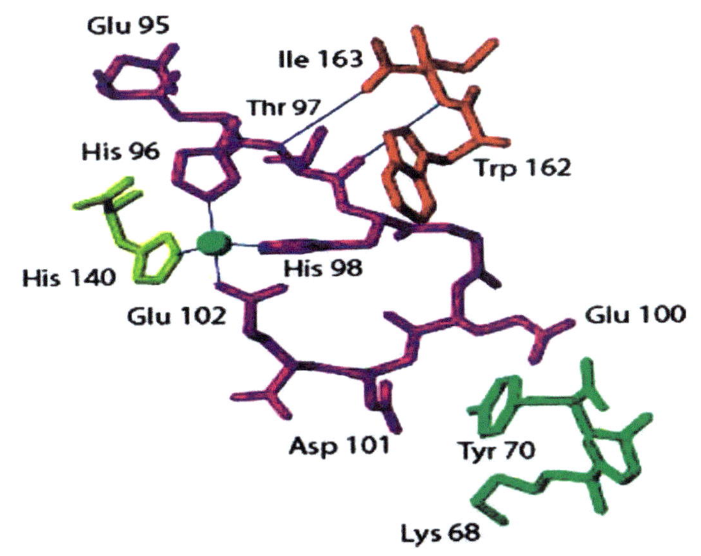

FIGURE 11.5 The structure of Ni^{II}ARD with Tyr residue in the second coordination sphere.[21]

Really, as mentioned above, we have found[2,3] that the inclusion of PhOH in complex $Ni(acac)_2 \cdot L^2$ (L^2 = *N*-methylpirrolidone-2), which is the primary model of Ni^{II}ARD, leads to the stabilization of formed triple complex $Ni(acac)_2 \cdot L^2 \cdot PhOH$. In this case, as we have emphasized above, ligand $(acac)^-$ is not oxygenated with molecular O_2. Also the stability of triple complexes $Ni(acac)_2 \cdot L^2 \cdot PhOH$ seems to be due to the formation of supramolecular macrostructures that are stable to oxidation with dioxygen. Formation of supramolecular macrostructures due to intermolecular (phenol–carboxylate) H-bonds and, possible, the other noncovalent interactions,[32–34] based on the triple complexes $Ni(acac)_2 \cdot L^2 \cdot PhOH$, that we have established with the AFM-method[4,5,35] (in the case of L^2=MP, HMPA, NaSt, LiSt), is in favor of this hypothesis (Fig. 11.6). Data of structures on the basis of complexes $\{Ni(acac)_2 \cdot MP \cdot PhOH\}$ that self-organized on the surface of the modified silicon (Fig. 11.6a) are in print.

Conclusive evidence in favor of the participation of tyrosine fragment in stabilizing primary Ni complexes as one of regulatory factors in mechanism of action of Ni-ARD has been obtained by AFM-microscopy, we observed at the first time the formation of nanostructures on Ni-based systems using L-tyrosine (Tyr) as an extraligand. The grow of self-assembly of supramolecular macrostructures due to intermolecular (phenol–carboxylate) H-bonds and, possible, the other noncovalent interactions,[32–34] based on the triple systems $\{Ni(acac)_2$ + MP + Tyr}, we observed at the apartment of a uterine H_2O solution of triple system $\{Ni(acac)_2$ + MP + Tyr} on surfaces of modified silicon (Fig. 11.7). Spontaneous organization process, that is, self-organization, of researched triple complexes (Figs. 11.6 and 11.7) at the apartment of a uterine hydrocarbon solution of complexes on surfaces of modified silicon are driven by the balance between intermolecular, and molecule–surface interactions, which may be the consequence of hydrogen bonds and the other noncovalent interactions.[36]

Histogram of volumes of the particles based on systems $\{Ni^{II}(acac)_2$ + MP + Tyr}, and also the empirical and theoretical cumulative normal probability distribution of volumes, and the empirical and theoretical cumulative Log normal distribution of volumes of the particles based on systems $\{Ni^{II}(acac)_2$ + MP + Tyr}, formed on the surfaces of modified silicon, are presented in Figure 11.8. As can be seen, distribution of volumes of the particles in this case is well described by a log–normal law.

But as one can see in Figure 11.9, in case of binary systems $\{Ni(acac)_2$ + Tyr}, we also observed formation of nanostructures due to H-bonds.

But these nanoparticles as well as particle based on {Ni(acac)$_2$·MP} complexes (Fig. 11.4) differ on form and high from the nanostructures on the basis of triple systems {Ni(acac)$_2$ + MP + Tyr} (compare Figs. 11.4, 11.7, and 11.8).

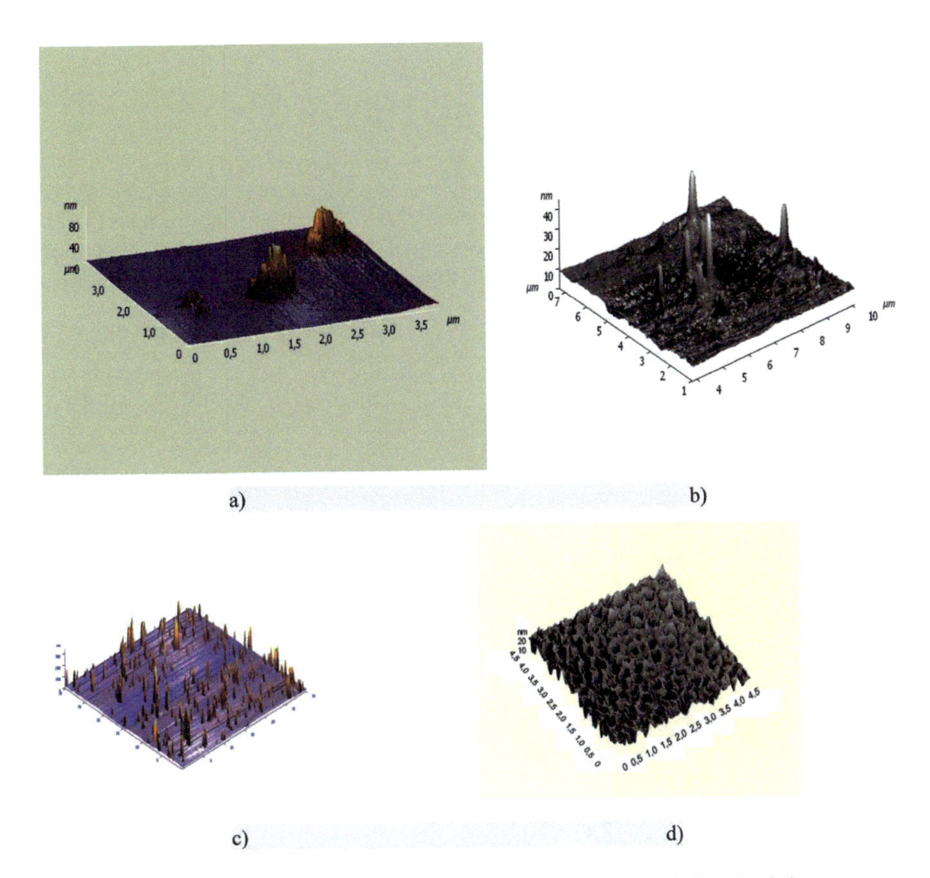

FIGURE 11.6 (a) The AFM three-dimensional image (5.0 × 5.0 (μm)) of the structures (h ~ 80–100 nm) formed on a surface of modified silicone on the basis of triple complexes Ni^{II}(acac)$_2$·MP·PhOH. (b) The AFM three-dimensional image (6.0 × 6.0 (μm)) of the structures (h ~ 40 nm) formed on a surface of modified silicone on the basis of triple complexes {Ni^{II}(acac)$_2$·HMPA·PhOH}. (c) The AFM three-dimensional image (30 × 30 (μm)) of the structures (h ~ 80 nm) formed on a surface of modified silicone on the basis of triple complexes Ni^{II}(acac)$_2$·NaSt·PhOH. (d) The AFM three-dimensional image (4.5 × 4.5 (μm)) of the structures (h ~ 10 nm) formed on a surface of modified silicone on the basis of triple complexes Ni^{II}(acac)$_2$·LiSt·PhOH.

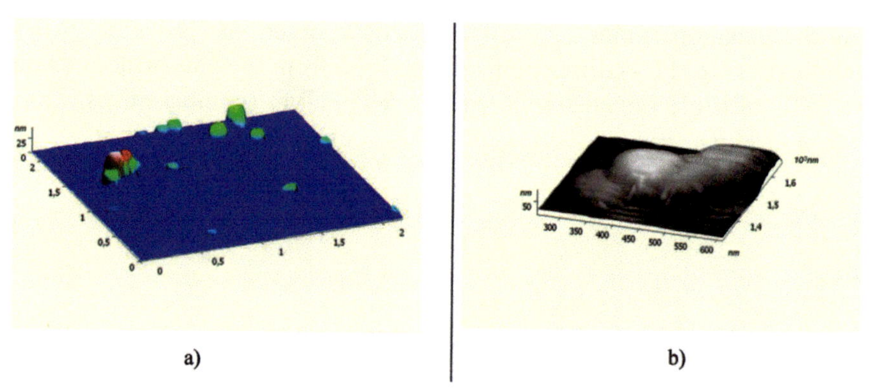

FIGURE 11.7 The AFM three-dimensional image (2.0 × 2.0 (μm)) of the structures (h ~25 nm) (a) and three-dimensional image (0.3 × 0.6 (μm)) of the structures (h ~50 nm) (b), formed on a surface of modified silicone on the basis of triple systems {$Ni^{II}(acac)_2$ + MP + Tyr}.

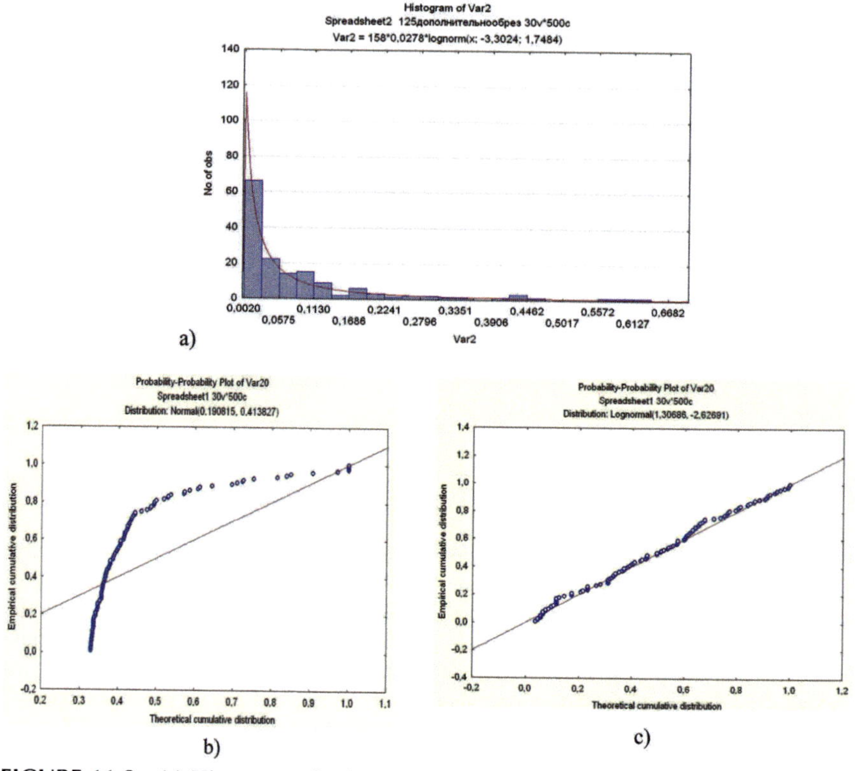

FIGURE 11.8 (a) Histogram of volumes of the particles based on systems {$Ni^{II}(acac)_2$ + MP + Tyr}, (b) the empirical and theoretical cumulative normal distribution of volumes of the particles based on systems {$Ni^{II}(acac)_2$ + MP + Tyr}, (c) the empirical and theoretical cumulative Log–normal distribution of volumes of the particles based on systems {$Ni^{II}(acac)_2$ + MP + Tyr}.

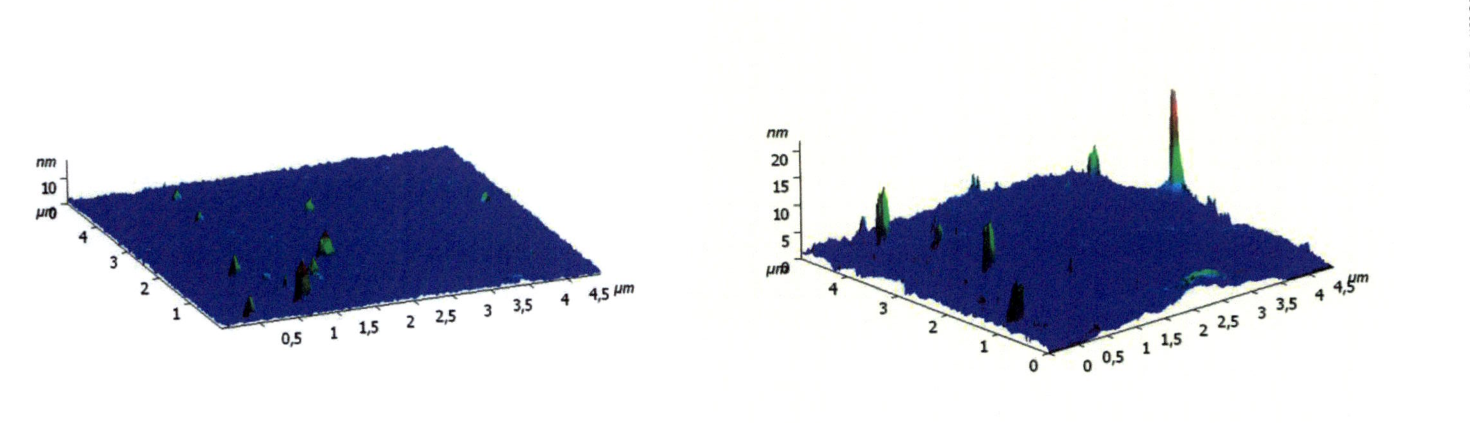

FIGURE 11.9 (a,b) The AFM three-dimensional image (4.5.0 × 4.5 (μm)) of the structures (h ~ 10–15 (20) nm) formed on a surface of modified silicone on the basis of binary systems $Ni^{II}(acac)_2$ + Tyr.

At the same time, it is necessary to mean that important function of Ni^{II}ARD in cells is established now. Namely, CO is formed as a result of action of nickel-containing dioxygenase Ni^{II}ARD. It was established that CO is a representative of the new class of neural messengers and seems to be a signal transducer like nitrogen oxide, NO.[12,21]

11.4 CONCLUSION

Usually in the quest for axial modifying ligands that control the activity and selectivity of homogeneous metal complex catalysts, the attention of scientists is focused on their steric and elcctronic properties. The interactions in the outer coordination sphere, the role of hydrogen bonds and also the other noncovalent interactions is less studied.

We have assumed that the high stability of heteroligand $M^{II}{}_xL^1{}_y(L^1{}_{ox})_z(L^2)_n(H_2O)_m$ (M = Ni, Fe, L^1=acac$^-$, $L^1{}_{ox}$=OAc$^-$, L^2 = electron-donating mono-, or multidentate activating ligands) complexes as selective catalysts of the ethylbenzene oxidation to PEH, formed during the ethylbenzene oxidation in the presence of $\{ML^1{}_n + L^2\}$ systems as a result of oxygenation of the primary complexes $(M^{II}L^1{}_2)_x(L^2)_y$ can be associated with the formation of the supramolecular structures due to the intermolecular H-bonds.

The supramolecular nanostructures on the basis of catalytic active iron $Fe^{III}{}_x(acac)_y18C6_m(H_2O)_n$, and nickel complexes $Ni^{II}{}_xL^1{}_y(L^1{}_{ox})_z(L^2)_n(H_2O)_m$ (L^1 = acac$^-$, $L^1{}_{ox}$ = OAc$^-$, L^2 = *N*-methylpirrolidone-2, $x = 2$, $y = 1$, $z = 3$, $m = 2$), $\{Ni(acac)_2 \cdot L^2 \cdot PhOH\}$ (L^2 = MP, HMPA, NaSt, LiSt), obtained with AFM method, indicate high probability of supramolecular structures formation due to H-bonds in the real systems, namely, in the processes of alkylarens oxidation. Since the investigated complexes are structural and functional models of $Ni^{II}(Fe^{II})$ARD dioxygenases, the data could be useful in the interpretation of the action of these enzymes.

Specific structural organization of iron complexes may facilitate the first step in Fe^{II}ARD operation: O_2 activation and following regioselective addition of activated oxygen to acireductone ligand (unlike mechanism of regioselective addition of no activated O_2 to acireductone ligand in the case of Ni^{II}ARD action), and reactions leading to formation of methionine.

The formation of multidimensional forms (in the case of Ni^{II}ARD) may be one way of controlling $Ni^{II}(Fe^{II})$ARD activity.

In this chapter, we at first time assumed the participation of Tyr-fragment which is in the second coordination sphere in mechanism of $Ni^{II}(Fe^{II})$ARD operation, as one of possible mechanisms of reduce in enzymes activity in $Ni^{II}(Fe^{II})$ARD enzymes operation and received experimental facts in favor this assumption. So we observed the formation of supramolecular macrostructures due to intermolecular (phenol–carboxylate) H-bonds and, possible, the other noncovalent interactions, based on the triple complexes $Ni(acac)_2 \cdot L^2 \cdot PhOH$ established by us with the AFM-method (in the case of L^2 = MP, HMPA, NaSt, LiSt), and self-assembly based on triple systems that included L-tyrosine as extraligand $\{Ni^{II}(acac)_2$ + MP + Tyr$\}$. Self-assembly based on triple systems that included L-tyrosine as extraligand $\{Ni^{II}(acac)_2$ + MP + Tyr$\}$, formed on a surface of modified silicone, which we observed first.

KEYWORDS

- **homogeneous and enzymatic catalysis**
- **dioxygen**
- **AFM method**
- **nanostructures based on complexes $Fe^{III}x(acac)_y18C6_m(H_2O)n$**
- **$Ni_xL^1_y(L^1_{ox})_z(L^2)_n(H_2O)_m$**
- **$\{Ni^{II}(acac)_2 \cdot L^2 \cdot PhOH\}$ [L^2 = MP**
- **HMPA**
- **MSt (M = Na, Li)]**
- **and system $\{Ni^{II}(acac)_2$ + Tyr + PhOH$\}$ (Tyr = L-tyrosine)**
- **models of Ni(Fe)ARD dioxygenases**

REFERENCES

1. Matienko, L. I. Solution of the Problem of Selective Oxidation of Alkylarenes by Molecular Oxygen to Corresponding Hydroperoxides. Catalysis Initiated by Ni(II), Co(II), and Fe(III) Complexes Activated by Additives of Electron-Donor Mono- or Multidentate Extra-ligands. In *Reactions and Properties of Monomers and Polymers*; D'Amore, A., Zaikov, G., Eds.; Nova Science Publ.: New York, NY, 2007; pp 21–41.
2. Matienko, L. I.; Mosolova, L. A.; Zaikov, G. E. *Selective Catalytic Hydrocarbons Oxidation. New Perspectives*. Nova Science Publ. Inc.: New York, NY, 2010; 150 p.
3. Matienko, L. I.; Binyukov, V. I.; Mosolova, L. A. Mechanism of Selective Catalysis with Triple System {Bis(acetylacetonate)Ni(II) + Metalloligand + Phenol} in Ethylbenzene Oxidation with Dioxygen. Role of H-Bonding Interactions. *Oxid. Commun.* **2014,** *37*, 20–31.

4. Matienko, L. I.; Mosolova, L. A.; Binyukov, V. I.; Mil, E. M.; Zaikov, G. E. The New Approach to Research of Mechanism Catalysis with Nickel Complexes in Alkylarens Oxidation. *Polymer Yearbook 2011*; Nova Science Publ.: New York, NY, pp 221–230.
5. Matienko, L. I.; Binyukov, V. I.; Mosolova, L. A.; Mil, E. M.; Zaikov, G. E. Supramolecular Nanostructures on the Basis of Catalytic Active Heteroligand Nickel Complexes and their Possible Roles in Chemical and Biological Systems. *J. Biol. Res.* **2012,** *1*, 37–44.
6. Borovik, A. S. Bioinspired Hydrogen Bond Motifs in Ligand Design: The Role of Noncovalent Interactions in Metal Ion Mediated Activation of Dioxygen. *Acc. Chem. Res.* **2005,** *38*, 54–61.
7. Holm, R. H.; Solomon, E. I. Biomimetic Inorganic Chemistry. *Chem. Rev.* **2004,** *104*, 347–348.
8. Tomchick, D. R.; Phan, P.; Cymborovski, M.; Minor, W.; Holm, T. R. Structural and Functional Characterization of Second-Coordination Sphere Mutants of Soybean Lipoxygenase-1. *Biochemistry* **2001,** *40*, 7509 7517.
9. Uehara, K.; Ohashi, Y.; Tanaka, M. Bis(acetylacetonato) Metal(II)-Catalyzed Addition of Acceptor Molecules to Acetylacetone. *Bull. Chem. Soc. Jpn.* **1976,** *49*, 1447–1448.
10. Lucas, R. L.; Zart, M. K.; Murkerjee, J.; Sorrell, T. N.; Powell, D. R.; Borovik, A. S. A Modular Approach toward Regulating the Secondary Coordination Sphere of Metal Ions: Differential Dioxygen Activation Assisted by Intramolecular Hydrogen Bonds. *J. Am. Chem. Soc.* **2006,** *128*, 15476–15489.
11. Nelson, J. H.; Howels, P. N.; Landen, G. L.; De Lullo, G. S.; Henry, R. A. Catalytic Addition of Electrophiles to β-Dicarbonyles. In *Fundamental Research in Homogeneous Catalysis,* vol. 3; Plenum: New York, London, 1979; pp 921–939.
12. Dai, Y.; Pochapsky Th. C.; Abeles, R. H. Mechanistic Studies of Two Dioxygenases in the Methionine Salvage Pathway of *Klebsiella pneumonia. Biochemistry* **2001,** *40*, 6379–6387.
13. Gopal, B.; Madan, L. L.; Betz, S. F.; Kossiakoff, A. A. The Crystal Structure of a Quercetin 2,3-Dioxygenase from *Bacillus subtilis* Suggests Modulation of Enzyme Activity by a Change in the Metal Ion at the Active Site(s). *Biochemistry* **2005,** *44*, 193–201.
14. Balogh-Hergovich, E.; Kaizer, J.; Speier, G. Kinetics and Mechanism of the Cu(I) and Cu(II) Flavonolate-Catalyzed Oxygenation of Flavonols, Functional Quercetin 2,3-Dioxygenase Models. *J. Mol. Catal. A: Chem.* **2000,** *159*, 215–224.
15. Straganz, G. D.; Nidetzky, B. Reaction Coordinate Analysis for β-Diketone Cleavage by the Non-Heme Fe^{2+}-Dependent Dioxygenase Dke 1. *J. Am. Chem. Soc.* **2005,** *127*, 12306–12314.
16. Ma, J. C.; Dougherty, D. A. The Cation–π Interaction. *Chem. Rev.* **1997,** *97*, 1303–1324.
17. Graham, J. D.; Buytendyk, A. M.; Wang, D.; Bowen, K. H.; Collins, K. D. Strong, Low-Barrier Hydrogen Bonds May be Available to Enzymes. *Biochemistry* **2014,** *53*, 344–349.
18. Stang, P. J.; Olenyuk, B. Self-Assembly, Symmetry, and Molecular Architecture: Coordination as the Motif in the Rational Design of Supramolecular Metallacyclic Polygons and Polyhedra. *Acc. Chem. Res.* **1997,** *30*, 502–518.
19. Drain, C. M.; Varotto-Radivojevic, A. I. Self-organized Porphyrinic Materials. *Chem. Rev.* **2009,** *109*, 1630–1658.
20. Beletskaya, I.; Tyurin, V. S.; Tsivadze, A. Yu.; Guilard, R.; Stem Ch. Supramolecular Chemistry of Metalloporphyrins. *Chem. Rev.* **2009,** *109*, 1659–1713.
21. Chai, S. C.; Ju, T.; Dang, M.; Goldsmith, R. B.; Maroney, M. J.; Pochapsky Th. C. Characterization of Metal Binding in the Active Sites of Acireductone Dioxygenase Isoforms from *Klebsiella* ATCC 8724, *Biochemistry* **2008,** *47*, 2428–2435.

22. Leitgeb, St.; Straganz, G. D.; Nidetzky, B. Functional Characterization of an Orphan Cupin Protein from *Burkholderia* Xenovorans Reveals a Mononuclear Nonheme Fe^{2+}-Dependent Oxygenase that Cleaves β-Diketones. *FEBS J.* **2009,** *276*, 5983–5997.
23. L. I. Matienko, L. A. Mosolova. The Modeling of Catalytic Activity of complexes Fe(II,III)$(acac)_n$ with R_4NBr or 18-Crown-6 in the Ethylbenzene Oxidation by Dioxygen in the Presence of Small Amounts of H_2O. *Oxid. Commun.* **2010,** *33*, 830–844.
24. Allpress, C. J.; Grubel, K.; Szajna-Fuller, E.; Arif, A. M.; Berreau, L. M. Regioselective Aliphatic Carbon–Carbon Bond Cleavage by Model System of Relevance to Iron-Containing Acireductone Dioxygenase. *J. Am. Chem. Soc.* **2013,** *135*, 659–668.
25. Radi, R. Protein Tyrosine Nitration: Biochemical Mechanisms and Structure Basis of Functional Effects. *Acc. Chem. Res.* **2013,** *46*, 550–559.
26. Koide, S.; Sidhu, S. S. The Importance of Being Tyrosine: Lessons in Molecular Recognition from Minimalist Synthetic Binding Proteins. *ACS Chem. Biol.* **2009,** *4*, 325–334.
27. Smirnov, V. V.; Roth, J. P. Tyrosine Oxidation in Heme Oxygenase: Examination of Long-Range Proton-Coupled Electron Transfer. *J. Biol. Inorg. Chem.* **2014,** *19*, 1137–1148.
28. Mbughuni, M. M.; Meier, K. K.; Münck, E.; Lipscomb, J. D. Substrate-Mediated Oxygen Activation by Homoprotocatechuate 2,3-Dioxygenase: Intermediates Formed by a Tyrosine 257 Variant. *Biochemistry* **2012,** *51*, 8743–8754.
29. Zhang, J.; Klinman, J. P. Enzymatic Methyl Transfer: Role of an Active Site Residue in Generating Active Site Compaction that Correlates with Catalytic Efficiency. *J. Am. Chem. Soc.* **2011,** *133*, 17134–17137.
30. Horowitz, S.; Dirk, L. M. A.; Yesselman, J. D.; Nimtz, J. S.; Adhikari, U.; Mehl, R. A.; Scheiner St.; Houtz, R. L.; Al-Hashimi, H. M.; Trievel, R. C. Conservation and Functional Importance of Carbon–Oxygen Hydrogen Bonding in AdoMet-Dependent Methyltransferases. *J. Am. Chem. Soc.* **2013,** *135*, 15536–15548.
31. Bond, J. S.; Beynon, R. J. The Astacin Family of Metalloendopeptidases, *Prot. Sci.* **1995,** *4*, 1247–1261.
32. Dubey, M.; Koner, R. R.; Ray, M. Sodium and Potassium Ion Directed Self-Assembled Multinuclear Assembly of Divalent Nickel or Copper and L-Leucine Derived Ligand. *Inorg. Chem.* **2009,** *48*, 9294–9302.
33. Basiuk, E. V.; Basiuk, V. V.; Gomez-Lara, J.; Toscano, R. A. A Bridged High-Spin Complex Bis-[Ni(II)(rac-5,5,7,12,12,14-hexamethyl-1,4,8,11-tetraazacyclotetradecane)]-2,5-pyridinedicaboxylate Diperchlorate Monohydrate. *J. Incl. Phenom. Macrocycl. Chem.* **2000,** *38*, 45–56.
34. Mukherjee, P.; Drew, M. G. B.; Gómez-Garcia, C. J.; Ghosh, A. (Ni_2), (Ni_3), and $(Ni_2 + Ni_3)$: A Unique Example of Isolated and Cocrystallized Ni_2 and Ni_3 Complexes. *Inorg. Chem.* **2009,** *48*, 4817–4825.
35. Matienko, L.; Binyukov, V.; Mosolova, L.; Zaikov, G. The Selective Ethylbenzene Oxidation by Dioxygen into α-Phenyl Ethyl Hydroperoxide, Catalyzed with Triple Catalytic System $\{Ni^{II}(acac)_2 + NaSt(LiSt) + PhOH\}$. Formation of Nanostructures $\{Ni^{II}(acac)_2 \cdot NaSt \cdot (PhOH)\}_n$ with Assistance of Intermolecular H Bonds. *Polym. Res. J.* **2011,** *5*, 423–431.
36. Gentili, D.; Valle, F.; Albonetti, C.; Liscio, F.; Cavallini, M. Self-Organization of Functional Materials in Confinement. *Acc. Chem. Res.* **2014,** *47*, 2692–2699.

CHAPTER 12

THE SILICA–POLYMER COMPOSITES OF THE SYSTEM OF HEMA–TEOS: THE SYNTHESIS, THERMOMECHANICAL PROPERTIES, AND THE FEATURES OF THE KINETIC OF THE PROCESS

G. KHOVANETS[1], YU. MEDVEDEVSKIKH[1], V. ZAKORDONSKIY[2], T. SEZONENKO[1], and G. ZAIKOV[3*]

[1]*Department of Physical Chemistry of Fossil Fuels InPOCC, National Academy of Sciences of Ukraine, Naukova Str. 3a, 79060 Lviv, Ukraine*

[2]*Ivan Franko National University of Lviv, Kyryla and Mefodiya Str. 6, 79005 Lviv, Ukraine*

[3]*N. M. Emmanuel Institute of Biochemical Physics RAS, Kosygin Str. 4, 19991 Moscow, Russia*

[*]*Corresponding author. E-mail: GEZaikov@yahoo.com*

CONTENTS

ABSTRACT

Hybrid organic–inorganic composites (HOIC), based on 2-hydroxyethyl-methacrylate–tetraethoxysilane system in a wide range of the composition system changes, were obtained by the methods of sol–gel synthesis and photoinitiated polymerization. The kinetics of photoinitiated polymerization of the obtained composites depending on time of gelation has been studied. The influence of the HOIC composition on their thermomechanical properties and parameters of their molecular structure was investigated.

Development of advanced polymer materials science requires a search of new methods for the synthesis of nanocomposites to obtain the new materials with improved properties. Especially effective in this direction is an approach that is based on the obtained hybrid nanocomposites based on organic–inorganic systems with two or a few phases, which are different by the chemical composition and structure, in which in some or other way manage "to compose" the properties of the individual components. This is explained of the ability of creation of functional materials to the multi-usage with directional control of performance specification wide range.[1–4] Such materials may have predefined functional properties, such as characterized by ionic conductivity, have biologically active, magnetic, anticorrosion, electrochromic, repellent properties, etc.[5] However, the general patterns of correlation properties of the initial components with the nanocomposite properties are not defined.

One of the wide methods of obtaining hybrid nanomaterials is sol–gel synthesis. The main advantages of the last one are the possibility of variation of the chemical nature and dimensions of organic and inorganic fragments, making directed synthesis of hybrid nanomaterials, the possibility of low-maintenance process, the high degree of homogeneity in multicomponent system, etc. According to this technology, the process of obtaining consists of the following of hydrolysis of precursor molecules, which are alcoholates or other derivatives Ti, Si, Al, Zr, Zn, Sr, Ge[6] and the subsequent reactions of polycondensation of generated products.

The present technology of sol–gel synthesis of composites allows to enter into the chemically inert and thermally stable silica matrix practically any organic monomers, oligomers, and polymers. The organic–inorganic hybrids that formed are used in a wide variety of technological forms such as solid block, thin films, fibers, and coatings on various substrates.

More perspective and cost-effective method to obtain hybrid nanomaterials is their forming in the process of the cooperative polymerization from the mixtures of liquid organic and inorganic components.[7–9] When using the organic and inorganic components of different chemical composition, the changing of their ratio and synthesis conditions allows to obtain the hybrid polymeric materials with a wide range of properties, including the high thermal stability, fire resistance, mechanical strength, improved deformation, and sorption and adhesion properties.[10–12] The structure of such hybrid organic–inorganic nanocomposites (HOIC) is a three-phase system including the phases of organic and inorganic components and phase, which is the product of their interaction.[5,10,12] Thus, the marked improvement of the functional characteristics and the change in the structure of the polymer matrix is achieved with the introduction in the polymer or monomer of the small quantity (3–5%) of the inorganic component.[13]

Polymers based on monofunctional methacrylates have the complex of performance important properties. For example, polymers based on 2-hydroxyethylmethacrylate (HEMA) are transparent, nontoxic, well compatible with the tissues of a living organism, and have high adhesion to the various substrates. That's why the purpose of our study was to synthesize the HOIC based on the monomethacrylates and the tetraethoxysilane (TEOS) by the method of sol–gel technology and photoinitiated polymerization and to investigate the effect of composition on the kinetics of polymerization process to the deep conversion and the thermomechanical properties and parameters of the molecular structure of the composites that were obtained.

For studies we have used: monomer of HEMA mark "chem. pure" (Aldrich); photoinitiator of 2,2-dimethoxy-1,2-diphenylethane-1-on (IRGACURE 651) mark "chem. pure" (Fluka); TEOS $Si(OC_2H_5)_4$ (for "ECOS-1," Russia, TU 2637-059-444493179-04); ethanol mark "chem. pure;" orthophosphoric acid mark "chem. pure." Monomer purification was performed by means of its mixing with the pre-activated Al_2O_3 powder and followed by centrifugation.

The synthesis was carried out as follows. At first, the composition of monomer and photoinitiator, 2 mol%, was prepared. Separately, the system in the ratio (ml) $TEOS:H_2O:C_2H_5OH:H_3PO_4$ = 2.2:0.36:4.08:0.0072 by sol–gel method was prepared. Then, the obtained systems were mixed, mixing them using a magnetic stirrer with a duration of 20 min at the room temperature. The liquid composite was selected by doser 0.04 ml from the obtained output system of HEMA–TEOS, and it was subjected to photoinitiated polymerization (see Fig. 12.1, curve 1), the output system was placed in the oven at 40°C to pass the sol–gel processes in the composition until the gelation. The same quantity of photocomposition was selected again from the system that was in the oven in some period of time (Fig. 12.1, curve 2), and the photoinitiated polymerization was carried out investigating process kinetics. Such procedure was repeated until the gelation of the output composition which was found in oven at 40°C without UV radiation.

The kinetics of the stationary photopolymerization of the system HEMA–TEOS at various times of the prior gelation was studied in thin films, closed against access of oxygen air by cover glass using a laser interferometer with a ratio of HEMA:TEOS 97.5:2.5, 95:5, 90:10, and 80:20 (vol%), when the UV-radiation of the lamp DRT-400 radiates. The obtained composition was polymerized at the room temperature and the intensity of UV radiation 48 W/m^2 to the deep conversion. The relative integral conversion P was estimated by means of the ratio of the running concentration and the threshold contraction layer to the boundary contraction (at $t \to \infty$). The experimental error of measurement value of the composition linear shrinkage is 2.24×10^{-7} m. The results of experimentation were presented in the form of the integral kinetic curves, the relative conversion P—time t (s) and their differential anamorphosis. The statistical analysis of experimental kinetic curves was carried out using the program ORIGIN 5.0.

To describe the kinetics of photoinitiated polymerization of HEMA–TEOS to the deep conversion, there was used the concept of microheterogeneous model of radical polymerization, which is based on the concept of system microheterogeneous which is polymerized, and its reaction areas, in each of which, polymerization process takes place with own regularities.[14]

It should be noted that the experimental error at the time of constructing a polymerization single kinetic curve is sufficiently small. This is observed by the nature of the curves in Figure 12.1a. However, the scatter of the kinetic curves at the same conditions of experimentation is much higher than the error of the individual kinetic curve. The poor reproducibility of kinetic measurements is a result of fluctuation sensitivity of the polymerization process, especially at the stage of auto-acceleration.[14] Therefore, for each condition of the experiment, three to six kinetic curves were received which then averaged to form of one kinetic curve (see Fig. 12.1b) that increased the reliability of the obtained estimates.

The comparison of some averaged integral kinetic curves of polymerization of HEMA–TEOS systems at the different duration of the previous gelation in depending on the composition of the system that is polymerized is shown in Figure 12.2.

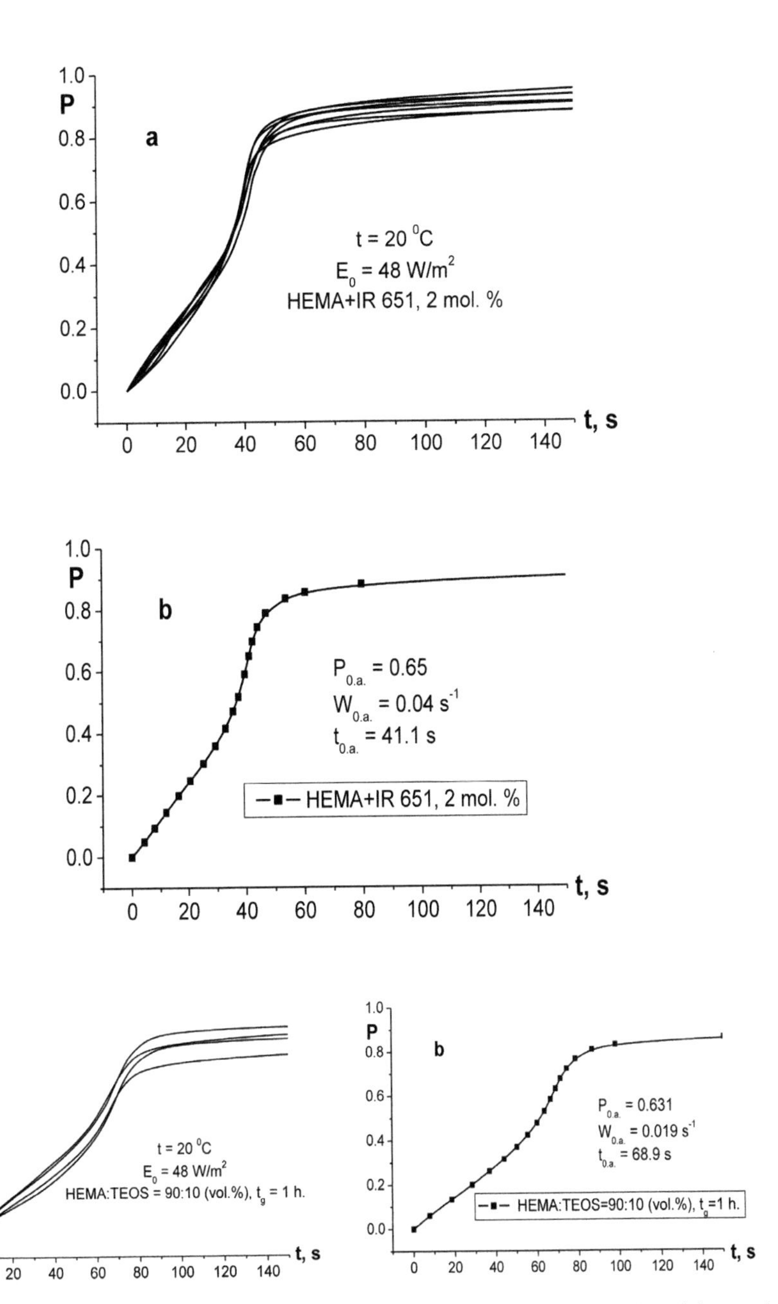

FIGURE 12.1 Output kinetic curves of polymerization of HEMA–TEOS (a) and the result of their averaging (b).

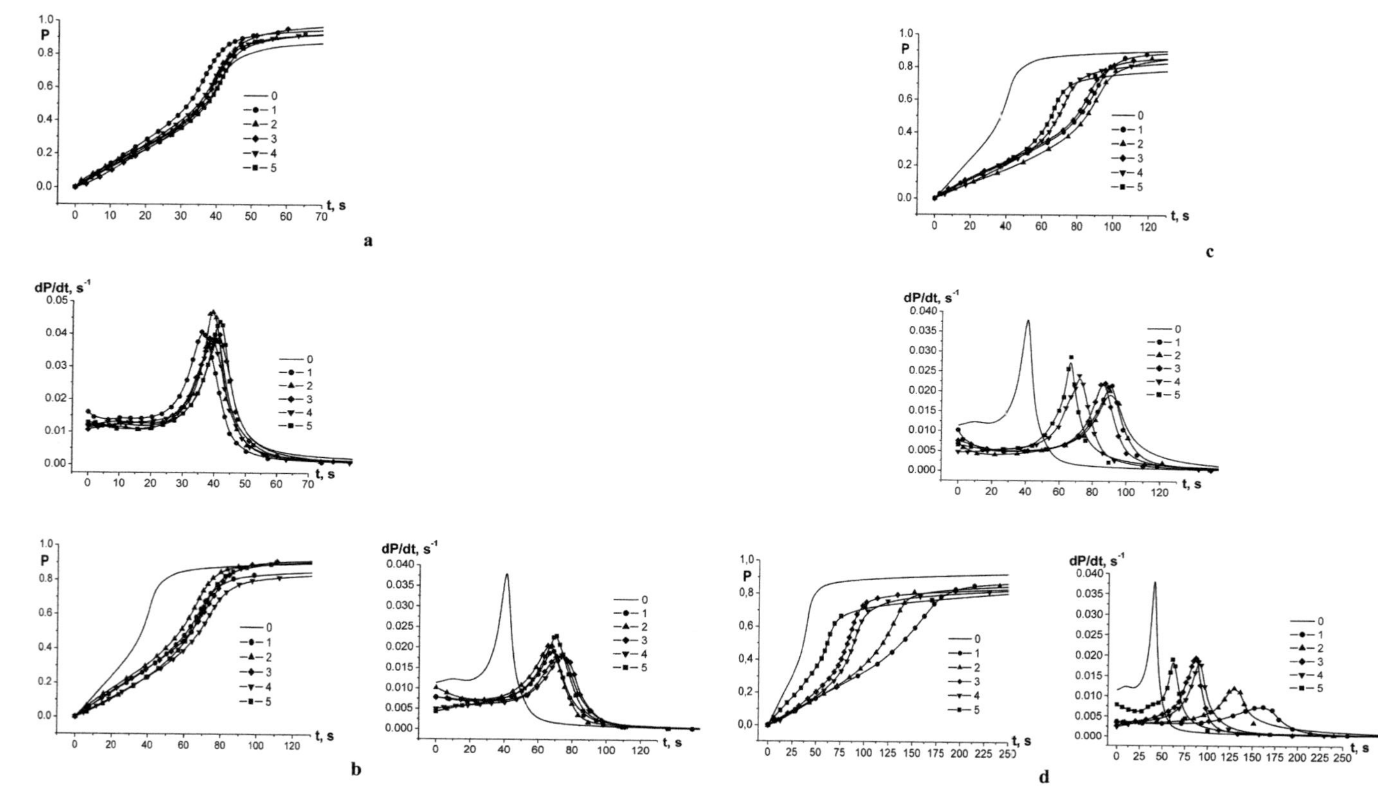

FIGURE 12.2 The integral and the differential kinetic curves of photoinitiated polymerization of HEMA + TEOS systems in the ratio (vol%) 97.5:2.5 (a), 95:5 (b), 90:10 (c), and 80:20 (d) depending on the different times of the prior gelation t_g: **0**—pure HEMA + IR 651, 2 mol%; **1**—1 h; **2**—24 h; **3**—48 h; **4**—72 h; **5**—96 h.

All integral kinetic curves, apart from the composition of the systems, are characterized by a typical S-like shape at the homopolymerization of HEMA and are consisted of prolonged, almost linear initial brief plot, but intense plot of auto-acceleration and long and slow plot of autodeceleration. The maximum speed W_0 of the polymerization on stage of autoacceleration and its corresponding conversion P_0 and time of its achievement for all investigated systems t_0 were defined by the method of numerical differentiation of integral kinetic curves (see Table 12.1).

From the experimental results, the appending of the inorganic components into the monomer system is accompanied by two effects is followed. The first effect consists of the fact that at the appending of the inorganic component, the velocity of the photoinitiated polymerization of monomer phase decreases significantly greater than it would be at its neutral dilution. It is visible by the sharp (two to four times) increasing of time t_0 to achieve the maximum speed of polymerization (compare the corresponding values of the curves of 0 and 1 in Table 12.1—HEMA:TEOS = 95:5, 90:10, and 80:20 [vol%]). Obviously, the inorganic component affects both on the stage of initiation and on the stage of chain growth as a result reducing them. The second effect consists of the fact that at the time increasing of gelation, the velocity of the photoinitiated polymerization is increased and stabilized at the end of gelation. It is visible by a decreasing t_0 and with time increasing of gelation, compare the corresponding values of the curves 1 and the followed ones in Table 12.1. This effect, however, is visible only at the high concentrations of inorganic components, that is, in TEOS–HEMA systems in the ratio of components 90:10 and 80:20 (vol%). This effect is, obviously, similar to the gel effect, which is the accelerating of polymerization process when the new polymer phase appears in the form of micrograins.[15]

The obtained polymeric composites, regardless of the ratio of components, are transparent, durable, elastic, homogeneous in their structure, have the high adhesion to different substrates and electrical conductivity (after doping of samples of solution 10% H_3PO_4 during hour), characterized by the simplicity and the ease of material manufacturing of specified form and can be used in different fields (from medicine and biotechnology to the telecommunications systems and the fuel cells of new generation).[1–5,10–13].

The modern thoughts concerning the nanomaterials allow to assert the possibility to control the structure of polymer composite during the synthesis or formation and to explain mechanical properties of the created composites improvement of the due to the formation of certain its morphology and the corresponding nanostructure.

TABLE 12.1 The Kinetic Parameters of Polymerization of HEMA–TEOS Systems at the Different Ratios of Components Depending on the Time of Gelation.

Number of the curve	97.5:2.5			95:5			90:10			80:20		
	W_0 (s^{-1})	P_0	t_0 (s)	W_0 (s^{-1})	P_0	t_0 (s)	W_0 (s^{-1})	P_0	t_0 (s)	W_0 (s^{-1})	P_0	t_0 (s)
1	0.041	0.64	36	0.02	0.63	69	0.022	0.69	91	0.007	0.67	162
2	0.047	0.65	39	0.021	0.64	65	0.02	0.6	90	0.012	0.62	130
3	0.04	0.71	42	0.018	0.62	71	0.022	0.66	87	0.019	0.59	87
4	0.039	0.67	40	0.018	0.6	74	0.024	0.57	72	0.018	0.52	90
5	0.044	0.66	42	0.023	0.64	71	0.029	0.55	67	0.019	0.5	61

Note: For pure HEMA+IR 651, 2 mol% (curve—**0**), $W_0 = 0.04\ s^{-1}$, $P_0 = 0.65$, $t_0 = 41$ s.

It is known that thermomechanical properties reflect the structural and the molecular features of polymers.[16] The samples for thermomechanical studies were prepared from the output liquid HOIC, based on the basis of HEMA and TEOS systems in the same ratios of the output components. Then, samples were subjected to the photoinitiated polymerization of a duration of 2.5 h for completion of the limited conversion of composites. The resulting samples were kept after polymerization in an oven at $t = 40°C$ during the week for passing the sol–gel process and were crushed to a powder-like state.

The samples for conducting of TM analysis were made by pressing of the powdered organic–inorganic composite, which was prewarmed in a mold to $T = 110°C$ at the pressure of 150 atm. The samples were kept in a mold under pressure until they cooled to the room temperature. The samples were kept in a cupboard at $T = 80°C$ for 10 h for relieving internal stresses after pressing.

TMA curves were filmed on the modified device for determining the heat resistance of polymers (Heckert, GDR) in mode of uniaxial compression under the load of 5.3×10^5 N/m^2 and at heating rate of 1.5 K/min. The sample for measurement had the form of a cylinder with diameter of 9.0 mm and night of 10.0 mm. The deformation of the sample was determined using the null indicator with accuracy of 0.01 mm. The temperature of the sample during the experiment was measured using the standard little inertial thermocouple THK.

The results of thermomechanical analysis of the composites HEMA–TEOS in view of dependence on relative deformation ε ($\varepsilon = \Delta h/h_0$, where Δh is the deformation of the sample, h_0 is the initial size of the sample) from temperature is shown in Figure 12.3.

The typical for polymers is the initial steep ascending plot which is clearly allocated on the TMA curves of the investigated organic–mineral composites, which corresponds to a structural transition of the polymer materials from the glassy state in highly elastic state (α-relaxation process). This plot concerns the disinhibition of the mobility of kinetic segments of the polymer matrix under the influence of temperature.[17] By extrapolation of this plot of TMA curve on T axis, one can define one of the important structural relaxation characteristics of the polymer. It is the temperature of glassing T_{glass}. The temperature of transition in highly elastic condition (T_{hel}) was defined as a point of crossing of the straight-line interval, which coincides with the interval of the thermomechanical curve that corresponds to the transition of the polymer from the glassy state to a highly elastic one with the straight-line interval of the highly elastic area. Accuracy of estimate T_{glass} and T_{hel} is ±2 K.

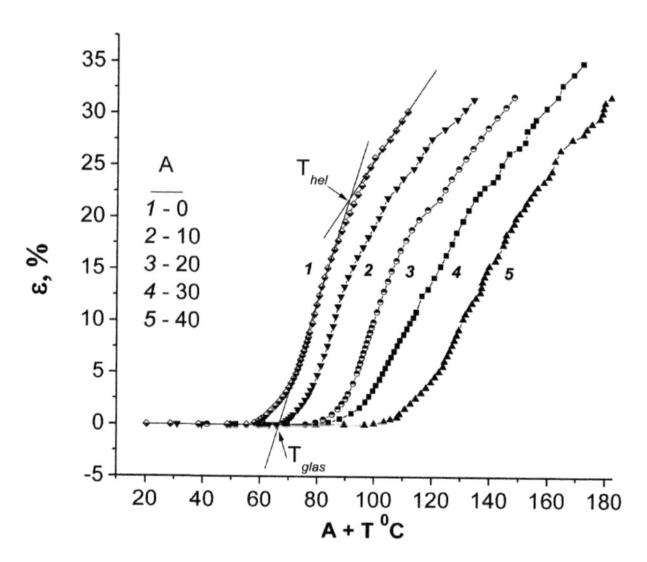

FIGURE 12.3 The thermomechanical curves: **1**—pure HEMA + IRGACURE 651, 2 mol%; the composites of HEMA:TEOS in ratios of components (vol%): **2**—97.5:2.5; **3**—95:5; **4**—90:10; **5**—80:20. For greater visibility, the curves 2–5 are offset on the *X*-axis concerning the curve 1 on the value *A*.

The study of the behavior of polymer in the highly elastic state allows to estimate the structural and molecular parameters of the spatial grid: the equilibrium modulus of highly elastic E_∞ and molecular weight of internodal segment M_c. Within the framework of the statistical theory of elasticity of the molecular grid of space crosslinked polymers, the dependence of module highly elastic E_∞ from density of crosslinks (value M_c) is described by the equation[13]:

$$E_\infty = \frac{3\gamma\rho RT\,\upsilon}{M_c\,\upsilon_0}, \tag{12.1}$$

where E_∞ is the modulus of high elastic; γ is the structural factor whose value depends on the nature and topology of the grid; ρ is the density of polymer; υ_o is the total number of cross-links; υ is the effective number of physical connections, which form the spatial structure; M_c is the molecular mass of the kinetic segment; and R is the gas constant; T is the temperature (K).

The quantitative inspection of this ratio is rather complicated as the proportion of functional groups that take part in the formation of physically active connections is unknown. There is even greater uncertainty connected with the estimate of the structural factor γ. If there is $E_\infty \leq 6 \times 10^7$ N/m^2, it can be accepted that $\upsilon = \upsilon_0$, $\gamma = 1$.[18] In this case, the calculation according to eq 12.1 is significantly simplified.

Module of highly elastic E_∞ was calculated by the equation:

$$E_\infty = \frac{P}{F \cdot \varepsilon}, \tag{12.2}$$

where P is the load on the sample (N); F is the cross-sectional area of the sample to which are efforts applied (m^2); and ε is the relative deformation of the sample in area of highly elastic.

The value M_c is a measure of the crosslinking density, which, consequently, is the factor that determines the full complex of physical and mechanical properties of polymer (strength, hardness, relaxation properties, etc.). The value of M_c allows to estimate the concentration of effective nodes of crosslinking of the polymer matrix υ (mol/cm^3):

$$\upsilon = \frac{\rho}{M_c} \tag{12.3}$$

The obtained values of the characteristic parameters of the studied organic–mineral composites are listed in Table 12.2.

TABLE 12.2 Thermomechanical Properties and Structural Parameters of Molecular Composites Synthesized on the Basis of HEMA–TEOS Systems.

Compositions (vol%)	T_{glass} (°C)	T_{hel} (°C)	$E_\infty \times 10^{-6}$ (N/m^2)	$M_c \times 10^{-3}$ (g/mol)	υ (mol/cm^3)
HEMA + IR 651, 2 mol%	67	91	2.72	3.5	0.353
HEMA:TEOS = 97.5:2.5	65	86	3.12	3.01	0.410
HEMA:TEOS = 95:5	65	88	3.34	2.83	0.436
HEMA:TEOS = 90:10	64	105	2.43	4.07	0.303
HEMA:TEOS= 80:20	74	110	2.22	4.62	0.267

The analysis of TM curves and studying of dependence of thermomechanical properties of the obtained HOIC from the content of inorganic component in the HEMA–TEOS systems (see Fig. 12.2, Table 12.2) gives the inconsistent results. At low concentrations of inorganic components (2.5 and 5 vol%), the equilibrium modulus of the highly elastic of composite E_∞ increases, which is connected with the decrease of the molecular weight of kinetic segment M_c and increasing concentration of the nods of crosslink υ. However, at the higher concentrations of inorganic components (10 and

20 vol%) in the composite, the opposite situation is observed. This contradiction clearly indicates that at the small additions of inorganic component (2.5 and 5 vol%) its phase, which appears in the form of nanoparticles, is more compatible with the phase of organic component, that is, confirmed by authors of work.[13] Although in the case of larger concentrations of phase inorganic components of (10 and 20 vol%) that are formed, get the less compatible, in consequence of which the microheterogeneity of the composition with formation of two interpenetrating grids of organic (polymeric) and mineral nature occur.[19]

Thus, the study of the composition effect of the systems on the kinetics of polymerization process to the deep conversion showed that if there is appending of the inorganic component in the monomer system in high concentrations, the velocity of photoinitiated polymerization of HEMA–TEOS systems is significantly reduced, which is shown evident as observed by the abrupt growth (two to four times) of the increasing time t_0 to reach the maximum velocity of polymerization. And, the time of gelation increasing the depth of photoinitiated polymerization also increases and stabilizes at the end of gelation. Therefore, the improved mechanical properties of the obtained composites are identified only at the small additions of inorganic component.

KEYWORDS

- **organic–inorganic composite**
- **photoinitiated polymerization**
- **sol–gel synthesis**
- **thermomechanical analysis**
- **sol–gel synthesis**
- **molecular structure**

REFERENCES

1. Roko, M. K.; Uilyamsa, R. S.; Alivisaosa, P. M. *Nanotekhnolohiya v blizhayshem desyatiletii*, Pod red. Mir, 2002.
2. Sergeev, G. B. *Nanokhimiya. M.*: MHU, 2003.
3. Shpaka, A. P. Nanosystemy, Nanomaterialy, Nanotekhnolohiyi, Pid red. K: Akademperiodyka **2003,** *1*, 1.

4. Shpak, A. P.; Kunitskiy, Y. A.; Karbovskiy, V. L. Klasterniye i nanostrukturnye materialy. *K: Akademperiodyka* **2003,** *1*, 1.
5. Myshak, V. D.; Semynoh, V. V.; Homza, Yu. P.; Nesin, S. D.; Klepko, V. V. Epoksydni nanokompozyty. Struktura ta vlastyvosti. *Polim. Zhurn.* **2008,** *30* (2), 146–153.
6. Pomahaylo, A. D.; Rozenberg, A. S.; Uflyand, Y. E. Nanochastitsy metallov v polimerakh. *M.: Khimiya*, 2000.
7. Veselovskiy, R. A.; Ishchenko, S. S.; Novikova, T. Y. Formirovaniye orhanomineralnoy kompozitsii na osnove poliizotsianata i zhidkogo stekla. *Ukr. Khim. Zhurn.* **1988,** *54* (3), 315–319.
8. Bronstein, L. M.; Karlinsey, R. L.; Ritter, K.; Joo, C. G.; Stein, B.; Zwanziger, J. W. Design of Organic–Inorganic Solid Polymer Electrolytes: Synthesis, Structure and Properties. *J. Mater. Chem.* **2004,** *14*, 1812–1820.
9. Lebedev, E. V.; Ishchenko, S. S.; Pridatko, A. B.; Babkina, N. V.; Lebedev, E. V. Polimernye orhanosilikatnye sistemy. *Kompoz. Polim. Mater.* **1999,** *21* (1), 3–12.
10. Ishchenko, S. S.; Rosovitskiy, V. F.; Pridatko, A. B.; Babkina, N. V.; Lebedev, E. V. Vliyanye organicheskikh modifikatorov na formirovanie organosilikatnykh polimernykh kompozitsiy. *Zhurn. Prikl. Khimii.* **1998,** *11*, 1929–1933.
11. Prudatko, A. B.; Ishchenko, S. S.; Lebedyev Ye. V. Fizyko-khimichni osoblyvosti formuvannya orhanosylikatnykh polimernykh system. *Fizyka kondensovanykh vysokomolekulyarnykh Syst.* **1998,** *4*, 31–33.
12. Mamunya Ye. P.; Yurchenko, M. V.; Lebedyev Ye. V.; Ishchenko, S. S.; Parashchenko, I. M. Sorbtsiyni vlastyvosti hibrydnykh orhano-neorhanichnykh system na osnovi uretanovykh olihomeriv i sylikatu natriyu. *Polim. Zhurn.* **2008,** *30* (1), 37–42.
13. Ivanchev, S. S.; Mesh, A. M.; Reichelt, N.; Khaykin, S. Y.; Hesse, A.; Myakin, S. V. Polucheniye nanokompozitov hidrolizom alkoksisilanov v matritse polipropilena. *Vysokomol. Soed. Ser. A* **2002,** *44* (6), 996–1002.
14. Berlin, A. A.; Korolev, H. V.; Kefeli, T. Y.; Sivergin, Y. M. Akrilovye olygomery i materialy na ikh osnove. *M.: Khimiya*, 1983.
15. Medvedevskikh, Y.; Khovanets, G.; Yevchuk, I. Kinetic Model of Photoinitiated Copolymerization of Monofunctional Monomers till High Conversions. *Chem. Chem. Technol.* **2009,** *3* (1), 1–6.
16. Teytelbaum, B. Y. Termomekhanicheskiy analiz polimerov. *M.: Nauka*, 1979.
17. Gul, V. E.; Kuleznev, V. N. Struktura i mekhanicheskiye svoystva polimerov. *M.: Vysshaya shkola*, 1972.
18. Trostyanskaya, E. B.; Babaevsiy, P. G. Formirovaniye setchatykh polimerov. *Uspekhi khimii* **1971,** *40* (1), 117–141.
19. Lipatova, Y. S. K. Fiziko-khimiya mnogokomponentnykh polimernykh sistem, v dvukh tomakh, Pod red: *Naukova Dumka* **1986,** *2*, 137–228.

CHAPTER 13

PHOTORESPONSIVE MATERIALS CONTAINING AZOMOIETIES—A FACILE APPROACH IN MOLECULAR IMPRINTING

T. SAJINI[1,2], BEENA MATHEW[2], and SAM JOHN[1*]

[1]*Research and Post Graduate Department of Chemistry, St Berchmans College, Kottayam, India*

[2]*School of Chemical Sciences, Mahatma Gandhi University, Kottayam, India*

[*]*Corresponding author. E-mail: samthanicken@yahoo.com*

CONTENTS

ABSTRACT

Stimuli-responsive molecularly imprinted polymers (MIPs) have recently received a significant attention because they represent a new generation of intelligent and self-regulated artificial receptors and have shown great potential in various applications. Photoresponsive materials exhibit unique advantages over systems that rely on other stimuli because light stimulus can be imposed instantly and delivered in specific amounts with high accuracy. In this chapter, we briefly report recent developments in photoresponsive MIPs especially those formed from functional monomer bearing azobenzene moiety.

13.1 INTRODUCTION

Molecularly imprinted synthetic polymers were discovered 40 years back. Molecular imprinting is a template-directed technique that allows the design and synthesis of polymers with well-defined artificially generated recognition sites that are intentionally engineered and specific for a target analyte or class of analytes.[1,2] In the imprinting process, the template (a small molecule, a biological macromolecule, or a microorganism) interacts with a polymerizable monomer that contains complementary functional groups or structural elements of the template through reversible covalent bond(s), electrostatic interactions, hydrogen bonding, van der Waals forces, hydrophobic interactions, or coordination with a metal center.[3,4] A schematic representation of the molecular imprinting process is shown in Scheme 13.1.

The incorporation of stimuli-responsive molecular functionalities within the receptor sites of rigid polymer materials generated by molecular imprinting is a new approach for the fabrication of stimuli-responsive materials (SRMs).[5] SRMs are able to alter volume and properties in response to environmental stimuli such as pH, temperature, ionic strength, electric field, and photo-irradiation.[6] Photo-irradiation is one of the most frequently adopted external stimuli for SRMs because it is convenient to apply and easy to control. Photoresponsive materials have attracted much attention because of their potential use in various optical applications such as nonlinear optics, erasable memory storage and processing, and electro-optical displays.[7]

Photoresponsive materials can be synthesized by functionalizing the material with photosensitive molecules such as cinnamic acid, cinnamylidene, or azo compounds. Out of these azobenzene is the most widely used photosensitive molecule due to its fast response on exposure to appropriate wavelength of light.[8] Azobenzene molecules are known to change their geometry upon photon absorption. Azobenzene is composed of two aromatic rings, where an azo linkage (−N=N−) joins the two phenyl rings.[9] Different type of azo compounds can be obtained by substituting an aromatic ring with various substituents to change geometry and electron donating/withdrawing mechanism. The light-induced configurational change renders azobenzenes good candidates for various photoresponsive applications. Due to their anisotropic shape, azobenzenes also contain directional information and are polarization sensitive. The phenomena arising from the photoisomerization reaction have applications not only in optics and photonics but also in the interfaces between light and surface science, information storage, imaging, biology, energy storage, and actuation.

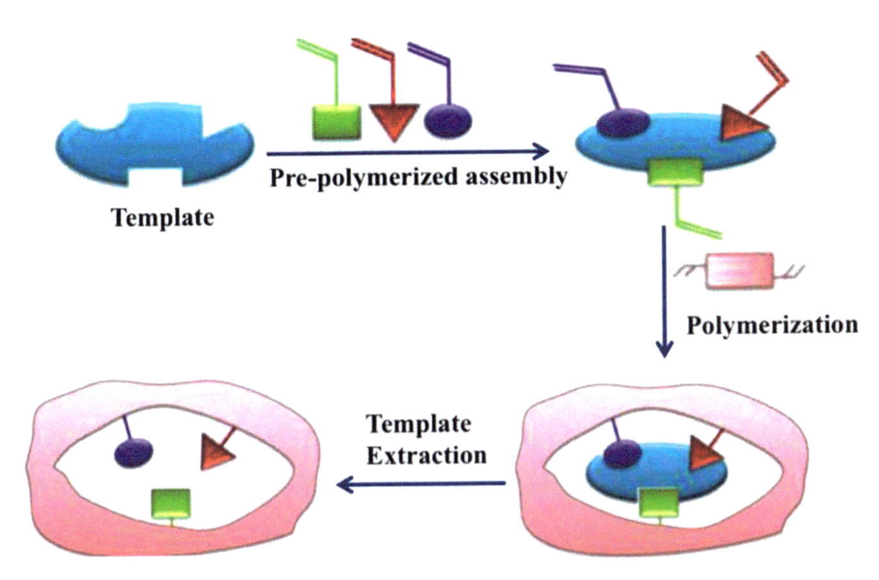

SCHEME 13.1 Schematic representation of molecular-imprinting process.

13.2 PHOTOISOMERIZATION OF AZOBENZENE DERIVATIVES

Azobenzenes, recognized by their nitrogen–nitrogen double bond, undergo clean and reversible conformational changes upon photon absorption, which is shown in Figure 13.1. This reaction, called photoisomerization, leads to large alterations in the physical and chemical properties of the molecules. There are two possible isomerization mechanisms in azobenzene: via rotation of the phenyl ring about the N=N bond or in-plane inversion around one of the nitrogen atoms. The *trans* form is thermodynamically more stable by 12 kcal mol^{-1} than the *cis* form, and the *cis* form is kinetically stabilized by an activation barrier of isomerization.[10] Under UV irradiation, the *trans* azobenzenes will be efficiently converted to the *cis* form with decrease in molecular size.[11] This *cis* form will thermally revert to the more stable *trans* form as the light source is switched off or switching back by illumination with visible light. Hence, azobenzenes reversibly change their geometry from a planar one to nonplanar upon irradiation with a drastic decrease in the distance between the *para* carbon atoms from 9 to 5.5 Å and a corresponding increase in the dipole moment from 0.5 to 3.1 D. This extremely clean photochemistry gives rise to the numerous remarkable photo-switching and photoresponsive behaviors observed in these systems.[12–15]

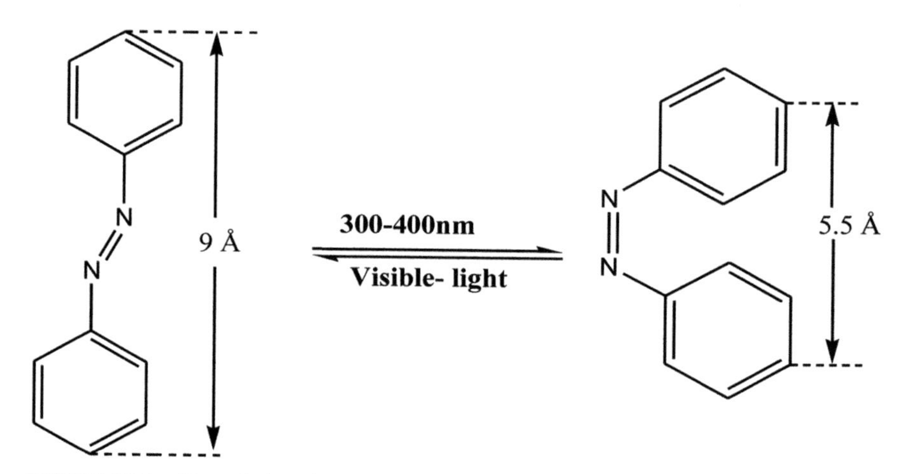

FIGURE 13.1 Photo-induced structural difference in azobenzene molecule.

As different geometries, polarities and electrical properties affect the two isomers, several functions can be photo-controlled including membrane dimensions, membrane potential, adsorption, solubility of polymer, wettability, swelling, enzyme activity, sol–gel transition of polymer, permeability, ion permeability, ion binding, photomechanical cycle,[16] etc.

13.3 PHOTORESPONSIVE MOLECULARLY IMPRINTED POLYMERS

The preparation of photoresponsive molecularly imprinted polymers (P-MIPs) involves replacing the commonly used functional monomers, such as methacrylic acid, acrylamide, and vinyl pyridine, by photoresponsive functional monomers, which always contain azobenzene. A schematic representation of reversible photoregulated substrate release and uptake process in P-MIPs are shown in Scheme 13.2.

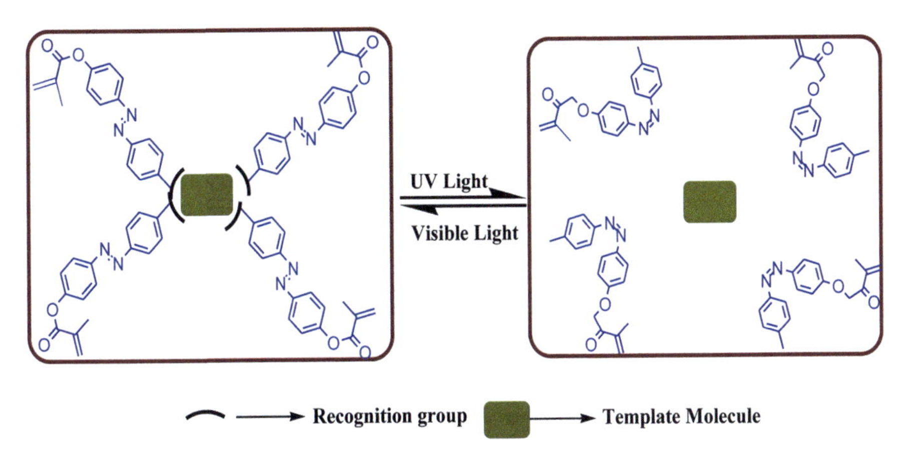

SCHEME 13.2 Schematic representation for reversible photoregulated substrate release and uptake process in P-MIPs.

The internal cavities of the P-MIPs no longer match with the template molecules in shape, size, and chemical functionality when irradiated by UV light, which would result in the release of the template. When irradiated by visible light, the recognition site would return to its original state, which can selectively adsorb template again. Thus, photoregulated release and uptake of substrates can be achieved by the incorporation of reversibly photo-switchable chromophores into the imprinted binding sites. The photoresponsive functional monomer is the key feature for preparation of P-MIPs. A photoresponsive functional monomer is composed of three groups. One is the photoresponsive group, the second is the recognition group, such as a carboxyl group or amino group, and the third is the polymerizable group, such as a vinyl double bond or silicon hydroxyl. To date, the most commonly investigated photoresponsive functional monomers contain azobenzene, and a series of azobenzene-containing functional monomers have been designed, as listed below. Azobenzene-containing functional monomers, similar to the functional monomers commonly used in molecular imprinting such as methacrylic acid, acrylamide, and 4-vinyl pyridine, have been synthesized by simply incorporating azobenzene into these molecules.

1. *p*-Phenylazoacrylanilide (PhAAAn).[17]

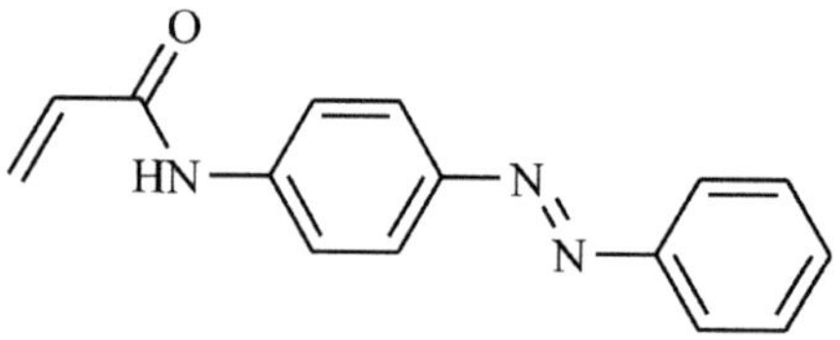

2. 4-[(4-Methacryloyloxy)phenylazo]benzoic acid (MPABA).[18]

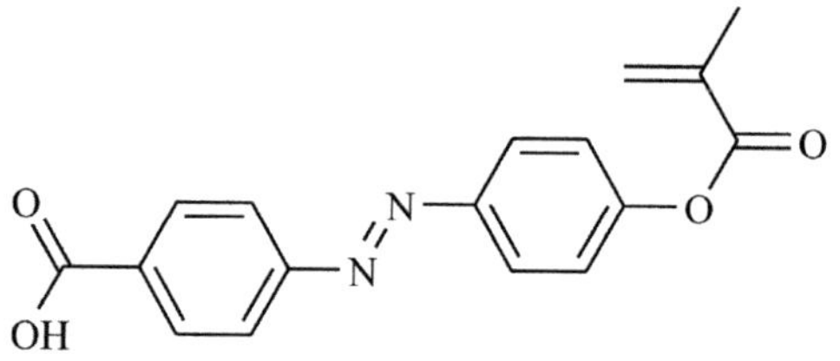

3. 4-{4-[2,6-Bis(*n*-butylamino)pyridine-4-yl]-phenylazo}-phenyl methacrylate.[21]

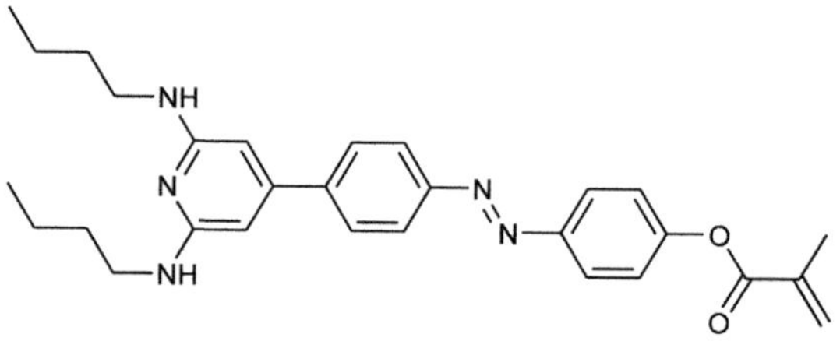

4. Di(ureidoethylenemethacrylate)azobenzene (Schmitzer et al., 2007).

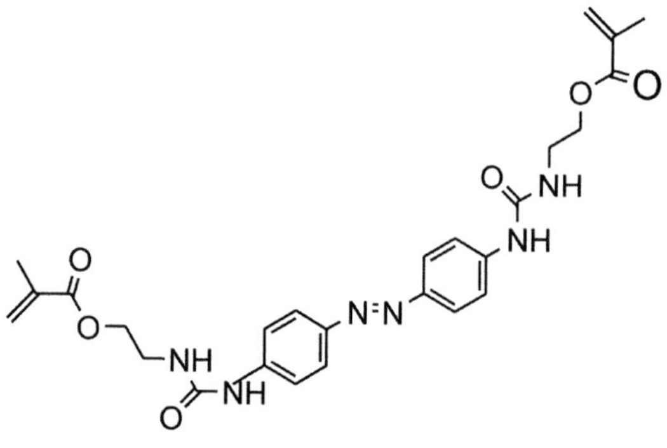

5. 4-[(4-Methacryloyloxy)phenylazo]benzene sulfonic acid (MAPASA).[19]

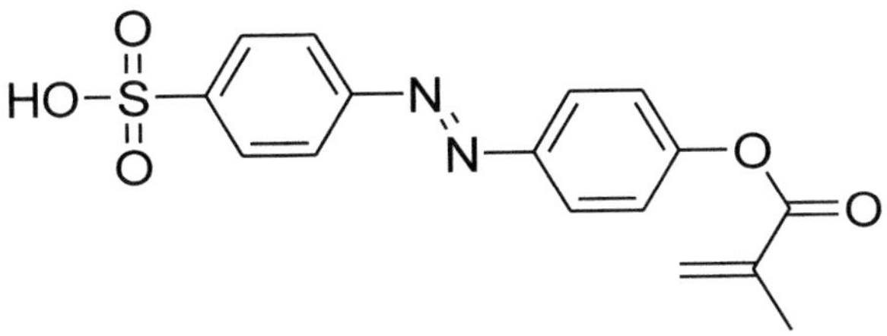

O
HO–S–
O
N
N–
O
O

6. 4-Amino-4-methacrylatylazobenzene (AMAAB) (Tang et al., 2010).

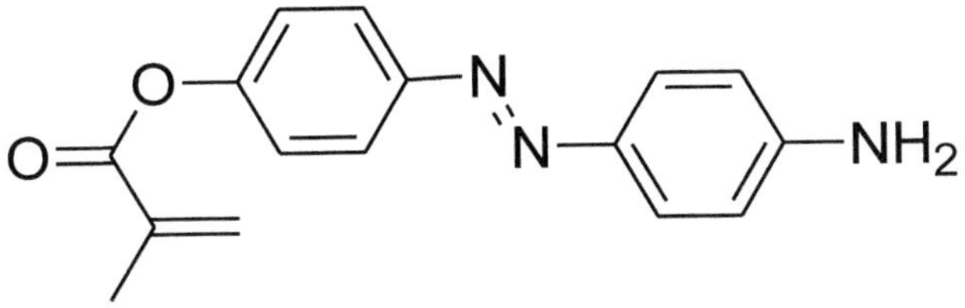

O
N
O
N–
–NH2

7. 4-{[4-(3-(Trimethoxysilyl)propoxy)phenyl]diazenyl}phenyl 2-(2,4) acetate.[23]

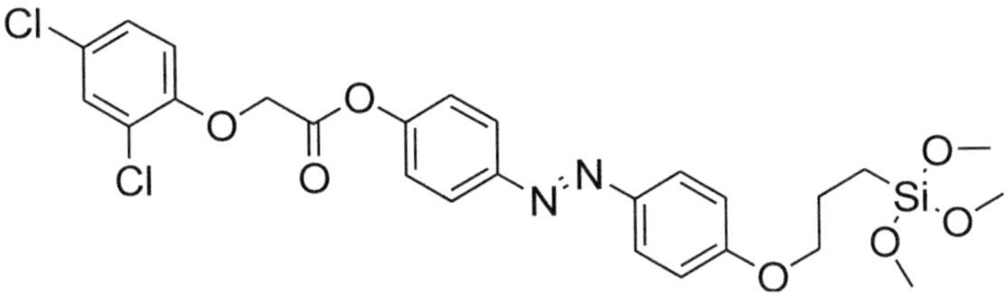

Cl
O
O
Cl
O
N N
O–
Si
O
O
O

8. 4-Hydroxyl-4-[(triisopropoxysilyl)propyloxy]azobenzene.[24]

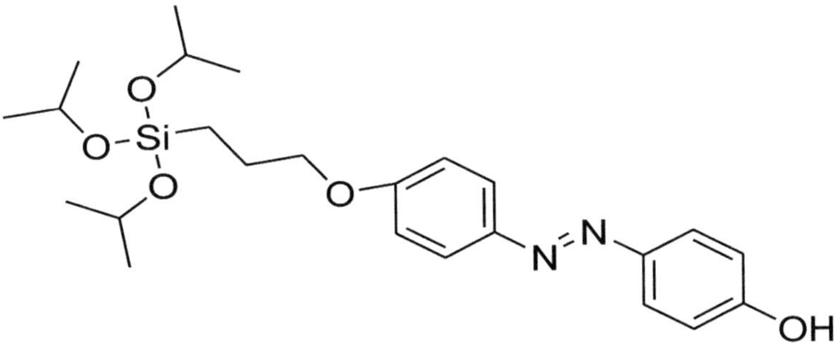

O
O–Si
O
O
N N
OH

9. (4-Methacryloyloxy)nonafluoroazobenzene.[24]

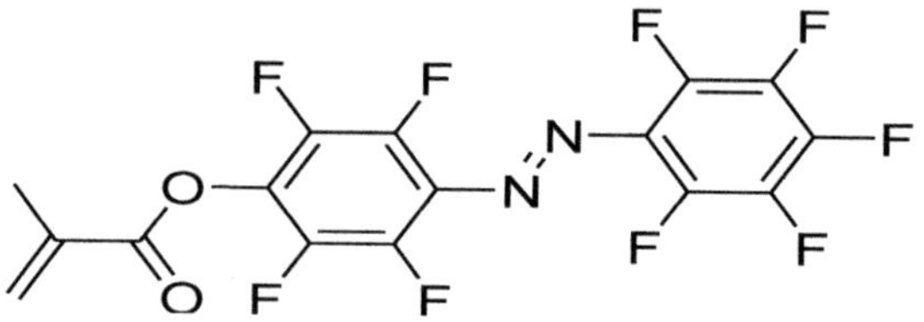

10. 4-((4-Methacryloyloxy)phenylazo)pyridine.[22]

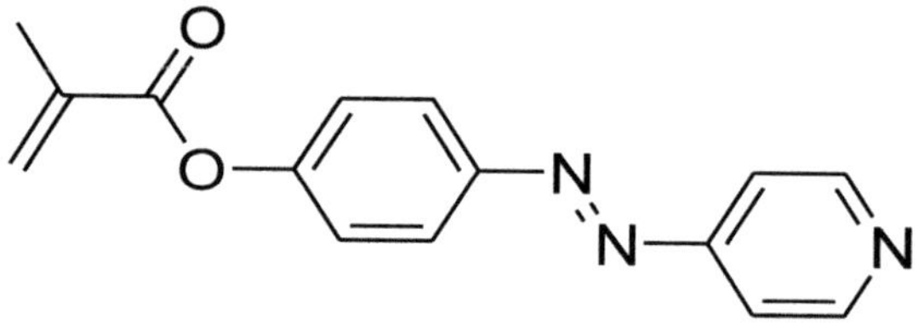

11. (4-Chloro-2-methylphenoxy)acetyloxy-4′-[(trimethoxysilyl)propyloxy]azobenze.[28]

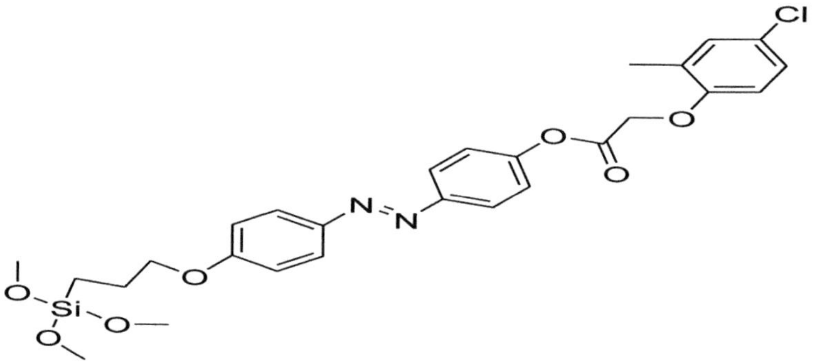

12. 2-Hydroxy-5-{4-[3-(4-trimethoxysilyl)propyloxy-phenyl)-acryloyl]-phenylazo} benzoic acid.[29]

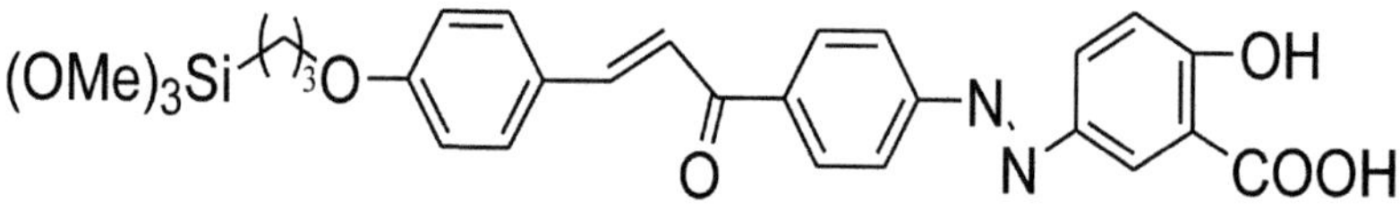

With the development of the molecular imprinting technique, various methods have been developed and used to prepare P-MIPs. Bulk polymerization is the most popular and general method to prepare molecularly imprinted polymers (MIPs) due to its attractive properties, such as rapidity and simplicity of preparation, and purity in the produced MIPs. Unquestionably, bulk polymerization has also been used to prepare P-MIPs.

Minoura and coworkers described the first preparation of azo-containing MIP membranes with photoregulated template-binding properties using *p*-phenylazoacrylanilide and dansylamide as the functional monomer and template molecule, respectively, via bulk polymerisation technique.[17]

Gong et al. synthesized methacrylic acid like azobenzene-containing functional monomer, 4-[(4-methacryloyloxy)phenylazo]-benzoic acid (MPABA) and is the most used azobenzene-containing functional monomer.[18] The synthesis route for the preparation of MPABA is shown in Scheme 13.3. Rate constants for the *trans–cis* and *cis–trans* isomerization of MPABA were found to be 7.59×10^{-4} and 14.68×10^{-4} s^{-1}, respectively. The adsorption selectivity in each case of the *trans*-form and the *cis*-form was not fully studied.

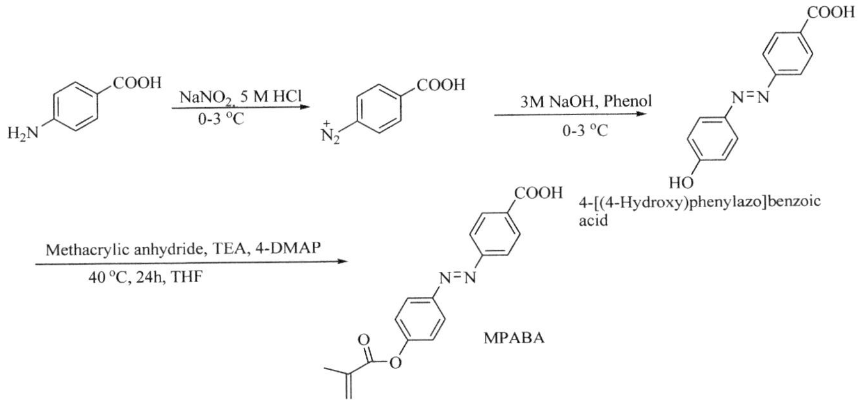

SCHEME 13.3 Synthesis of MPABA.

MPABA can be only dissolved in highly polar solvents, such as DMF and DMSO, which restricts its biomedical applications (e.g., drug-delivery systems) because of its low water compatibility. Water-soluble azobenzene-containing functional monomers are urgently needed. To meet this requirement, a kind of water-soluble azobenzene-containing functional monomer 4-[(4-methacryloyloxy)phenylazo] benzenesulfonic acid (MAPASA), with benzenesulfonic acid as the recognition element, has been developed for the fabrication of a P-MIPs hydrogel material that can function in the biocompatible aqueous media via precipitation polymerization (Scheme 13.4).[19] Their studies revealed that the MAPASA-containing polyacrylamide hydrogel fabricated from the cross-linker *N,N'*-hexylenebismethacrylamide was found to afford good optical transparency in the aqueous media, reasonable substrate binding affinity, and the fastest photoresponse rate. In this, specific and nonspecific binding strength of the resultant imprinted hydrogel toward the template paracetamol to be 1.96×10^5 and $747.0\ M^{-1}$, respectively.

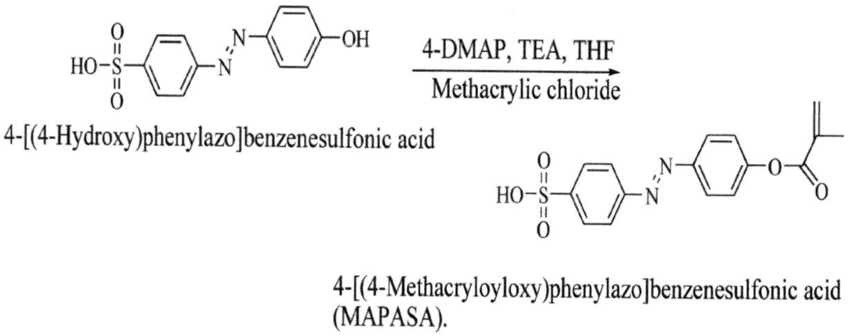

4-DMAP, TEA, THF

Methacrylic chloride

4-[(4-Hydroxy)phenylazo]benzenesulfonic acid

4-[(4-Methacryloyloxy)phenylazo]benzenesulfonic acid (MAPASA).

SCHEME 13.4 Synthesis of MAPASA.

Recently, a simple and quick detection method for trace bisphenol A (BPA) was developed by synthesizing P-MIPs on silica microspheres by surface polymerization using a water-soluble azobenzene-containing 4-[(4-methacryloyloxy)-phenylazo]benzenesulfonic acid as the functional monomer.[20] The SMIP microspheres displayed good photoresponsive properties and specific affinity toward BPA with high recognition ability (maximal adsorption capacity: 6.96 mmol g^{-1}) and fast binding kinetics (binding constant: $2.47 \times 10^4\ M^{-1}$) in aqueous media. Upon alternate irradiation at 365 and 440 nm, the P-MIPs microspheres could quantitatively bind and release BPA.

Schmitzer and coworkers demonstrated the utility of a new azo monomer di(ureidoethylenemethacrylate) azobenzene (Scheme 13.5) and successfully prepared a photoresponsive MIP for methotrexate recognition using it as both the functional monomer and cross-linker. They revealed the fact that this new cross-linking monomer combines interactive monomer functionality with a cross-linking format. This new cross-linking agent is readily copolymerizable under mild conditions and has been used for noncovalent MIPs preparation with potential improved performance, that more functionality can be introduced without suffering performance losses due to reduced cross-linking.

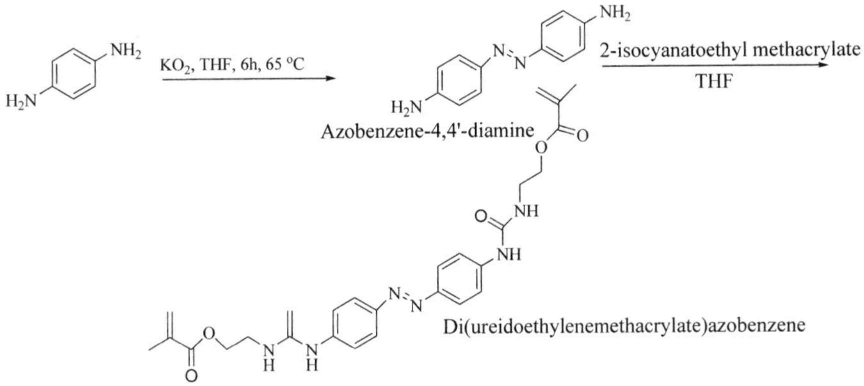

SCHEME 13.5 Synthesis of di(ureidoethylenemethacrylate).

Takeuchi et al. designed a photoresponsive functional monomer having diaminopyridine and azobenzene moieties, 4-{4-[2,6-bis(*n*-butylamino)pyridine-4-yl]-phenylazo}–phenyl methacrylate (Scheme 13.6) for preparing photoresponsive imprinted polymers for porphyrin derivatives with carboxylic acids. They found that the multiple hydrogen bonds could be formed between the template and functional monomer, facilitating the assembly of functional monomer with the template in appropriate positions

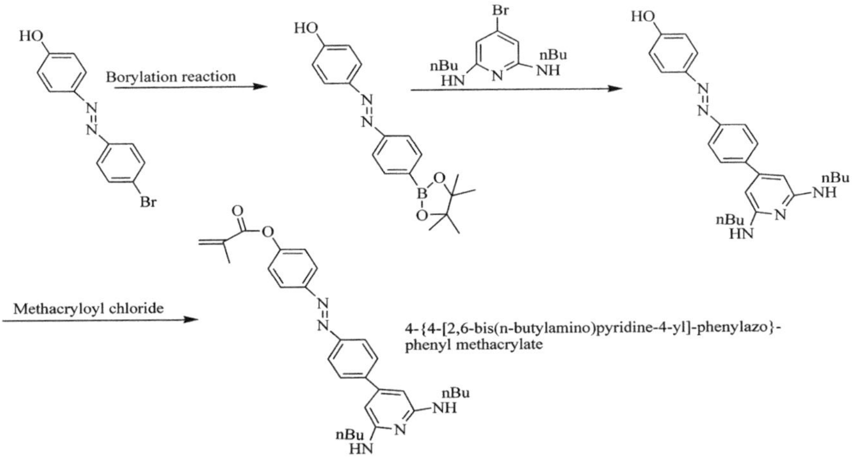

SCHEME 13.6 Synthesis of 4-{4-[2,6-bis(*n*-butylamino)pyridine-4-yl]-phenylazo}-phenyl methacrylate.

by polymerization, yielding selective imprinted cavities complementary to the target molecule.[21] The monolithic polymers obtained by bulk polymerization have to be crushed, grounded, and sieved to an appropriate size, which results in particles with irregular shapes and sizes, and some high-affinity binding sites are destroyed and changed into low-affinity sites. These rather limited physical formats (i.e., bulk polymer membranes, bulk monoliths) have significantly restricted their application. Therefore, precipitation polymerization, enjoying some advantages for synthesizing spherical particles, such as not requiring a surfactant, containing a single preparative step and allowing excellent control over the particle size, is needed to produce high-quality, uniform, and spherical P-MIPs.

An azo-containing MIP microspheres with photoresponsive template-binding properties via precipitation polymerization using an acetonitrile-soluble azo-functional monomer with a pyridine group, 4-{(4-methacryloyloxy)-phenylazo}pyridine (Scheme 13.7) was developed by Fang.[22] The resulting azo-containing MIP microspheres had a number–average diameter of 1.33 mm and a polydispersity index of 1.15, and they showed obvious molecular imprinting effects toward the template, 2,4-dichlorophenoxyacetic acid (2,4-D).

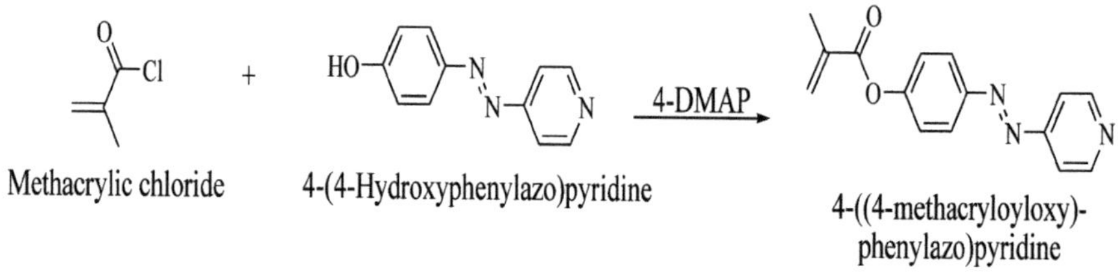

Methacrylic chloride + 4-(4-Hydroxyphenylazo)pyridine —4-DMAP→ 4-((4-methacryloyloxy)-phenylazo)pyridine

SCHEME 13.7 Synthesis of 4-{(4-methacryloyloxy)-phenylazo}pyridine.

As a convenient and versatile method, the sol–gel process has already been developed by a number of researchers and has been utilized to prepare MIPs. Generally, sol–gel based materials are prepared through acid-catalyzed or base-catalyzed hydrolysis of silanes, which is followed by polycondensation of the silanols into a polysiloxane network. The amorphous sol–gel molecularly imprinted materials have some advantages. For example, template removal from sol–gel materials can be more thorough under much stronger removal conditions, such as hydrolysis in strong acid and strong base, or even combustion. Also, the control of porosity, thickness, and surface area of sol–gel materials is more convenient.

A novel photoresponsive functional monomer which bearing a siloxane polymerizable group and azobenzene moieties, that is, an organic–inorganic polymerizable monomer, was synthesized by Jiang et al.[23] This azobenzene monomer was used to prepare photoresponsive molecularly imprinted polymers (Scheme 13.8), which have specific binding sites for 2,4-D through hydrogen-bonding interaction. The concentration of the 2,4-D can be quantitatively determined through the *trans*-to-*cis* photoisomerization rate of the azobenzene chromophore.

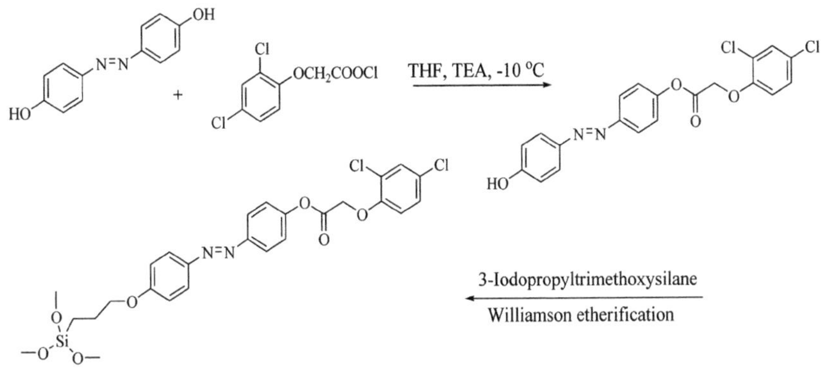

SCHEME 13.8 Synthesis of organic–inorganic polymerizable monomer.

Tang et al. designed and synthesized fluorine-substituted photoresponsive functional monomer, (4-methacryloyloxy) nonafluoroazobenzene (MANFAB) (Scheme 13.9).[24] The release and uptake of PAF (2,3,4,5,7,8,9,10-octafluorophenazine) from toluene is photoregulated by alternate irradiation at 315 and 440 nm, indicating that photoresponsive molecular recognition directed by fluorine–fluorine interaction is possible.

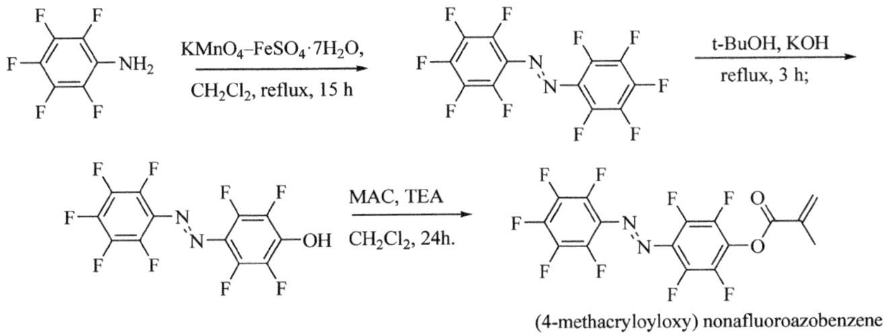

SCHEME 13.9 Synthesis of MANFAB.

Molecularly imprinted polymers (DR-MIPs) with photonic and magnetic dual responses were prepared by combination of stimuli-responsive polymers and a molecular imprinting technique.[25] Photoswitchable functional monomer of 4-[(4-methacryloyloxy)phenylazo]benzoic acid (MPABA) was used as functional monomer. The resultant DR-MIPs of Fe_3O_4@MIPs exhibited specific affinity for caffeine and photoisomerization induced reversible uptake and release of caffeine upon alternate UV and visible-light irradiation. The novel DR-MIPs were used as a sorbent for the enrichment of caffeine from real water and beverage samples. The idea of photonic and magnetic DR-MIPs makes possible the photoregulated uptake and release of pollutants from environment samples, and the magnetic property allows for magnetic separation. The photonic and magnetic DR-MIPs could also be used in a drug controlled-release system.

A general, facile, and highly efficient approach to obtain azobenzene (azo)-containing molecularly imprinted polymer (MIP) microspheres with both photo and thermoresponsive template binding properties in pure aqueous media is reported.[26] This involves the first synthesis of "living" azo-containing MIP microspheres with surface-immobilized alkyl halide groups via atom transfer radical precipitation polymerization. Subsequent modification of these microspheres via surface-initiated atom transfer radical polymerization of *N*-isopropylacrylamide (NIPAAm) would result in the MIP. This spherical azo-containing MIP particle with surface-grafted PNIPAAm brushes should be of tremendous potential in such applications as smart separation, extraction, and assays, as well as intelligent drug-delivery and bioanalytical analysis.

Photoresponsive surface molecularly imprinted poly(ether sulfone) microfibers were synthesized via nitration reaction, the wet-spinning technique, surface nitroreduction reaction, and surface diazotation reaction for the selectively photoregulated uptake and release of 4-hydrobenzoic acid (4-HA).[27] The prepared molecularly imprinted microfibers show selective binding to 4-HA under irradiation at 450 nm and release under irradiation at 365 nm. The simple, convenient, effective, and productive method for the preparation of azo-containing photoresponsive material is also applied to the modification of polysulfone and poly(ether ether ketone) (Scheme 13.10). All three benzene-ring-containing polymers show significant photoresponsibility after the azo modification.

A novel photoresponsive functional monomer bearing a siloxane polymerizable group and azobenzene moieties were reported by Zhou.[28] Photoresponsive molecularly imprinted sol–gel polymers were successfully fabricated from the synthesized functional monomer, using (4-chloro-2-methylphenoxy) acetic acid as a molecular template. They employed computational molecular modeling to study the hydrogen bond interactions between template molecules and functional monomer. The data indicate that the design of the MIP is rational.

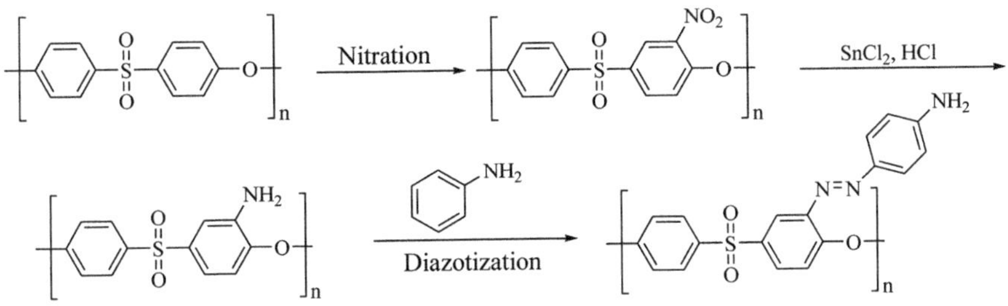

SCHEME 13.10 Synthesis of poly(ether sulfone).

Azobenzene containing photoresponsive molecule-imprinted silica microspheres were developed by Li et al.[29] This presynthesized azobenzene-based monomer contains a carboxyl and a hydroxyl group for substrate interaction (Scheme 13.11). The resulting uniform azo-containing MIP-silica particles were mechanically stable and showed obvious molecular imprinting effects toward the template, ibuprofen.

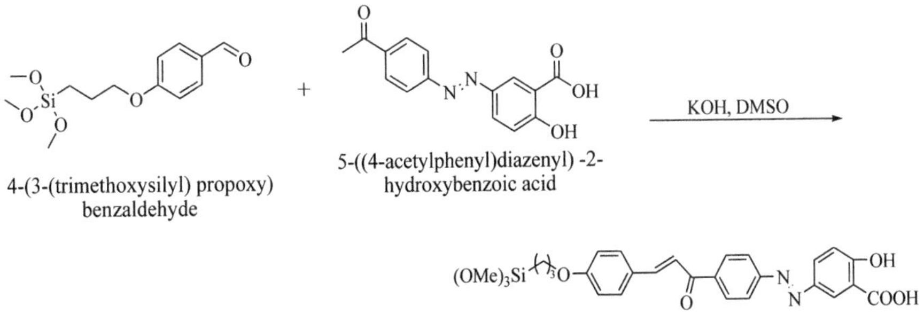

SCHEME 13.11 Synthesis of azobenzene-based monomer with carboxyl and hydroxyl groups.

13.5 CONCLUSIONS

P-MIP particles have wide range potential applications in many fields. However, the applications of photoresponsive MIPs are limited because of two reasons: (1) the photoinduced *trans–cis* isomerization of azobenzene functional monomers in highly cross-linked MIPs is slow, which cannot meet the requirements of rapid analysis, so the application of photoresponsive MIPs in environmental analysis is restricted; (2) visible light has a limited ability to penetrate human tissue, which in turn restricts its applications in drug-delivery systems. Only a few research groups have reported applications of P-MIPs. One of the examples is that of Gong and coworkers, who developed a reliable method to detect the content of melamine in dairy products with the advantages of the simple pretreatment of samples, a quick detection step, good sensitivity, and no need for expensive instruments. This kind of work revealed new potential applications for P-MIPs in chemosensing, food safety, and environmental analysis. Long-term efforts are still needed to further progress and development of the versatile functions of P-MIPs.

KEYWORDS

- **photoresponsive materials**
- **molecular imprinting**
- **azo monomer**
- **stimuli-responsive**
- **self-regulated artificial receptors**

REFERENCES

1. Garcia, R.; Cabrita, M. J.; Freitas, C. *Am. J. Anal. Chem.* **2011,** *2*, 16.
2. Nasrullah, S.; Mazhar, U.; Haneef, M.; Park, J. K. *J. Pharm. Res.* **2012,** *5*, 3309.
3. Moreno, M. C.; Navarro, F.; Pena, E. B.; Urraca, J. L. *Curr. Anal. Chem.* **2008,** *4*, 316.
4. Cheong, W. J.; Yang, S. H.; Ali, F. *J. Separat. Sci.* **2013,** *36*, 609.
5. Shoufang, X.; Hongzhi, L.; Xiuwen, Z.; Chen, L. *J. Mater. Chem. C* **2013,** *1*, 4406.
6. Theato, P.; Brent, S.; Rachel, K.; Thomas, H. *Chem. Soc. Rev.* **2013,** *42*, 7055.
7. Xie, S.; Natansohn, A.; Rochon, P. *Chem. Mater.* **1993,** *5*, 403.
8. Barrett, C. J.; Mamiya, J.; Yager, K. G. *Soft Matter* **2007,** *3*, 1249.
9. Iqbal, D.; Samiullah, M. H. *Materials* **2013,** *6*, 116.
10. Mina, H.; Takumu, H. *Sci. China Chem.* **2011,** *54*, 1955.
11. Yu, Y. L.; Nakano, M.; Ikeda, T. *Nature* **2003,** *425*, 145.
12. Angeloni, A. S.; Caretti, D.; Carlini, C.; Chiellini, E.; Galli, G. *Liq. Cryst.* **1989,** *4*, 513.
13. Ikeda, T.; Mamiya, J.; Yu, Y. L. *Angew. Chem. Int. Ed.* **2007,** *46*, 506.
14. Mamiya, J.; Yamada, M.; Naka, Y.; Kondo, M. *Kobunshi Ronbunshu* **2009,** *66*, 79.

15. Yu, Y.; Yin, R.; Xu, J.; Cheng, F.; Kondo, M. Liquid-Crystalline Polymers. *Proc. SPIE* **2008,** *7050*. DOI:10.1117/12.794183.
16. Nicoletta, F. P.; Cupelli, D.; Formoso, P.; Filpo, G. D.; Colella, V.; Gugliuzza, A. *Membranes* **2012,** *2*, 134.
17. Minoura, N.; Idei, K.; Rachkov, A.; Uzawa, H.; Matsuda, K. *Chem. Mater.* **2003,** *15*, 4703.
18. Gong, C.; Lam, M. H.; Yu, H. *Adv. Funct. Mater.* **2006,** *16*, 1759.
19. Gong, C.; Wong, K.; Lam, M. H. *Chem. Mater.* **2008,** *20*, 1353.
20. Yang, Y.; Tang, Q.; Ma, X.; Penga, J.; Lamb, M. *New J. Chem.* **2014,** *38*, 1780.
21. Takeuchi, T.; Akeda, K.; Murakami, S.; Shinmori, H.; Inoue, S.; Lee, W.; Hishiya, T. *Org. Biomol. Chem.* **2007,** *5*, 2368.
22. Fang, L.; Chen, S.; Zhang, Y.; Zhang, H. *J. Mater. Chem.* **2011,** *21*, 2320.
23. Jiang, G. S.; Zhong, S.; Chen, L.; Blakey, I.; Whitaker, A. *Radiat. Phys. Chem.* **2011,** *80*, 130.
24. Tang, Q.; Gong, C.; Lam, M. H.; Fu, X. *Sens. Actuat.* **2011,** *156*, 100.
25. Xu, S.; Li, J.; Song, X.; Liu, J.; Lub, H.; Chen, L. *Anal. Methods* **2013,** *5*, 124.
26. Fang, L.; Chen, S.; Guo, X.; Zhang, Y.; Zhang, H. *Langmuir* **2012,** *28*, 9767.
27. Wang, D.; Zhang, X.; Nie, S.; Zhao, W.; Lu, Y.; Sun, S.; Zhao, C. *Langmuir* **2012,** *28*, 13284.
28. Zhou, X.; Zhong, S.; Jiang, G. *Polym. Int.* **2012,** *61*, 1778.
29. Li, C.; Zhong, S.; Li, X.; Guo, M. *Colloid Polym. Sci.* **2013,** *291*, 2049.

CHAPTER 14

GREEN NANOTECHNOLOGY: AN APPROACH TOWARD ENVIRONMENT SAFETY

ANAMIKA SINGH*

Department of Botany, Maitreyi Collage, University of Delhi, New Delhi, India

**E-mail: arjumika@gmail.com*

CONTENTS

ABSTRACT

Green nanotechnology commonly used in the development of clean technologies. In this chapter, different aspects of this technology are reviewed.

14.1 INTRODUCTION

Past few decades were extremely dedicated for the novel research in science especially nanotechnology. Nanotechnology expands itself not only in few areas of applied science, but it is also helping in the field of energy, environment, sewage technologies, pollution control, etc. Nowadays, new nanotechnological materials have been designed which are having high impact on biodiversity, conservation, and human health benefits. One of the important aspect of nanotechnology is green nanotechnology, which is actually the use of nanotechnology to conserve and protect the environment. It also includes the use of green nanoproducts or using nanoproducts in support of sustainability.

Green nanotechnology commonly used in the development of clean technologies (pollution-free environment). It uses the nanotechnological products or encourages the replacement of existing products with a new nanoproduct. These new nanoproducts are actually environment friendly and cause no damage to the environment as compare to the previous one.[1]

Green nanotechnology has two main motives:

1. Production of nanomaterial and products that are eco-friendly to human health. The basis of green technology is green chemistry and green engineering.[2] So basically, green nanotechnology is nothing but is conversion of non-nanotechnological products to nanotechnological products that are biodegradable and eco-friendly. The nanoproducts used in green nanotechnology are nontoxic and renewable. Due to this, nanotechnology plays an important role in environment protection, conservation, and human health.
2. The second aim of green nanotechnology involves development of products that are benefiting environment directly or indirectly. Nanomaterials or its products can clean hazardous waste, desalinate water, treat pollutants, or sense and monitor environmental pollutants directly. In spite of that indirectly, it can also help in automobile and other means of transportations which are useful to save fuel. Reduction of pollution is one of the major uses of nanotechnology.[1]

> Green nanotechnology takes a wide range of view as it ensures unforeseen consequences of pollutants and help to minimize its impacts on environment.[3]

There are so many questions related to nanotechnology that how is nanotechnology used in biodiversity and how it helps in human health and conservation? Many researches were performed in different areas (Fig. 14.1) in the past many years related to nanotechnology. Here, we are going to discuss few aspects of nanotechnology in different fields.

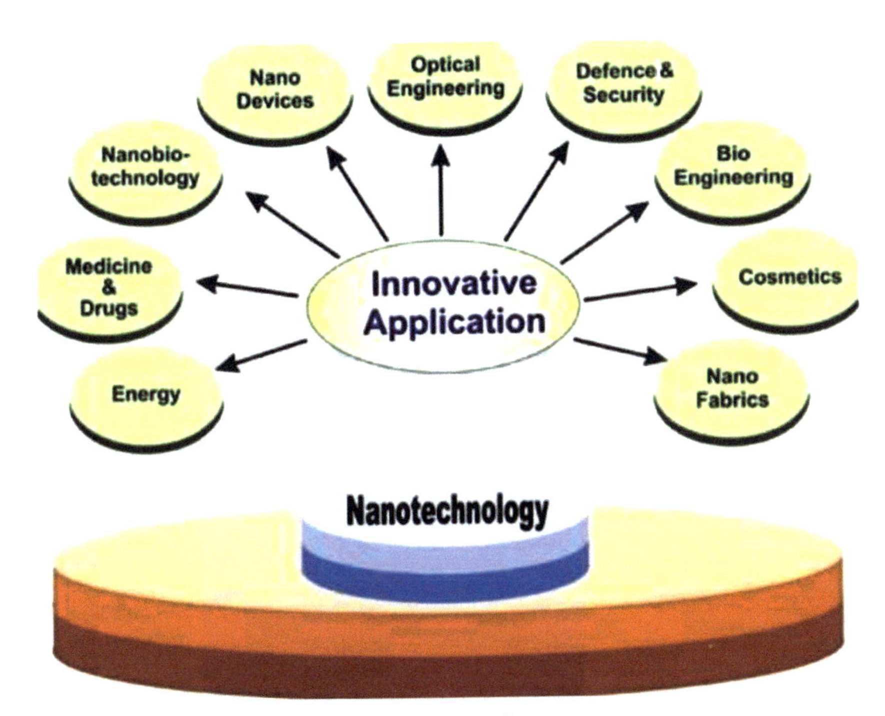

FIGURE 14.1 Areas of nanotechnological research.

14.2 NANOTECHNOLOGY: ENERGY APPLICATION

To make a step forward toward current science and toward energy generation technology, scientists have developed a new area on nanotechnology which solves the problems related to energy generation and its utility.

Nanofibrication: One of the important subfield of nanotechnology is nanofabrication, which is designing and creating devices on nanoscale, that helps in capturing, storing, and transferring energy from one system to another. Through the use of nanofabrications, many problems can be solved related to current (electricity) generation. Nanotechnology in energy field is used to develop those products which are beneficial for a consumer. Several benefits have been analyzed in this field; some of them are like high efficiency of heat and light, electrical storage capacity, and decreases in environment pollution. These are a few points that prove the involvement of nanotechnology in the field of energy.

Solar cells: These are an electrical device which can convert the light energy directly into electrical energy by photovoltaic (PV) effect. These solar cells are also known as PV cells. The quality and quantity of current, voltage, and resistance vary, and it depends upon the type of light absorbed.[4] In general, it is refer a generation of electricity from light. The cells known as PV cell even light source is different to sun light, like lamp light, artificial light, etc. PVs is the field of technology and research and its application is PV cells which produces electricity from light, though it is often used specifically to refer to the generation of electricity from sunlight. Cell is sometimes used as a photodetector (e.g., infrared detectors), detecting light or other electromagnetic radiation near the visible range or measuring light intensity.

The operation of a PV cell requires three basic attributes:

1. The absorption of light, generating either electron–hole pairs or excitons.
2. The separation of charge carriers of opposite types.
3. The separate extraction of those carriers to an external circuit.

Similar to this, solar thermal power absorbed sun light either by direct heating or indirect electrical power generation. In this method, nanoparticles like titanium dioxide, silver, quantum dots, and cadmium telluride are used in the preparation of thin solar cells.

There are also types of nanoparticle spray available which have capacity to increase the efficiency of solar light and help in more instant transformation of solar.[5]

Fuel cell: It is a device that converts the chemical energy from a fuel into electricity through a chemical reaction with oxygen or another oxidizing agent.[6] Hydrogen is the most common fuel, but hydrocarbons such as natural gas and alcohols like methanol are sometimes used. Fuel cells are different from batteries in that they require a constant source of fuel and oxygen/air to sustain the chemical reaction; however, fuel cells can produce electricity continuously for as long as these inputs are supplied.

There are many types of fuel cells, but they all consist of an anode, a cathode, and an electrolyte that allows charges to move between the two sides of the fuel cell. Electrons are drawn from the anode to the cathode through an external circuit, producing direct current electricity.[7] In addition to electricity, fuel cells produce water, heat, and, depending on the fuel source, very small amounts of nitrogen dioxide and other emissions. The energy efficiency of a fuel cell is generally between 40% and 60%, or up to 85% efficient in cogeneration if waste heat is captured for use.[8]

Nanobatteries are fabricated batteries employing technology at a nanoscale, a scale of minuscule particles that measure less than 100 nm or 100×10^{-9} m. In comparison, traditional Li-ion technology uses active materials, such as cobalt oxide or manganese oxide, with particles that range in size between 5 and 20 μm (5000 and 20,000 nm—over 100 times nanoscale). It is hoped that nanoengineering will improve many of the shortcomings of present battery technology, such as recharging time and battery memory.

14.3 NANOTECHNOLOGY: WATER TREATMENT

Two-fifth of the world population lack proper sanitation.[9] Although there are so many technologies available but due to high cost and availability to a common man, it is quite tough to provide a proper sanitation and clean water throughout the world. Conventional water-treatment technologies are available and are used from thousands of years at household level.[10] These conventional methods are very effective for poorer group of people and these methods are filters (ceramic, activated carbon, granular media, fiber, and fabric), desalination (reverse osmosis, distillation, etc.), chemical, and radiation treatment.[11]

Monitoring devices: By the use of nanotechnology, sensor devices have been designed and are able to sense the quality and quantity of water resource and it can also detect the contaminations. Due to the use of nanotechnology, the water can be treated which is cost efficient, cheap, and durable compared to the conventional household methods.[12] Commercially, membrane, meshes, filters, ceramics, clay and absorbent, zeolites, and catalysts are available in which nanomaterials are used. Zeolite nanoparticles can be prepared by laser-induced fragmentation of zeolite LTA microparticles using a pulsed laser or by hydrothermal activation of fly ash. Zeolites are used as an ion-exchange media for metal ions and effective sorbents for removal of metal ions. Zeolites have been reportedly used in the removal of heavy metals such as Cr(III), Ni(II), Zn(II), Cu(II), and Cd(II) from metal electroplating and acid-mine wastewaters.

Nanoscale membrane: It helps to separate the useful chemical product from the chemical reaction of waste materials.

Nanoscale catalyst: By the use of small amount of nanomaterial as catalyst makes a chemical reaction more efficient and it also decreases the waste product of the reaction.

Nanoscale sensor: It helps to control the process of water treatment and it will go smooth step by step.

Nanofiltration is the new membrane filters which are used to soften and remove the decomposition byproduct of natural organic synthetic organic matter.[13,14] Nanofiltration based on cross flow technology which is in between ultrafilters and reverse osmosis.

Carbon nanotube membranes: It provides a high surface of filtration with high permeability and very strong mechanical and thermal stability. They can remove bacteria, virus, sediments, and organics contaminations. It is very cost effective and does not need frequent maintenance.[15] But one of the most important aspects to be discussed is the use of asbestos in these nanofilters, which are harmful for the flora and fauna.[16]

14.4 NANOTECHNOLOGY: REMEDIATION

It is a method used to decontaminate the ground water by the use of nanomaterials. Especially, zero-valent metals were used in this method. It involves the injection of nanoparticles in contaminated water and these nanoparticles get transported to the source of contamination by the ground water flowed and degrade the contaminants. The degraded products are less harmful as compared to the previous one.[17] In general, nano-irons are used in this process. The less harmful product may be trichloroethylene and dichloroethene, while vinyl chloride may also form, which is more harmful as compared to the parent compound.[17] There is no need to remove the nanoparticle from the water as it is frequently transported and decontaminates the source and forms a less harmful waste product.[18,19]

Nanopollutions are pollutants which are generated from nanomaterials during their production. These waste products are very minute in size so cannot be detected easily. Due to small size and light weight, it can easily enter animals through penetration of epidermis and can also float in the environment. This kind of waste may be very dangerous because of its size. In general, only these nanoparticles were generated by humans which are not found in nature. It is having a very high impact on human health either positive or negative.[20] Nanofilterations are used to separate the nanomaterials of smaller than 10-nm size. Similar to this, magnetic nanoparticles are used to separate the heavy particle from the wastewater. Traditional and normal filtration methods are nowadays replaced by nanoscale particles which is highly efficient in contamination removal from wastewater as compared to the previous two. Nanotechnology is also used to clean the environment by removing nanomaterials from the environments solar cell, fuel cells, and environmental friendly batteries.

14.5 CONCLUSION

Previously, one of the major area of nanotechnology was drug delivery, but new areas have also emerged and they prove that nanotechnology can also support for a better human life. In spite of this, it can also protect and conserve our environment. It also helps to decontaminate water and provide clean water. Nanotechnologies will solve many efficient issue related energy generation. Application of nanotechnology involves cost effective and innovative methods for environmental remediation and waste management. Materials created using nanotechnologies are lighter and stronger generates less waste material.

KEYWORDS

- **nanotechnology**
- **pollution control**
- **energy**
- **environment**
- **nanofabrication**
- **sustainability**
- **nanoproducts**
- **human health**

REFERENCES

1. Nowack, B. Pollution Prevention and Treatment Using Nanotechnology. In *Environmental Aspects*; Krug, H., Ed.; WILEY VCH, Verlag GmbH & Co. KGaA: Weinheim, 2008, vol. 2.
2. http://www.epa.gov/opptintr/greenengineering/pubs/whats_ge.html.
3. Zhuang, J.; Gentry, R. W. Environmental Application and Risks of Nanotechnology: A Balanced View Biotechnology and Nanotechnology Risk Assessment: Minding and Managing the Potential Threats around Us. *ACS Symp. Ser.* **2011,** *3*, 41–67.
4. Smee, A. *Elements of Electro-biology: or the Voltaic Mechanism of Man; of Electro-pathology, Especially of the Nervous System; and of Electro-therapeutics*; Longman, Brown, Green, and Longmans: London, 1849; p 15.
5. http://news.nationalgeographic.com/news/2005/01/0114_050114_solarplastic.html.
6. Paull, J. Nanotechnology: No Free Lunch. *Platter* **2010,** *1* (1), 8–17.
7. Paull, J. Nanomaterials in Food and Agriculture: The Big Issue of Small Matter for Organic Food and Farming. In *Proceedings of the Third Scientific Conference of ISOFAR*; Neuhoff, D., Halberg, N., Rasmussen, I. A., Hermansen, J. E., Ssekyewa, C., Sohn, S. M., Eds.; ISOFAR, Bonn, 2011; vol. 2, pp 96–99.
8. United States National Institute for Occupational Safety and Health. Occupational Exposure to Titanium Dioxide. *Current Intelligence Bulletin 63*. United States National Institute for Occupational Safety and Health, 2012 (retrieved 19.2.2012).
9. Barlow, M. *Blue Covenant—The Global Water Crisis and the Coming Battle for the Right to Water*. Black Inc.: Melbourne, 2007.
10. Hillie, T.; Munasinghe, M.; Hlope, M.; Deraniyagala, Y. *Nanotechnology, Water and Development*, 2007.
11. Cenihr, S. *Risk Assessment of Products of Nanotechnologies*, 2009.
12. UNESCO. Water—A Shared Responsibility (Executive Summary). UN-WATER/WWAP/2006/3, 2006.
13. Letterman, R. D., Ed. *Water Quality and Treatment*, fifth ed. American Water Works Association and McGraw-Hill: New York, 1999. ISBN 0-07-001659-3.
14. Hillie, T.; Hlophe, M. Nanotechnology and the Challenge of Clean Water. *Nat. Nanotechnol.* **2007,** *2* (11), 663–664.
15. Diallo, M. S.; Christie, S.; Swaminathan, P.; Johnson, J. H.; Goddard, W. A. Dendrimer Enhanced Ultra-filtration Recovery of Cu(II) from Aqueous Solutions Using Gx-NH2-PAMAM Dendrimers with Ethylene Diamine Core. *Environ. Sci. Technol.* **2005,** *39*, 1366–1377.
16. Lindsten, D. C. Technology Transfer: Water Purification, U.S. Army to the Civilian Community. *J. Technol. Transf.* **1984,** *9* (1), 57–59.
17. Lowry, G. V. Nanotechnology for Ground Water Remediation. In *Environmental Technology*; Wiesner, M. R., Bottero, J., Eds.; The McGraw-Hill Companies: New York, NY, 2007; pp 297–336.
18. Zhang, W.; Cao, J.; Elliot, D. Iron nanoparticles for site remediation. In *Nanotechnology and the Environment: Applications and Implications*; Karn, B.; Masciangioli, T.; Zhang, W.; Colvin, V.; Alivisatos, P., Eds.; Oxford University Press, Washington, DC, 2005; pp 248–261.
19. http://www.nanorev.in/nanorev—nanotechnology.html.
20. Mueller, N.; Nowack, B. Exposure Modeling of Engineered Nanoparticles in the Environment. *Environ. Sci. Technol.* **2008,** *42*, 4447–4453.

CHAPTER 15

A NOTE ON PREPARATION OF NANOFILTER FROM CARBON NANOTUBES

M. ZIAEI and S. RAFIEI*

University of Guilan, Rasht, Iran

*Corresponding author. E-mail: saeedeh.rafieii@gmail.com

CONTENTS

ABSTRACT

Carbon nanofibers (CNFs) possess properties that are rarely present in any other types of carbon adsorbent, including a small cross-sectional area, combined with a multitude of slit-shaped nanopores that are suitable for adsorption of certain types of molecules. Because of their unique properties, these materials can be used for the selective adsorption of organic molecules.

On the other hand, activated carbon fiber (ACF) has been widely applied as an effective adsorbent for micropollutants in recent years. ACF effectively adsorbs and removes a full spectrum of harmful substances. Although there are various methods of fabricating CNFs, electrospinning is perhaps the most versatile procedure. This technique has been given great attention in current decades because of the nearly simple, comfortable, and low cost. Spinning process control and achieve optimal conditions are important to effect on its physical properties, absorbency, and versatility with different industrial purposes. Modeling and simulation are suitable methods to obtain this approach. In this chapter, activated CNFs were produced during electrospinning of polyacrylonitrile solution. Stabilization, carbonization, and activation of electrospun nanofibers in optimized conditions were achieved, and mathematical modeling of electrospinning process was done by focusing on governing equations of electrified fluid jet motion (using FeniCS software). Experimental and theoretical results will be compared with each other to estimate the accuracy of the model. The simulation can provide the possibility of predicting essential parameters, which affect the electrospinning process.

15.1 INTRODUCTION

Water and air represent two environmental systems where the most pressing environmental issues remain. Water pollution and deteriorating freshwater supplies are frequently mentioned as critical global problems.[1] Various agricultural processes and industrial sectors are the major sources of pollutants into the environment. These include oil and petrochemical production, mining and mineral processing, battery manufacture, printing and photographic industry, electroplating processes, textile industries and agricultural activities, burning of fossil fuels, incineration of wastes, and automobile exhausts processes.[2] Dyes are one of the most complex pollution that is entering the water sources by industries. The conventional water treatment process of flocculation–filtration–disinfection is effective to remove suspended solid in raw water and has been practiced for over century for water supply. All the same, it is disabled to remove micropollutants such as dye particles that may cause serious problem of water safety in raw water.[1,3] On the other hand, the growing environmental concerns of worldwide heating and climate change have motivated significant research activities, such as separation, capture, and storage of greenhouse gases, especially carbon dioxide (CO_2), which will be more and more significant in the future world economic system.[4]

Activated carbon nanofibers (ACNFs) has been shown to be an effective adsorbent for the removal of a spacious sort of organic and inorganic pollutants from aqueous or gaseous media. It is widely practiced due to its exceptionally high surface area, well-developed internal micro- or nanoporosity, and broad spectrum of surface functional groups.[5] ACF effectively adsorbs and removes a full spectrum of harmful substances. Introducing of nanosized porosity has been proven as an effective scheme to achieve better adsorption. On the other hand, fibrous materials have filled requests in lots of areas due to their intrinsically high surface and porous structures. In concept, coupling nanoporosity and ultrafine form of fibrous materials would lead to the highest possible specific surface area.[1,6] An activated form of carbon nanofibers (CNFs) is one of the most famous porous ultrafine fibers with significant flexibility.[1,7] Although there are several methods of fabricating CNFs, such as a traditional vapor-growth method or plasma-enhanced chemical vapor deposition method, electrospinning is the most versatile and useful operation. A simple and affordable approach is to utilize electrospinning to prepare carbon precursor polymer nanofibers and subsequent thermal treatment to produce carbon nanostructure.[8,9] Among the various precursors for producing CNFs, polyacrylonitrile (PAN) is the most commonly used polymers, primarily due to its high carbon yield, flexibility for tailoring the structure of the final CNF products, and the easiness of obtaining stabilized products due to the organization of a ladder structure via nitrile polymerization.[10,11] Electrospinning PAN followed by stabilization and carbonization has become a straightforward and convenient path to make continuous CNFs.[12,13]

Electrospinning that first patented by Cooly in 1902 has been used to produce nanofibers from polymers or composites. This method is seen in terms of its versatility, flexibility, and simplicity of fiber yield. At a laboratory level, a typical electrospinning set-up only requires a high-voltage power supply (up to 30 kV), a syringe, a flat-tip needle, and a conducting collector.[14] Schematic diagram of electrospinning is shown in Figure 15.1.

PAN nanofibers are converted into CNFs by the sequenced processes of stabilization, carbonization, and graphitization.[15] It is necessary to select the optimum conditions of temperature, heating rate, and oxidation period. Too low temperatures lead to slow reactions and incomplete stabilization, whereas too high temperatures can fuse or even burn the fibers.[1,10] A significant component of our information of the electrospinning process comes from experimental observations while the complexity of the process makes empirical determination of affected parameters difficult. By employing a suitable theoretical model, the effects of parameters can be assessed. It will help researchers in predicting the influence of variables and controlling the process. Therefore, without conducting any experiments, one can easily estimate features of the product under unknown conditions.[16,17]

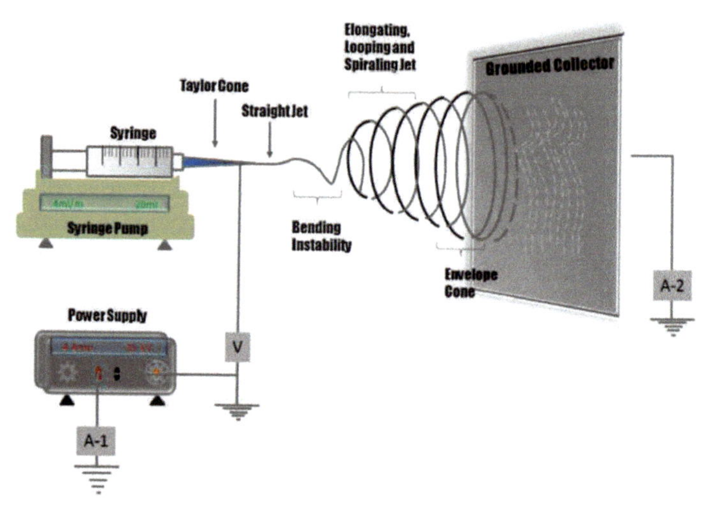

FIGURE 15.1 Schematic diagram of electrospinning showing details of the jetting processes, the whipping instability, as well as the fiber morphologies that can be obtained.

In the current work, ACNFs were produced during electrospinning of PAN solution, stabilization, carbonization, and activation of electrospun nanofibers in optimized conditions, and then mathematical modeling of electrosinning process will be done by focusing on governing equations of electrified fluid jet motion by using FEniCS software.

15.2 EXPERIMENTAL

15.2.1 MATERIALS

Poly(acrylonitrile-ran-venylacetate) (94.6%) was received from Polyacryle Co. (Isfahan, Iran). Dimethylformamide (DMF) was purchased from Merck

Co. as appropriate solvent due to its solubility parameters[18] of the PAN powder dissolving and was used without further purification.

15.2.2 METHODS

PAN was dissolved in DMF in a concentration of 11 wt% and was prepared by using magnetic stirrer for 24 h. The solution then was electrospinned using a syringe pump with pumping rate of 8 μl/min, gamma high voltage provided 18 kV potentials, in 12 cm spinning distances where provided the optimum conditions. Scanning electron microscopy (SEM) was used to select the best electrospinning conditions, according to the appearances, uniformity, and diameters of electrospun nanofibers[1,19] Micrographs to obtain the optimal conditions are shown in Figure 15.2. The obtained webs were converted to ACNFs during two stages. (1) Stabilization in a chamber furnace Nabertherm Controller Co. at temperatures 240–270°C and time ranging from 60 to 120 min, in heating rate of 2°C/min, to find the best stabilization conditions. (2) Simultaneous carbonization and activation in a furnace with pure nitrogen (99.99%) atmosphere at temperatures ranging from 800 to 1200°C for 1 h. The properties of obtaining ACNF were examined. Then studies were conducted on modeling and simulation. In electrospinning, the jet is elongated by electrostatic forces and gravity, while pressure, viscosity, and density also play a part. As the jet thins, the surface charge density σ varies, which in turn affects the electric field E and the pulling force. We assume the "leaky dielectric model"[20] and the slender-body approximation apply. The jet can now be represented by four steady-state equations: the conservation of mass and electric charges, the linear momentum balance and Coulomb's law for the E field, with all quantities depending only on the axial position z.[21]

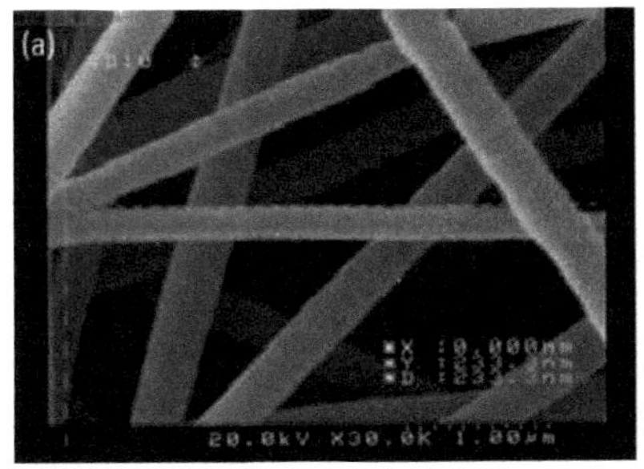

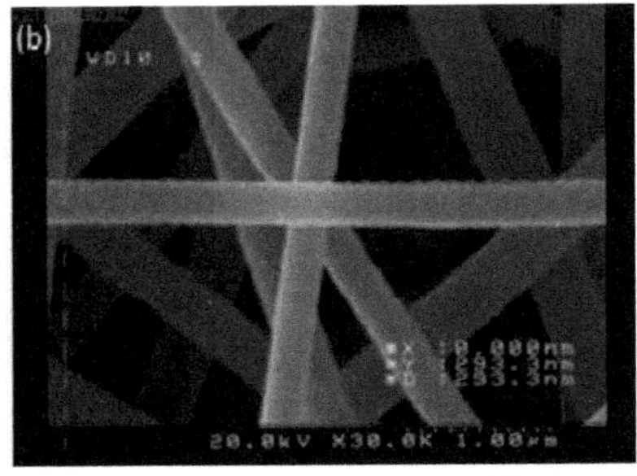

FIGURE 15.2 The micrographs (a) and (b) show electrospun polyacrylonitrile nanofibers in optimized conditions (in different voltages). (a) concentration (11 wt%), pumping rate (8 μl/min), spinning distance (12 cm), and voltage (18 kV), (b) concentration (11 wt%), pumping rate (8 μl/min), spinning distance (12 cm), and voltage (16 kV).

TABLE 15.1 Symbols Employed and Their Definitions.

Symbol	Definition	Units
τ_{prr}	Radial polymer normal stress	N/m^2
λ	Relaxation time	s
v	Velocity of the jet	m/s
α	Mobility factor	–
η_p	Viscosity of the solution due to the polymer	Pa s
τ_{pzz}	Axial polymer normal stress	N/m^2

15.3 MATHEMATICAL MODELING PROCEDURE

There are parameters in the model that effect on the morphology and diameter of electrospun nanofibers. The governing parameters are diameter, velocity fluid jet, the electric field (E, v, d). Once the jet flows away from the Taylor cone in a nearly straight line, the traveling liquid jet is subjected to a variety of forces, such as an electrostatic force, a viscoelastic force, a surface tension force, a gravitational force, and an inertia force.[22,23]

Using the equation of these forces operating on the fluid jet and eqs 15.1 and 15.2 that called Giesekus constitutive equations, are considered here to translate the nonuniform uniaxial extension of viscoelastic polymer,[24] the model was thought. The notations are provided in Table 15.1.

$$\tau_{prr} + \lambda\left(v\tau'_{prr} + v'\tau_{prr}\right) + \alpha\frac{\lambda}{\eta_p}\tau^2_{prr} = -\eta_p v' \quad (15.1)$$

$$\tau_{pzz} + \lambda\left(v\tau'_{pzz} + v'\tau_{pzz}\right) + \alpha\frac{\lambda}{\eta_p}\tau^2_{pzz} = 2\eta_p v' \quad (15.2)$$

15.4 RESULTS AND DISCUSSION

The PAN nanofiber and its stabilized forms were evaluated using SEM, FTIR, and DSC. As can be seen in Figure 15.2, by using SEM, the optimal conditions were obtained. It was observed that PAN nanofibers lose some of their diameters during stabilization and carbonization processes.

Comparing the FTIR spectra of untreated and stabilized PAN nanofiber in different temperature and measure, the extent of stabilization reaction (EOR) refer to (15.3) was achieved stabilization optimal conditions (Fig. 15.3).

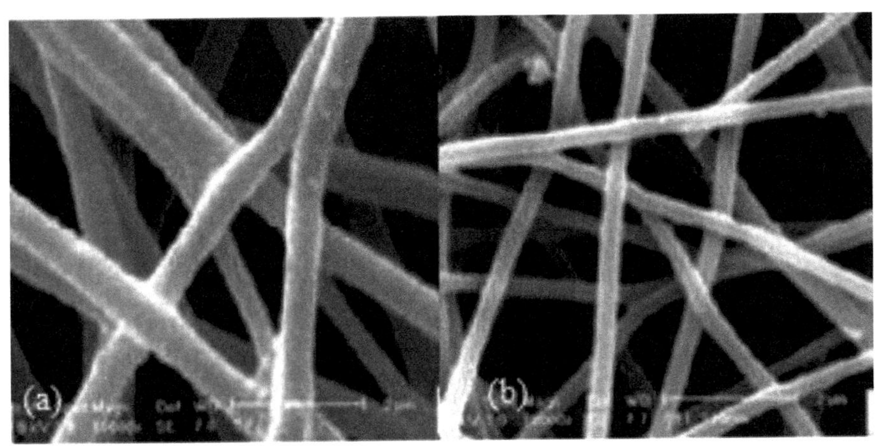

FIGURE 15.3 Micrograph of (a) stabilized and (b) carbonized nanofibers in optimum condition.

$$\text{ÅÏ R} \, (\%) = \frac{I_{1600}}{I_{2240} + I_{1600}} \qquad (15.3)$$

The optimal conditions (times and temperatures) are chosen according to higher calculated EOR.

By using DSC analysis, the thermal behavior of untreated and stabilized PAN nanofibers was studied. It is known that the quality of resulting carbon fibers depends strongly upon the level of stabilization process during CNF production. According to the consistent confirmed results obtained by employing analyzing techniques, the conditions that contain ranges of temperature and time of 270°C and 2 h using heating rate below 2°C/min had the maximum stabilization efficiency, so they were selected as the suitable conditions for stabilization stage prior to carbonization and activation process. Simultaneous carbonization and activation were carried out under pure nitrogen (99.99%) atmosphere at temperatures ranging from 800 to 1200°C for 1 h to achieve ACNF.

The model capability to predict the behavior of the process parameters was demonstrated using simulation. The plot obviously showed the changes of R versus axial position. The plot shows similar behavior to those reported by other researchers and experimental data. According to the figures as the jet becomes thinner downstream, the increase in jet speed reduces the surface charge density and thus electric force. The rate of R is maximum at the beginning of motion and then relaxes smoothly downstream toward zero (Figs. 15.4 and 15.5).

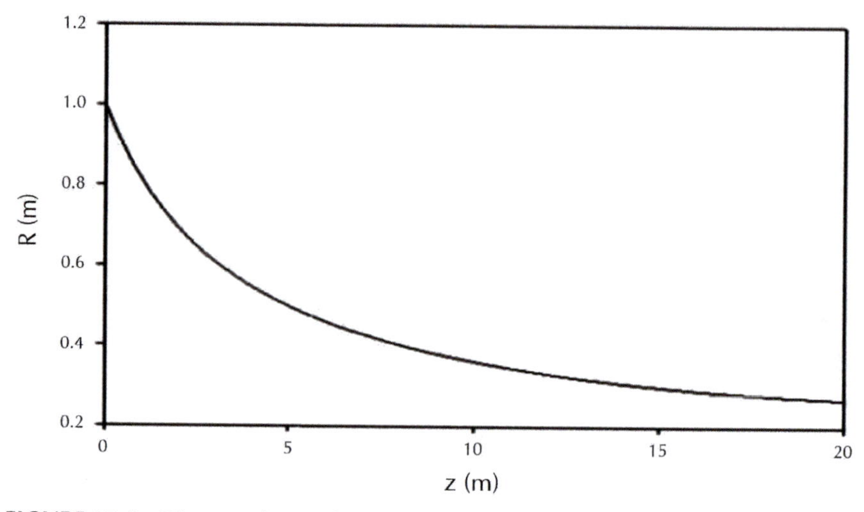

FIGURE 15.4 Diameter changes in jet axis direction.

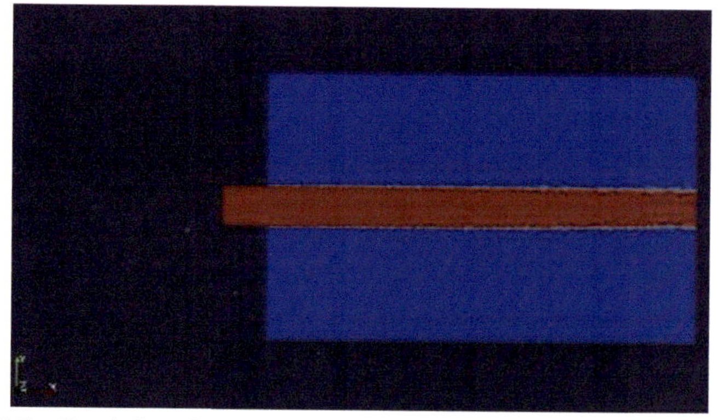

FIGURE 15.5 Simulation of diameter changes in jet axis direction using FEniCS software.

15.5 CONCLUSION

By applying optimum conditions of electrospinning, stabilization, and simultaneous carbonization and activation, using SEM, FTIR, DSC, and BET techniques, ACNFs with the surface area of 850 m^2/g and the porosity of 65% with micropores (average width pores of 0.7 nm) were produced in this study. Besides, it is claimed that the production parameters can be arranged in the manner which is leading to the appropriate pore volume and pore size distribution for the ACNF application as molecular sieving in water contamination control.

A regular adsorption experiment was used to evaluate the dye removal potential of produced ACNF. Adsorption experiment carried out with 200 mg/l of basic blue 41 dye (a cationic dye) with different kind of adsorbents such as activated carbon, chitin, and ACNF. Dye removal efficiency of ACNF was much higher that other adsorbents (around 52%) that is related to higher porous structure of ACNFs providing physical entrapping of adsorbate molecules.

KEYWORDS

- **carbon nanofibers**
- **electrospinning**
- **electrospinning modeling**
- **simulation**
- **fluid jet**

REFERENCES

1. Rafiei, S.; Noroozi, B.; Arbab, S.; Haghi Rafiei, A. K. Characteristic Assessment of Stabilized Polyacrylonitrile Nanowebs for the Production of Activated Carbon Nanosorbents. *Chin. J. Polym. Sci.* **2014,** *32*, 449–457.
2. Gupta, V. K.; Saleh, T. A. Sorption of Pollutants by Porous Carbon, Carbon Nanotubes and Fullerene—An Overview. *Environ. Sci. Pollut. Res.* **2013,** *20*, 2828–2843.
3. Mostafavi, S.; Mehrnia, M.; Rashidi, A. Preparation of Nanofilter from Carbon Nanotubes for Application in Virus Removal from Water. *Desalination* **2009,** *238*, 271–280.
4. Lee, S.-Y.; Park, S.-J. Determination of the Optimal Pore Size for Improved CO_2 Adsorption in Activated Carbon Fibers. *J. Colloid Interface Sci.* **2013,** *389*, 230–235.
5. Rivera-Utrilla, J.; Sanchez-Polo, M.; Gomez-Serrano, V.; Alvarez, P. M.; Alvim-Ferraz, M. C. M.; Dias, J. M. Activated Carbon Modifications to Enhance its Water Treatment Applications. An Overview. *J. Hazard. Mater.* **2011,** *187*, 1–23.
6. Thavasi, V.; Singh, G.; Ramakrishna, S. Electrospun Nanofibers in Energy and Environmental Applications. *Energy Environ. Sci.* **2008,** *1*, 205–221.
7. Oh, G. Y.; Ju, Y. W.; Jung, H. R.; Lee, W. J. Preparation of the Novel Manganese-Embedded PAN-Based Activated Carbon Nanofibers by Electrospinning and their Toluene Adsorption. *J. Anal. Appl. Pyrol.* **2008,** *81*, 211–217.
8. Ji, L.; Lin, Z.; Medford, A. J.; Zhang, X. Porous Carbon Nanofibers from Electrospun Polyacrylonitrile/SiO_2 Composites as an Energy Storage Material. *Carbon* **2009,** *47*, 3346–3354.

9. Li, D.; Xia, Y. Electrospinning of Nanofibers: Reinventing the Wheel? *Adv. Mater.* **2004,** *16*, 1151–1170.
10. Nataraj, S.; Yang, K.; Aminabhavi, T. Polyacrylonitrile-Based Nanofibers—A State-of-the-Art Review. *Progr. Polym. Sci.* **2012,** *37*, 487–513.
11. Litmanovich, A. D.; Platé, N. A. Alkaline Hydrolysis of Polyacrylonitrile. On the Reaction Mechanism. *Macromol. Chem. Phys.* **2000,** *201*, 2176–2180.
12. Zhang, L.; Aboagye, A.; Kelkar, A.; Lai, C.; Fong, H. A Review: Carbon Nanofibers from Electrospun Polyacrylonitrile and their Applications. *J. Mater. Sci.* **2014,** *49*, 463–480.
13. Esrafilzadeh, D.; Morshed, M.; Tavanai, H. An Investigation on the Stabilization of Special Polyacrylonitrile Nanofibers as Carbon or Activated Carbon Nanofiber Precursor. *Synth. Met.* **2009,** *159*, 267–272.
14. Teo, W.; Ramakrishna, S. A Review on Electrospinning Design and Nanofibre Assemblies. *Nanotechnology* **2006,** *17*, R89.
15. Arshad, S. N.; Naraghi, M.; Chasiotis, I. Strong Carbon Nanofibers from Electrospun Polyacrylonitrile. *Carbon* **2011,** *49*, 1710–1719.
16. Thompson, C. J.; Chase, G. G.; Yarin, A. L.; Reneker, D. H. Effects of Parameters on Nanofiber Diameter Determined from Electrospinning Model. *Polymer* **2007,** *48*, 6913–6922.
17. Yarin, A. L.; Koombhongse, S.; Reneker, D. H. Taylor Cone and Jetting from Liquid Droplets in Electrospinning of Nanofibers. *J. Appl. Phys.* **2001,** *90*, 4836–4846.
18. P. Heikkilä, Harlin, A. Electrospinning of Polyacrylonitrile (PAN) Solution: Effect of Conductive Additive and Filler on the Process. *Express Polym. Lett.* **2009,** *3*, 437–445.
19. Ziabari, M.; Mottaghitalab, V.; Haghi, A. K. A New Approach for Optimization of Electrospun Nanofiber Formation Process. *Korean J. Chem. Eng.* **2010,** *27*, 340–354.
20. Saville, D. Electrohydrodynamics: the Taylor–Melcher Leaky Dielectric Model. *Annu. Rev. Fluid Mech.* **1997,** *29*, 27–64.
21. Feng, J. The Stretching of an Electrified Non-Newtonian Jet: A model for Electrospinning. *Phys. Fluids (1994–present)* **2002,** *14*, 3912–3926.
22. Reneker, D. H.; Yarin, A. L.; Fong, H.; Koombhongse, S. Bending Instability of Electrically Charged Liquid Jets of Polymer Solutions in Electrospinning. *J. Appl. Phys.* **2000,** *87*, 4531–4547.
23. Angammana, C. J.; Jayaram, S. H. A Theoretical Understanding of the Physical Mechanisms of Electrospinning. In *Proceedings of ESA Annual Meeting on Electrostatics*, 2011.
24. Peters, G.; Hulsen, M.; Solberg, R. *A Model for Electrospinning Viscoelastic Fluids.* Department of Mechanical Engineering, Eindhoven University of Technology, 2007; p 26.

CHAPTER 16

CONTROL OF FLUIDIC JET REPULSION IN THE ELECTROSPINNING PROCESS

M. ZIAEI and S. RAFIEI*

University of Guilan, Rasht, Iran

Corresponding author. E-mail: saeedeh.rafieii@gmail.com

CONTENTS

ABSTRACT

Physical reasoning problems, like electrospinning phenomena, have usually required a representational apparatus that can deal with the vast amount of physical knowledge that is used in reasoning tasks. Mathematical and theoretical modeling and simulating procedure will permit to offer an in-depth prediction of electrospun fiber properties and morphology. Utilizing a model to express the effect of electrospinning parameters will assist researchers in making an easy and systematic way of presenting the influence of variables, and by means of that, the process can be controlled. Considering the fact that electrospinning like most of the scientific problems is inherently nonlinear so the governing equations that affect fluid motions do not have analytical solution and should be solved by using other methods. Although there are several numerical methods, which have been applied in literatures to solve such problems, the main novelty in this research will be the application and presentation of a new mathematical model with a minimum applicable time for running simulation program.

16.1 INTRODUCTION

Electrospinning is an easy but relatively inexpensive procedure, in which an electrical charge is applied to draw really fine (typically in the micro- or nanoscale) fibers from polymer solution or melt.[1] This technique is able to manufacture high-volume production of fibers from a vast variety of materials including polymers, composites, and ceramics.[2,3] Electrospinning technology was first developed and patented by John Francis Cooley in the 1902s,[4] and a few years later, the actual developments were triggered by Reneker and coworkers.[5] In this method, nanofibers are produced by solidification of a polymer solution stretched by an electric field.[6,7] There are commonly two standard electrospinning setups, vertical and horizontal, that is shown in Figure 16.1. By expanding this technology, several researchers have developed more intricate systems that can fabricate more complex nanofibrous structures in a more controlled and efficient style.[8]

In the electrospinning process, a high voltage is applied to produce an electrically charged jet of a polymer solution or melt, which dries or solidifies to leave a polymer fiber. This technique consists of three stages that correspond to the behavior of the electrospun jet: the formation of the Taylor cone, the ejection of the straight jet, and the unstable whipping jet region, as shown in Figure 16.2.[9]

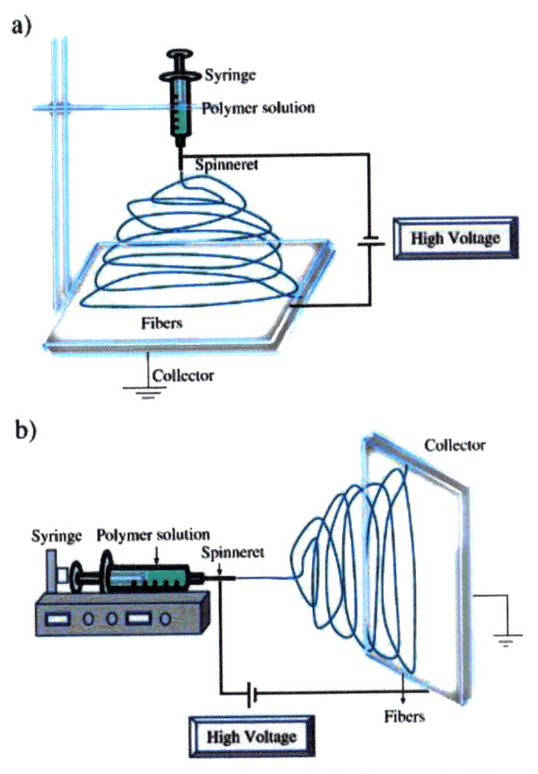

FIGURE 16.1 Electrospinning set-ups: (a) vertical alignment of the electrodes (top-to-bottom design) and (b) horizontal alignment (bottom-to-top design).

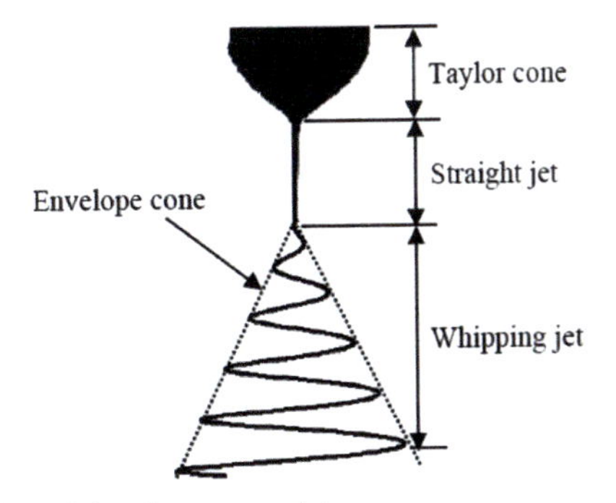

FIGURE 16.2 Behavior of the electrospun jet.

An important stage of nanofibers formation in electrospinning includes fluid instabilities such as whipping instabilities.[10] Therefore, the fluid jet and electrospinning method are out of control. The efficiency of this method can be improved by applying modeling and simulating. Mathematical and theoretical modeling and simulating procedure will help in presenting an in-depth insight into the physical comprehension of complex phenomena during electrospinning and might be very applicable to manage contributing factors to increase the production rate.[11] Although electrospinning technology is simple, industrial applications of this method are still relatively rare, mainly due to the significant problems of very low fiber production rate and difficulties in controlling the process.[12] Modeling and simulation yield appropriate information about how something will act without actual testing in real. Modeling is a representative of a real object or system of objects to visualize its appearance or analyze its behavior. Simulation is transitioning from a mathematical or computational model for the description of the system behavior based on lots of input parameters.[13,14]

Moreover, in experimental situations when the possibility of error is high, simulation can be even more realistic than experiments, as they let the free configuration of environmental and operational parameters and are able to be run faster than in actual time.[1] It is necessary for the expansion theoretical and numerical models of electrospinning, because each material demands a different optimization procedure. Utilizing a model to express the effect of electrospinning parameters will assist researchers in making an easy and systematic way of offering the influence of variables and by means of that, the process can be controlled. Additionally, predicting the results under a novel combination of parameters becomes possible. Therefore, without conducting any experiments, one can easily estimate features of the product under unknown conditions.[1]

In this research, assignment will be the application and presentation of novel mathematical model with a minimum applicable time for running simulation program.

16.2 MODEL DEVELOPMENT

In electrospinning, the jet is elongated by a variety of forces, such as a Coulomb force, an electric force imposed by the external electric field, a viscoelastic force, a surface tension force, a gravitational force, and an air-drag force.[15] It is assumed the "leaky dielectric model"[16] and the slender-body approximation apply. Therefore, the electrospun jet can be represented by four steady-state equations as follows.[17] The notations are provided in Table 16.1.

TABLE 16.1 Symbols Employed and Their Definitions.

Symbol	Quantity	Conversion from Gaussian and CGS EMU to SI
R	Jet radius	m
Q	Volume flow rate	m^3/s
E	z Component of the electric field	V/m
v	Velocity of the jet	m/s
K	Conductivity of the solution	S/m
σ	Surface charge density	C/m^2
I	Jet current	A
ρ	Fluid density	kg/m^3
p	Pressure	N/m^2
γ	Surface tension of the solution	N/m
R'	Slope of the jet surface	–
τ_{zz}	Viscous normal stress in the axial direction	N/m^2
t_t^e	Tangential stress exerted on the jet surface due to the electric field	N/m^2
t_n^e	Normal stress exerted on the jet surface due to the electric field	N/m^2
E_∞	External electric field	V/m
τ_{prr}	Radial polymer normal stress	N/m^2
τ_{pzz}	Axial polymer normal stress	N/m^2
λ	Relaxation time	s
α	Mobility factor	–
η_p	Viscosity of the solution due to the polymer	Pa s
η_0	Viscosity of the solution at zero shear rate	Pa s
$\bar{\varepsilon}$	Dielectric constant of the ambient air	–
ε	Dielectric constant of the solution	–
R_0	Jet radius at the origin	m

Conservation of mass

$$\pi R^2 = Q \tag{16.1}$$

Conservation of charge

$$\pi R^2\, KE + 2\pi R\, v\sigma I \tag{16.2}$$

The momentum equation

$$\frac{\mathrm{d}}{\mathrm{d}z}\left(\pi R^2 \rho v^2\right) = \pi R^2 \rho g + \frac{\mathrm{d}}{\mathrm{d}z}\left[\pi R^2\left(-P + \tau_{zz}\right)\right] + \frac{\gamma}{R}\cdot 2RR' + 2\pi R\left(t_t^e - t_n^e R'\right) \tag{16.3}$$

Coulomb's law equation for electric field

$$E(z) = E_{\infty}(z) - \ln \chi\left(\frac{1}{\overline{\varepsilon}}\frac{\mathrm{d}(\sigma R)}{\mathrm{d}z} - \frac{\beta}{2}\frac{\mathrm{d}^2\left(ER^2\right)}{\mathrm{d}z^2}\right) \tag{16.4}$$

In addition, eqs 16.5 and 16.6 that called Giesekus constitutive equations, are considered here to represent the nonuniform uniaxial extension of visco-elastic polymer solutions.[16]

$$\tau_{prr} + \lambda\left(v\tau'_{prr} + v'\tau_{prr}\right) + \alpha\frac{\lambda}{\eta_p}\tau_{prr}^2 = -\eta_p v' \tag{16.5}$$

$$\tau_{prr} + \lambda\left(v\tau'_{prr} + v'\tau_{prr}\right) + \alpha\frac{\lambda}{\eta_p}\tau_{prr}^2 = -\eta_p v' \tag{16.6}$$

The equations can be converted to dimensionless form using the following characteristic scales and dimensionless groups.

Characteristics parameters

Length R_0

Velocity $V_0 = \frac{Q}{\pi R_0^2}$

Electric field $E_0 = \frac{I}{\pi R_0^2 K}$

Dimensionless groups employed

Froude number $Fr = \frac{V_0^2}{gR_0}$

Reynolds number $$Re = \frac{\rho v_0 R_0}{\eta_0}$$

Weber number $$We = \frac{\rho v_0^2 R_0}{\gamma}$$

Inserting these dimensionless parameters and groups into eqs 16.1–16.6 gives the dimensionless equations. Then applied boundary conditions. The mathematical model is formulated using the partial differential equation (PDE) module using FEniCS package software.

16.3 RESULT AND DISCUSSION

The model capability to predict the behavior of the process parameters was demonstrated using simulation. The plots obviously showed the changes of each parameter versus axial position. According to the figures as the jet becomes thinner downstream, the increase in jet speed reduces the surface-charge density and thus electric force. The rates of R are maximum at the beginning of motion and then relax smoothly downstream toward zero. Axial shear stress and, as a result, axial viscosity of the polymer solution increases versus z.

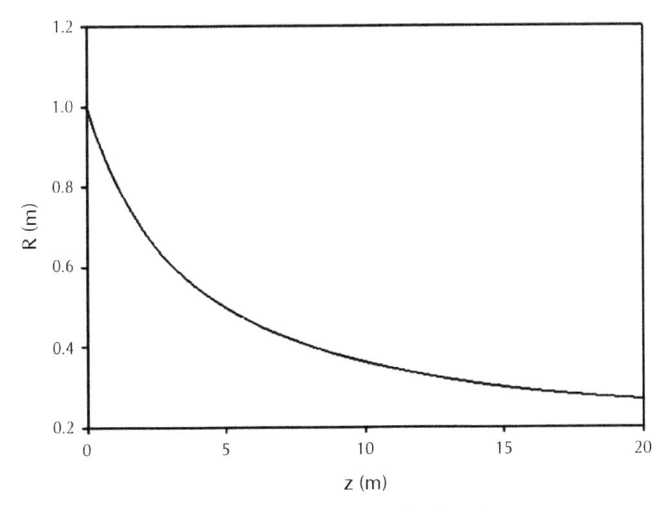

Diameter changes in jet axis direction.

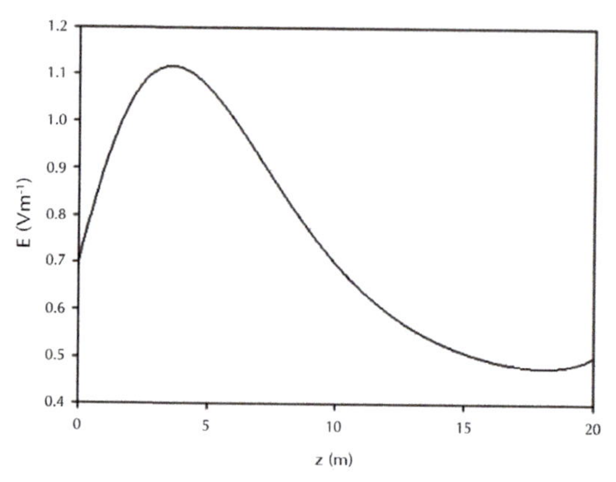

Electric force in jet axis direction.

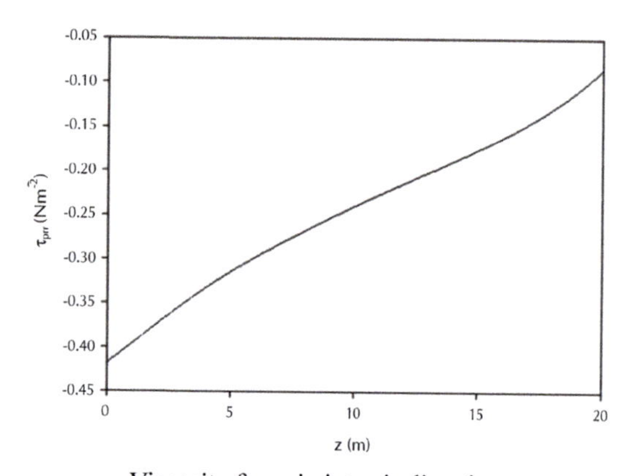

Viscosity force in jet axis direction.

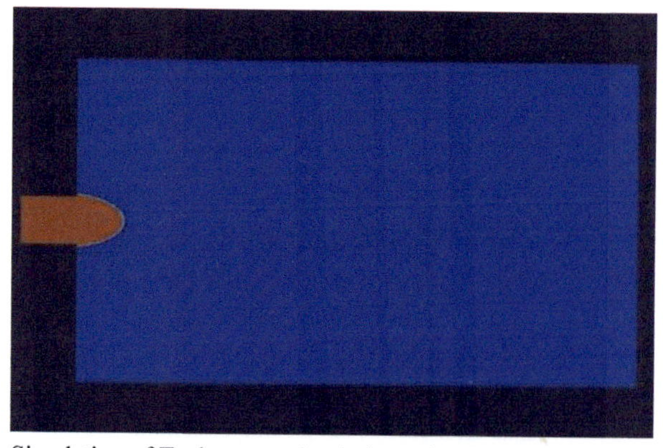

Simulation of Taylor cone simulation using FEniCS software.

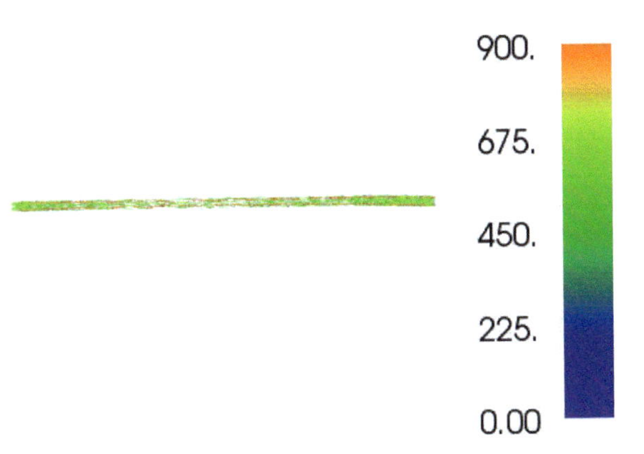

Simulation of electric field changes during electrospinning process using FEniCS software.

Simulation of diameter changes in jet axis direction using.

16.4 CONCLUSION

In this chapter, electrospun nanofibers which are produced during electrospinning of polymeric solution in optimized conditions are simulated using mathematical modeling of electrospinning process. This procedure is done by focusing on governing equations of electrified fluid jet motion by using FeniCS software. Comparing experimental and theoretical results represents good accuracy of the model. The simulation provides the possibility of predicting essential parameters which affect the electrospinning process such as electric field changes and diameter of final nanofiber.

KEYWORDS

- **electrospinning**
- **electrospinning modeling**
- **nanofibers**
- **simulation**
- **fluid motions**

REFERENCES

1. Rafiei, S.; Maghsoodloo, S.; Saberi, M.; Lotfi, S.; Motaghitalab, V.; Noroozi, B.; Haghi, A. K. New Horizons in Modeling and Simulation of Electrospun Nanofibers: A Detailed Review. *Cellulose Chem. Technol.* **2014,** *48*, 401–424.
2. Solberg, R. H. M. *Position-Controlled Deposition for Electrospinning*. Department Mechanical Engineering, vol. 67, 2007.
3. Chronakis, I. S. Micro-Nano-Fibers by Electrospinning Technology: Processing, Properties and Applications. *Micromanufacturing Engineering and Technology*; Elsevier: Boston, 2010; pp 264–286.
4. Tucker, N.; Stanger, J. J.; Staiger, M. P.; Razzaq, H.; Hofman, K. The History of the Science and Technology of Electrospinning from 1600 to 1995. *J. Eng. Fabrics Fibers* **2012,** *7*.
5. Reneker, D. H.; Chun, I. Nanometre Diameter Fibres of Polymer, Produced by Electrospinning. *Nanotechnology* **1996,** *7*, 216.
6. Theron, S. A.; Zussman, E.; Yarin, A. L. Experimental Investigation of the Governing Parameters in the Electrospinning of Polymer Solutions. *Polymer* **2004,** *45*, 2017–2030.
7. Xu, L. A Mathematical Model for Electrospinning Process under Coupled Field Forces. *Chaos, Solut. Fract.* **2009,** *42*, 1463–1465.
8. Bhardwaj, N.; Kundu, S. C. Electrospinning: A Fascinating Fiber Fabrication Technique. *Biotechnol. Adv.* **2010,** *28*, 325–347.
9. Angammana, C. J.; Jayaram, S. H. A Theoretical Understanding of the Physical Mechanisms of Electrospinning. In *Proceedings of ESA Annual Meeting on Electrostatics*, 2011; pp 1–9.
10. Shin, Y. M.; Hohman, M. M.; Brenner, M. P.; Rutledge, G. C. Experimental Characterization of Electrospinning: The Electrically Forced Jet and Instabilities. *Polymer* **2001,** *42*, 09955–09967.
11. Haghi, A. K. *Electrospinning of Nanofibers in Textiles*. CRC Press: Boca Raton, FL, 2011.
12. Frenot, A.; Chronakis, I. S. Polymer Nanofibers Assembled by Electrospinning. *Curr. Opin. Colloid Interface Sci.* **2003,** *8*, 64–75.
13. Fritzson, P. *Principles of Object-Oriented Modeling and Simulation with Modelica 2.1.* John Wiley & Sons: Hoboken, NJ, 2010.

14. Collins, A. J.; Meyr, D.; Sherfey, S.; Tolk, A.; Petty, M. *The Value of Modeling and Simulation Standards*; Virginia Modeling, Analysis and Simulation Center, Old Dominion University, 2011; pp 1–8.
15. Reneker, D. H.; Yarin, A. L.; Fong, H.; Koombhongse, S. Bending Instability of Electrically Charged Liquid Jets of Polymer Solutions in Electrospinning. *J. Appl. Phys.* **2000,** *87*, 4531–4547.
16. Peters, G. W. M.; Hulsen, M. A.; Solberg, R. H. M.. A *Model for Electrospinning Viscoelastic Fluids*. Department of Mechanical Engineering, Eindhoven University of Technology, vol 26, 2007.
17. Feng, J. J. The Stretching of an Electrified Non-Newtonian Jet: A Model for Electrospinning. *Phys. Fluids (1994–present)* **2002,** *14*, 3912–3926.

CHAPTER 17

MODIFICATION OF UREA–FORMALDEHYDE RESIN WITH COLLAGEN BIOPOLYMERS

JÁN SEDLIAČIK[1*], JÁN MATYAŠOVSKÝ[2], PETER JURKOVIČ[2], MÁRIA ŠMIDRIAKOVÁ[1], and LADISLAV ŠOLTÉS[3]

[1]*Technical University in Zvolen, Masaryka 24, 96053 Zvolen, Slovakia*

[2]*VIPO a.s., Partizánske, Gen. Svobodu 1069/4, 95801 Partizánske, Slovakia*

[3]*Institute of Experimental Pharmacology and Toxicology, Slovak Academy of Sciences, 84104 Bratislava, Slovakia*

**Corresponding author. E-mail: sedliacik@tuzvo.sk*

CONTENTS

17.1 INTRODUCTION

With the rapid development of science and technology in almost all industry branches, bonding systems combining different materials using adhesives are coming to the forefront. Recent years have been characterized by rapid development in the use of bonded joints in the engineering industry, the automotive and aerospace industry, in building construction, woodworking, and other industries. Not only does the application of adhesives provide a technological advantage, but, particularly, it also has a relatively large economic effect in the sectors where adhesive consumption is constantly increasing. In the application of all kinds of adhesives, it is also necessary to respect the technological processes given by the producer of the adhesive, as well as the safety rules when working with the adhesive, because some types of adhesives are classified as toxic, or they are otherwise harmful to human health.

Urea–formaldehyde (UF) adhesives are currently the most frequently used and most widespread adhesives for wood-based composite materials. Their industrial consumption is increasing quickly around the world. This growth in production is due to the increased production of agglomerating materials, mainly particle-board, fiberboard, and plywood. At the same time, it is also conditioned by their advantageous properties such as the possibilities for curing at a wide range of temperatures (from 10 to 150°C), they have a relatively short curing time and are used in the form of water solutions, are colorless, partially water resistant, etc.

A disadvantage of these adhesives is the release of formaldehyde (fd) in the production of board as well as during their storage and use. They are currently very intensive efforts being made to remove or at least reduce the release of formaldehyde. However, free formaldehyde and formaldehyde released via slow hydrolysis from amino-plastic bonds are highly reactive and easily bonds with proteins in the human organism. This may cause a painful inflammation of eye, nose, and mouth mucous. Even a low concentration of formaldehyde vapor in the air may cause unwanted irritation of the nose and eyes. However, this irritation usually disappears quickly and without causing permanent damage. There are also occasional allergic and anaphylactic reactions, and it is therefore necessary to prevent staying in such an environment. Emission of formaldehyde and harmful effect of formaldehyde is still a problem mainly for products used in interior. Adversely, affects the respiratory system, eyes, skin, genetic material, reproductive organs, it has a strong effect on the central nervous system. The International Agency for Research on Cancer (IARC) categorizes formaldehyde as carcinogen, which can cause allergies.

Improving UF adhesives to have increased resistance to humidity as well as a reduction in the release of free formaldehyde is currently being resolved by the addition of different modifiers, melamine formaldehyde resin, or by condensation of a mixture of urea and melamine with formaldehyde. However, the problem of the release of formaldehyde must be thoroughly resolved, since it could jeopardize their use, such as wood-based panels in wooden constructions. Some producers tested isocyanate adhesives in their operations—despite their high price—mainly due to concerns about sales due to the release of harmful formaldehyde when using UF adhesives.[5] Isocyanate adhesives found application in the production of oriented strand board. Mamiński et al.[14] developed a formaldehyde-free adhesive based on urea (U) and glutaraldehyde (GA). High reactivity of the U–GA mix at ambient temperature allows for its cold setting. The glues of GA/U molar ratio between 0.8 and 1.2 were examined. It was found that for the satisfactory performance of the system, blending with nano-Al_2O_3 was necessary.

Fibril proteins of leather, mainly collagen, keratin, glycoproteins, heteropolysaccharides as hyaluronic acid are the most significant from the view of application in different technical applications. Collagen is the most widespread animal protein component of skin, tendon, bones and ligament. Keratin is the main component of hair, fur, feathers, hooves, horns, and outer surface of the skin. Keratin is characterized by a high content of sulfur amino acid cysteine with a typical formation of disulfide bridges. Modified collagen is the basis for several types of test colloids, where agents for the regeneration of the skin, with a high degree of its hydration appear very promising. Originality of the research was ensured as well by the biopolymer keratin as natural antisolar protection of skin. This knowledge is based on physiological presence of keratinocytes in leather and connected protective mechanism against the effects of the sun. The aim of this work was to develop liposome colloid systems based on biopolymers with a multifunctional effect and to obtain higher benefit of cosmetic preparations, for example, increased hydration, regeneration, protection against ultraviolet radiation, barrier protection of the skin, etc. and to ensure microbiologic stability of these systems by the application of colloid silver. Samples of biopolymers, dispersions, emulsions, and liposomes were evaluated by the determination of their basic qualitative parameters as viscosity, dry-content matter, size of particles, and stability. Hydrolyzates of keratin lower surface tension from the value $\gamma = 72.8$ mN/m to the value approx. $\gamma = 55.5$ mN/m, and therefore, research was oriented to the possibility of lower dosing of synthetic emulsifier at keeping of required stability of hydrogels and hydrocreams. Results of testing samples confirmed increased hydration of the skin and protection mainly against UVB radiation. Evaluation of disinfectant efficiency of prepared colloidal silver confirmed the bactericidal, fungicidal, and sporicidal effect against a broad spectrum of bacteria, fungi, and microbes.[16]

Collagen is the organic matter, which is included among proteins–amines, and it is a characteristic compound of animal body. The most used sources of collagen from the leather tanning industry are nontanned (chemically not cross-linked) waste with following utilization in industries:

- food (collagen packs for food, food gelatin, pharmacy, cosmetics),
- adhesives (classical, included special properties—increasing of the joint elasticity, additives for lowering of formaldehyde emission, ensuring the stability of UF adhesives against humidity—in all cases after relevant modification), and
- agriculture (biodegradable foils, activator of growth, surface-active matters, filling preparations for different utilization).

Another source of collagen is chromium-tanned leather waste, which is the risk factor for environment due to presence of Cr^{3+}, which is washed from waste, for example, by acid rains at dumping. At waste-water treatment, Cr^{3+} is oxidized to Cr^{6+}, which is carcinogenic. Therefore, there is an effort to use waste as secondary raw material. Waste contains collagen fibrils and large amount of amino groups reactive with formaldehyde. This research was aimed at determining the possible usefulness (ecologic, efficiency, and economic) of collagen material, after dechroming, to modify polycondensation adhesives for application in the wood-working industry. Polycondensation adhesives based on UF and phenol–formaldehyde were modified with protein hydrolyzates. It is known that collagen hydrolyzate of chrome-tanned leather waste added to dimethylol urea clearly limits formation of relatively unstable oxy-methylene bridges that may be regarded as potential source of formaldehyde emissions. The aim of this study was to verify the change of glue joint properties in following ways: influence of collagen hydrolyzate on shear strength properties of glued joints under action of water and high humid conditions; increasing the water-resistance of glued materials, development of adhesive mixtures suitable for gluing of wood at higher moisture content, influence of collagen hydrolyzate on lowering the formaldehyde emissions from wood-based panels. The intended sources of protein hydrolyzates were solid waste from leather production (e.g., chromium shavings and chippings, hypodermic and adipose ligament, leather chippings, gelatin production, and food packing). These various analytical investigations confirmed significant reduction of formaldehyde emission from wood-based panels, increased water-resistance of glued materials, and conforming shear strength properties of glued joints. Application of collagen from dechromed waste allows improvement of ecologic and economic parameters of bonding processes in the woodworking industry (e.g., eliminating of harmful effects on the environment and lowering of costs at keeping required quality).[23]

Just the fact that polycondensation (formaldehyde) adhesives contain certain amount of free formaldehyde in the liquid state, and proteins contain amine groups, which are able to bond formaldehyde, this is the principle of their utilization for their modification.

The aim of the research is to decrease the formaldehyde content in UF adhesives by application of collagen colloid, hardener, and additives. In a laboratory and industrial conditions, the technology of collagen preparation and their modifications were optimized. The work describes possibility to lower formaldehyde emission from wood products glued with UF adhesives at keeping of required strength of glued joints.

17.2 EXPERIMENTAL PART

17.2.1 UREA–FORMALDEHYDE RESIN

UF adhesives are prepared by the condensation of urea and formaldehyde. The chemical reaction and properties of the adhesives obtained depend upon these main factors: the molar ratio of the initial components (influences the reactivity of the adhesive as well as the content of free formaldehyde), temperature, concentration, possibly a catalyst or inhibitor such as hydrogen or hydroxyl ions (an alkaline environment supports the addition of fd and urea which results in the creation of methylol compounds, whereas an acidic environment supports condensation and the creation of resin-like products). In the first phase, in a neutral or weak alkaline environment with a molar ratio of initial compounds of 1:1, monomethylol urea is created:

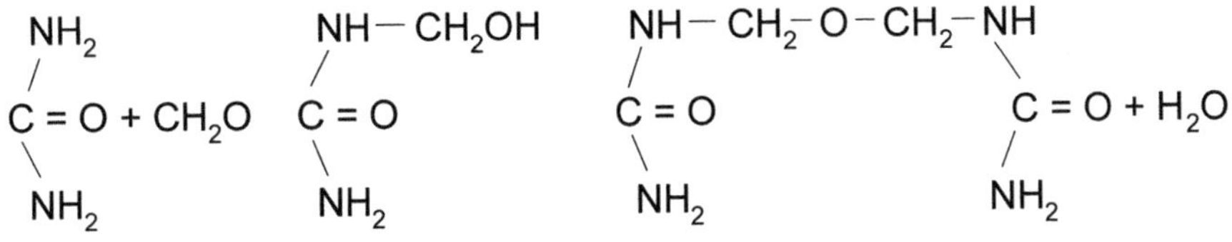

$$\begin{array}{l} NH_2 \\ / \\ C=O + CH_2O \\ \backslash \\ NH_2 \end{array} \quad \begin{array}{l} NH-CH_2OH \\ / \\ C=O \\ \backslash \\ NH_2 \end{array} \quad \begin{array}{lcr} NH-CH_2-O-CH_2-NH & & \\ / & & \backslash \\ C=O & & C=O + H_2O \\ \backslash & & / \\ NH_2 & & NH_2 \end{array}$$

monomethylol urea dimethylene ether bond

Cured resins contain methylol groups which had no reaction; it is assumed that these are groups which are not able to react since they are spatially trapped in a cross-linked macromolecule. Together with dimethylene ether bonds, these trapped methylol groups worse the properties of cured UF resins (they mainly decrease water resistance) and also cause the release of fd.

Curing agents for UF adhesives. UF adhesives cannot be used in the form in which they are supplied. A curing agent, that is, a compound which reduces the pH value, must be added. The optimum pH value for curing is 3–3.5. Generally, curing agents for UF adhesives are salts of strong acids and weak bases. At present, the most used curing agents are

- water solution of NH_4NO_3, modified with HCOOH or H_3PO_4 to pH 4–5,
- powder based on $(NH_4)_2SO_4$ with the addition of acidic compounds so the resulting pH will be between 4.1 and 5.0.

Catalytic influence of wood on the hardening behavior of UF adhesives used for wood-based panels has little or no effect on the curing progress of the resins.[24]

Fillers and extenders for UF adhesives. The cured film of UF adhesive in a glued gap should not be thicker than 0.1 mm. With increased thickness, it becomes more brittle, it ages more quickly and its mechanical strength decreases. After the adhesive is cured, the volume of adhesive film is reduced—film shrinks. The filler should not change the pH of an adhesive mixture. The fragility and quick aging of film with a greater thickness, as well as the contraction in volume during curing, are the main reasons for using fillers. The principle of the action of fillers is in dividing the thicker film of an adhesive into individual thin layers between particles of the filler. This will result in distribution of the internal strength of the adhesive film. Extenders thicken the adhesive and increase the stability of a glued bond and reduce production costs by lowering the consumption of adhesives. Wheat flour (wheat, rye, corn), potato starch, blood albumin, and others are used. The most suitable extenders are substances with a particular adhesive ability. If they are used up to an amount of approximately 20%, they do not really have a great influence upon strength and water resistance of a bond.

17.2.2 PROTEIN COLLAGEN MODIFIER

The basic components of adhesives of animal origin are proteins in the form of colloid solutions which have adhesive properties. This research of modification of adhesives for wood-working industry is based on natural nontoxic, biologically degradable, and cheap protein biopolymers. Market offers large amount of biopolymers (e.g., collagen waste from food and leather productions), which can be used as modifiers of adhesives for woodworking industry.[1,13,26] Proteins of amino acids with peptide bond are the source of large amount of amino groups $-NH_2$, which are reactive with formaldehyde:

$$\text{protein}-NH_2 + CH_2O \longrightarrow \text{protein}-N=CH_2 + H_2O$$

$$\text{protein}-NH_2 + CH_2O \longrightarrow \text{protein}-NH-CH_2-OH$$

$$2\ \text{protein}-NH_2 + CH_2O \longrightarrow \text{protein}-NH-CH_2-NH-\text{protein} + H_2O$$

Fibril character of collagen presents similar analogy with cellulose fibers and its structure can be stabilized with chemical bond, for example, formaldehyde, GA, etc. Another advantage of biopolymers is their nontoxicity and biodegradation ability to basic structural elements. For adhesives, modification reactions of proteins have the significance. Proteins lose their original solubility by affecting of formaldehyde. This property is used for lowering formaldehyde emission from UF adhesives, increasing water resistance of leather glue.

17.2.3 PLYWOOD PREPARATION

Plywood is wood-based board material usually used for testing of polycondensation glue compositions according to European standards. The technological process of plywood production for determination of strength properties and content of free formaldehyde with utilization of collagen hydrolyzate was following:

- beech-wood (*Fagus sylvatica* L.) veneers with the thickness of 1.7 mm,
- moisture content of veneers 4–6%, and
- spread of adhesive mixture with gluing roller in amount of 153 resp. 135 g/m^2.

Plywood was prepared with three or five layers. Plywood pressing was carried out in electrically heated press FONTIJNE at the temperature of 125°C and specific pressure of 1.8 MPa during the pressing time 10 min. After pressing and conditioning, samples for testing of mechanical and physical properties were prepared. The content of formaldehyde was determined by perforator method according to EN 120.[6] Dry content matter content was determined gravimetrically according to EN 322.[7]

17.3 RESULTS AND DISCUSSION

The research of modification of adhesives is aimed on utilization of products, which are easy accessible and their application save the costs for resin production. Leather and food industry produces amount of different biopolymer waste, which pollutes the environment.[4,17,21] Protein waste arising at the processing of leather consist approx. of 55% carbon, 21% oxygen, 7% hydrogen, 17% nitrogen, and/or sulfur and phosphorus and their properties are conditioned by the chemical composition, amino acid sequence, molecular size, and dimensional structure. Besides peptide bond, in proteins there are very often covalent disulfide bond (disulfide bridge) and other covalent bonds, for example, ester.[3]

For the preparation of adhesives, they have special importance modifying reactions of proteins. By affecting of formaldehyde, proteins lose their original solubility. This property is used to reduce formaldehyde emission of UF adhesives, increase the resistance of leather glue against moisture, and also the increase resistance of albumin glues, and at the manufacture of artificial horn. Fibril character of collagen is analogic with cellulose fibers and its structure can be stabilized by chemical bond, for example, formaldehyde, GA, etc. Substantial advantage of biopolymers is their nontoxicity and ability of biodegradation into their constituent elements.[11,18,20] Langmaier et al.[12] in experiments used hydrolyzate of chromium waste from leather industry obtained by enzymatic hydrolysis. Nonisothermal thermogravimetric method was used at investigation of condensation reactions of dimethylolurea and its mixtures with different weight content of urea, hydrolyzate, or acid hardener.

GA is chemical matter, which is often tested for modification of hardeners; there is the assumption, which is completely cross linked into the structure of the adhesive. Maminski et al.[15] investigated melamine–urea–formaldehyde adhesive, they added GA into the hardener in form of 50% water solution. Shear strength of birch samples glued with modified adhesive was significantly higher in comparison with the reference sample. Also, there is a direct bond of GA with chemical compounds of wood, what significantly increase the strength of glued joint. The percent of fiber destruction was much higher, modified adhesive proved stronger interaction adhesive wood.

In the experimental research, commercial UF resin KRONORES CB 1639F and hardener ammonium nitrate (R-60) was used. Natural modifiers of UF resins were raw materials based on collagen prepared from waste of leather industry. For modification of collagen hydrolyzate urea, di-aldehyde, and glycerol were applied. The composition of collagen hydrolyzates is in Table 17.1.

TABLE 17.1 Composition of Collagen Hydrolyzate Modifiers.

Collagen No. 1—collagen hydrolyzate prepared from waste of leather industry
Collagen No. 2—collagen hydrolyzate modified with urea
Collagen No. 3—collagen hydrolyzate modified with urea and glutaraldehyde

Experimental research was aimed on testing the influence collagen hydrolyzate prepared from leather waste and its modifications on adhesive properties—mainly on lowering of fd emission, viscosity, surface tension, life-time, and strength of glued joint. Parameters of collagen colloid for application into UF resins are in Table 17.2.

TABLE 17.2 Physical and Chemical Parameters of Dry Collagen Colloid.

Size of particles	60 mesh or 0.25 mm
Bloom at concentration 6.67% and temperature 10°C	191
Viscosity at concentration 6.67% and temperature 60°C	2.6 mPa s
pH	5.2
Dry content matter	90.8%
Ash	<2%

Compositions of UF resin and collagen colloid mixtures applied in further experiments are described in Table 17.3.

TABLE 17.3 Compositions of UF Adhesive Mixtures.

Standard—KRONORES CB 1639F + 3% hardener R-60
Modification 1—KRONORES CB 1639F + 3% hardener R-60 + (2%, 5%, 8%, 10%) substitution of resin with collagen
Modification 2—KRONORES CB 1639F + 3% hardener R-60 + (2%, 5%, 8%, 10%) substitution of resin with collagen modified with urea
Modification 3—KRONORES CB 1639F + 3% hardener R-60 + (2%, 5%, 8%, 10%) substitution of resin with collagen modified with urea and glutaraldehyde

17.3.1 THE INFLUENCE OF COLLAGEN CONCENTRATION ON THE CHANGE OF SURFACE TENSION

Studies on surface tension were carried out by drop number method using Traube's stalagmometer technique. This stalagmometric method is one of the most common methods for measuring surface tension. Traube's stalagmometer is an instrument for measuring surface tension by determining the exact number of drops in a given quantity of a liquid. The drop-number method is based on the principle that a fixed volume of a liquid delivered is free falling from a capillary tube held vertically approximately proportional to the surface tension of liquid. Surface tension is the result of the difference between attractions of molecule of the substance on the other side of the interface. The surface tension of water is created by van der Waals forces. Obtained results of measurements of the influence of collagen concentration on the change of surface tension are described in Table 17.4.

TABLE 17.4 The Influence of Collagen Concentration on the Surface Tension and pH Value.

Collagen concentration (%)	Surface tension (γ = mN/m)	pH
0	72.80	7.4
0.25	68.21	7.2
0.5	66.88	6.8
1	65.50	6.3
2	63.53	6.1
3	60.97	5.5
5	60.19	5.3
8	56.42	5.2
10	55.55	5.2
20	54.40	5.2

The reference value of the surface tension without the collagen biopolymer is γ = 72.8 mN/m. Collagen hydrolyzates lower the surface tension from the value of reference sample down to the value of γ = 54.40 mN/m, whereas this value depends on biopolymer concentration.

17.3.2 THE INFLUENCE OF ADDITION OF MODIFIED COLLAGEN COLLOIDS ON VISCOSITY OF UF ADHESIVE MIXTURE

The influence of addition of modified collagen colloids in different concentrations on the viscosity of UF adhesive mixture is presented in Table 17.5.

TABLE 17.5 Effect of Modification of Collagen to Viscosity UV Adhesive Mixture.

Sample	Mod. 1 (mPa s)	Mod. 2 (mPa s)	Mod. 3 (mPa s)
Standard (0%)	450	450	450
Concentration 2%	480	480	550
Concentration 5%	520	550	600
Concentration 8%	580	620	750
Concentration 10%	650	750	980

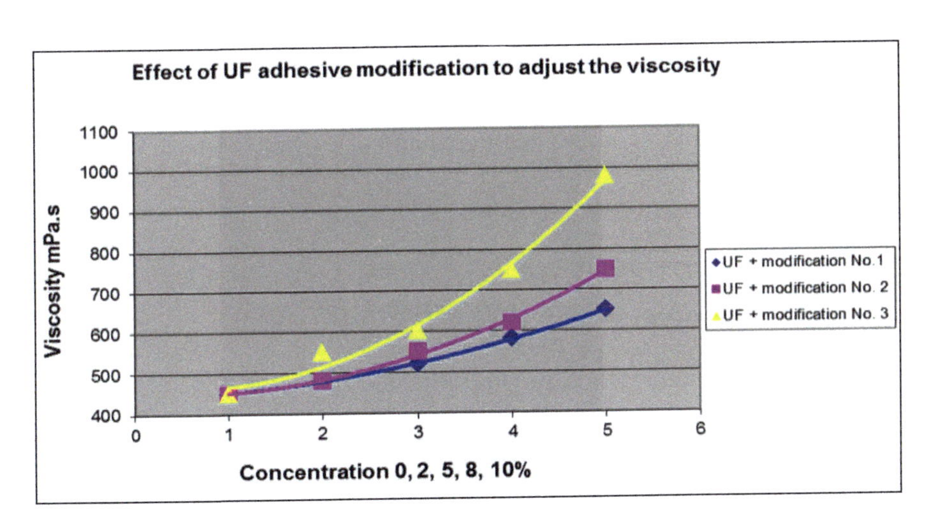

Modification 1 increases the viscosity of UF adhesive at 10% concentration up to 650 mPa s from the standard value of 450 mPa s. Modification 2 increases the viscosity of UF adhesive at 10% concentration up to 750 mPa s. Modification 3 most significantly increases the viscosity of UF adhesive at 10% concentration up to 980 mPa s.

Investigation the changes of viscosity of UF adhesive mixtures confirmed, that collagen jellies are suitable modifiers of UF adhesive viscosity.

17.3.3 THE INFLUENCE OF MODIFICATIONS OF COLLAGEN COLLOIDS ON THE LIFE-TIME OF UF ADHESIVE MIXTURE

The influence of addition of modified collagen colloids in different concentrations on the compositions life of UF adhesive mixture is presented in Table 17.6. The influence of collagen modification Nos. 1, 2, and 3 with the concentration (2%, 5%, 8%, 10%) on the life-time of UF adhesive mixtures was compared with the life-time of reference sample.

TABLE 17.6 Effect of Modification of Collagen to Life UV Adhesive Mixture.

Sample	Life-time of adhesive mixture at 20°C
Reference	>2 weeks
Modification 1	>2 weeks
Modification 2	>2 weeks
Modification 3	>1 week < 2 weeks

The life-time of adhesive mixtures—modification Nos. 1 and 2 are comparable with the reference sample of original UF resin. The modification No. 3—collagen modified with urea and GA—has shorter time of life of adhesive mixture.

17.3.4 THE INFLUENCE OF COLLAGEN MODIFICATIONS ON THE CURING TIME OF UF ADHESIVE MIXTURES

Results of the determination of the influence of the amount of collagen modification Nos. 1, 2, and 3 on the curing time of UF adhesive mixture are described in Table 17.7. Curing time was determined in a test tube at the temperature of 100°C.

TABLE 17.7 The Influence of Collagen Modification Nos. 1, 2, and 3 on Curing Time of UF Adhesive Mixtures.

Sample	Mod. 1 (s)	Mod. 2 (s)	Mod. 3 (s)
Standard (0%)	63.6	63.6	63.6
Concentration 2%	63.5	63.6	63.0
Concentration 5%	63.2	64.0	62.5
Concentration 8%	62.5	64.5	61.0
Concentration 10%	61.0	65.9	59.0

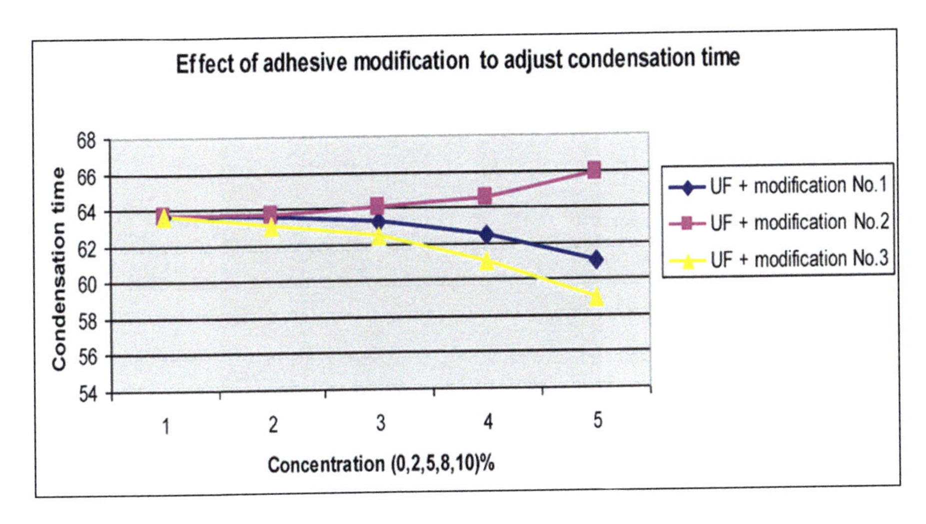

From the obtained results, follow that the curing time of UF adhesives is affected by the type of modification and the concentration of collagen modifiers in the adhesive mixture:

- urea slows the curing time and
- GA accelerates the curing time.

Modification 3—UF resin + 3% hardener R-60 + (2%, 5%, 8%, 10%) substitution of resin with collagen modified with urea and GA quickly accelerates the curing time from the value of 63.6–59.0 s.

17.3.5 THE INFLUENCE OF COLLAGEN MODIFICATION ON THE CONTENT OF FORMALDEHYDE IN HARDENED UF ADHESIVES

Results of the influence of UF resin modification on the content of free fd in 1 g hardened UF adhesive mixture are presented in Table 17.8.

TABLE 17.8 Formaldehyde Content in UF Hardened Modified Samples.

Sample	Mod. 1 fd (mg/g)	Mod. 2 fd (mg/g)	Mod. 3 fd (mg/g)
Standard (0%)	0.35	0.35	0.35
Concentration 2%	0.33	0.31	0.29
Concentration 5%	0.28	0.27	0.27
Concentration 8%	0.25	0.22	0.21
Concentration 10%	0.22	0.19	0.17

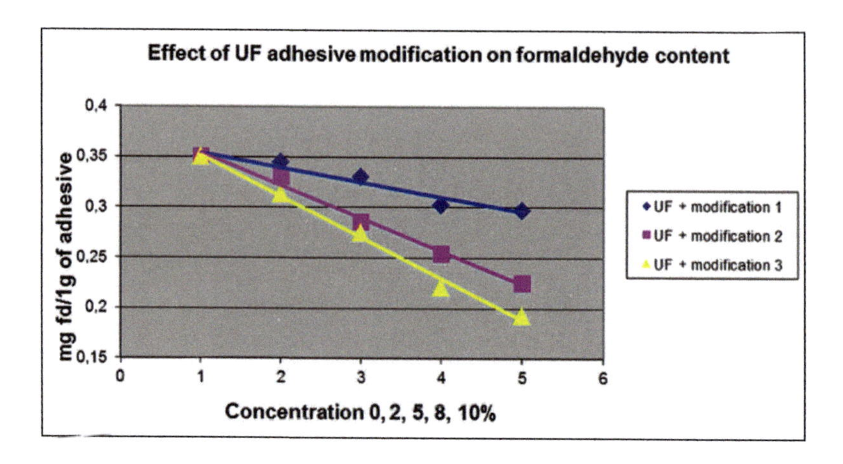

From obtained results, follow that collagen is suitable modifier for lowering of fd content in hardened UF adhesive mixtures. Collagen modified with urea and GA most significantly lowers fd content, and this effect is stronger with increasing concentration from the value of the reference sample 0.35 mg fd/g down to 0.17 mg fd/g of hardened adhesive.

17.3.6 THE INFLUENCE OF COLLAGEN MODIFICATIONS ON THE FORMALDEHYDE CONTENT IN PLYWOOD

The stability of amino-plastic thermoset adhesives is important hygienic parameter; therefore, the research effort is aimed on reduction and/or avoiding of formaldehyde release from glued material.[19,22] When curing adhesives, present acidic hardeners increase the rate of both types of cross-linking bonds of adhesive structure (nonstable dimethyl-ether and also more stable methylene bonds), but also influence the transformation of dimethyl-ether bonds to methylene. With lowering of amount of dimethyl-ether cross-link bonds in hardened adhesive film of amino plastic, we are able to reduce the emission of formaldehyde in hardened film.

The formaldehyde content in wood material was tested by the perforator method according to EN 120.[6] The addition of different concentrations of modified samples of collagen with reactive amino groups was tested on ecologic parameters of wood products. Obtained results confirmed the decrease of formaldehyde content in comparison with the reference sample. Results of measurements of the influence of modification and concentration of collagen colloids No. 1, 2, and 3 on the formaldehyde content in prepared plywood with UF adhesive mixtures are presented in Table 17.9.

TABLE 17.9 Effect of Collagen Colloid and its Modifications on fd Content in Plywood.

Sample	Fd content
Reference	3.57 mg fd/100g a.d. sample[a]
Modification 1	3.17 mg fd/100g a.d. sample
Modification 2	2.54 mg fd/100g a.d. sample
Modification 3	1.68 mg fd/100g a.d. sample

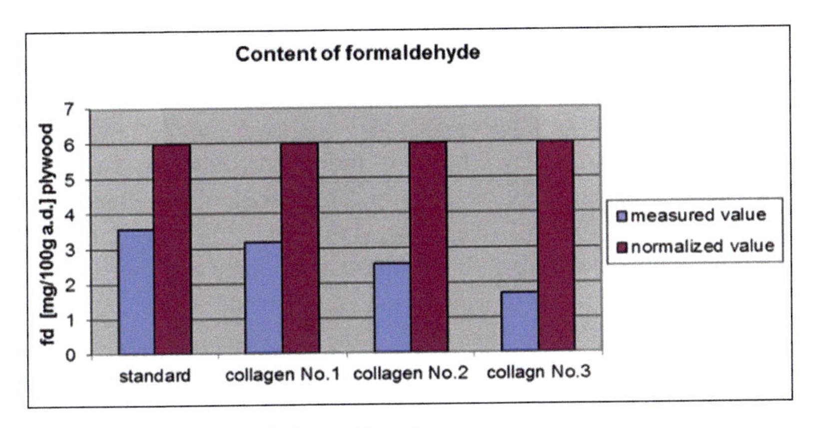

[a]100 g absolutely dry sample of plywood board.

Results of laboratory tests confirmed that collagen prepared from leather waste is suitable additive for lowering of formaldehyde emission from wood products glued with UF adhesive. Increased efficiency of collagen was obtained by modification with urea and di-aldehyde. Tests confirmed the decrease of formaldehyde content in comparison with the standard down to 50% according to perforator method.

17.3.7 THE INFLUENCE OF COLLAGEN MODIFICATIONS ON STRENGTH PROPERTIES OF PLYWOOD

The shear strength of glued joint directly depends on the resistance against humidity. Suitable modification of adhesive mixtures can reach better cross-linking of the structure of hardened adhesive, increase of durable chemical bonds and lowering of the hydrolysis of adhesive. The research aimed not only on the study of properties of wood and adhesives, but as glued products are also the subject exposed to the environment in which they are located, and also to study the interactions of UF wood adhesive system.[8,16,25] Water has an important role in wood aging process. In presence of water, the influence of radiation, oxidation, and heat is more intensive and more pronouncedly reflected in the wood surface degradation and in changes to its morphology. In case of water-free regimen, the wood surface is subject to chemical changes due to UV radiation and heat-induced effects, but there need not be major changes to the wood morphology.[9]

The influence of collagen modifications on strength properties of plywood was tested according the standards EN 314-1 and 314-2 (pretreatment for humid conditions: 6 h boiling in water, 1 h cooling in water 20°C). It was subsequently tested, if better wetting of the veneer surface of plywood bonded with modified UF adhesive mixture with collagen enables to reduce the amount of adhesive spread at keeping all required quality parameters. There were prepared reference and testing samples of three-layer plywood. The pressing process was carried out in the electrically heated press at the temperature of 125°C, pressing time 10 min, at the spread of reference UF adhesive mixture of 153 g/m^2, and at lowered spread of UF adhesive mixtures with collagen to 135 g/m^2, considering the better wetting of the veneer wood surface.[2,10] Obtained results of shear strength properties of plywood are presented in Table 17.10.

TABLE 17.10 Strength Properties of Plywood Prepared with Modified UF Adhesive Mixtures.

Sample	Mod. 1 (MPa)	Mod. 2 (MPa)	Mod. 3 (MPa)
Standard (0%)	2.55	2.55	2.55
Concentration 2%	2.50	2.40	2.65
Concentration 5%	2.35	2.24	2.81
Concentration 8%	2.30	2.17	2.84
Concentration 10%	2.24	2.06	2.95

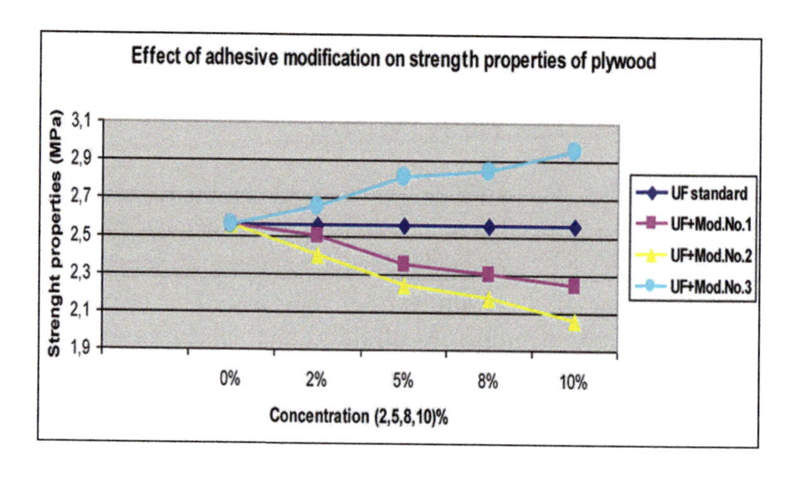

Testing of samples proved that collagen in UF adhesive mixtures ensured uniform wetting of the veneer surface even at about 10% lowered adhesive spread in comparison with the reference sample at parallel improving physical and mechanical properties of plywood. From obtained results, follow that those collagen modifications are suitable modifiers of UF adhesive mixtures. Modification Nos. 1 and 2 lower the strength of bonded joint in comparison with standard UF adhesive from the value of reference sample 2.55 MPa down to 2.24 resp. 2.06 MPa at their 10% concentration. On the other side, modification No. 3 increases the strength of the glued joint in comparison with the standard UF adhesive from the value of reference sample 2.55 MPa up to 2.95 MPa at 10% concentration.

Taking into account all obtained results, for the further testing of formaldehyde emission and shear strength of plywood, 5% concentrations of modified collagen colloids 1, 2, and 3 were applied with a spread of reference UF adhesive mixture 153 g/m^2 and a spread of 135 g/m^2 modified UF adhesive compositions with collagen.

17.3.7 THE INFLUENCE OF COLLAGEN MODIFICATIONS ON STRENGTH PROPERTIES AND HYGIENIC PROPERTIES OF PLYWOOD—INDUSTRIAL TESTING

Industrial experiments were done in the company for plywood production. Reference and modified samples of plywood were prepared for testing of qualitative parameters. The pressing process was carried out in the steam-heated press within the temperature range of 125–130°C, pressing time 10 min, at the spread of adhesive mixture with collagen 135 g/m^2. The aim of the experiment was to obtain required shear strength of plywood for classification for intended use in humid conditions (6 h boiling, 1 h cooling in water 20°C). Prepared five-layer plywood was tested in an industrial company laboratory and in a laboratory at TU in Zvolen. Obtained results shear strength of plywood from industrial company laboratory are presented in Table 17.11.

Sampling:

Standard, industrial production—reference sample,
No. 1—5% concentration of collagen modification 1,
No. 2—5% concentration of collagen modification 2, and
No. 3—5% concentration of collagen modification 3.

TABLE 17.11 Shear Strength of Plywood—Results of Industrial Company Laboratory.

Sample	Standard value (MPa)	x (MPa)	min (MPa)	max (MPa)
Standard	1	1.91	1.32	2.26
Mod. 1	1	1.98	1.33	2.31
Mod. 2	1	1.52	1.26	1.91
Mod. 3	1	2.15	1.52	2.46

Obtained results of shear strength of plywood from the university laboratory are presented in Table 17.12.

TABLE 17.12 Shear Strength of Plywood—Results of TU in Zvolen.

Classification for humid conditions (6 h boiling, 1 h cooling in water 20°C)				
Sample	**Standard**	**Mod. 1**	**Mod. 2**	**Mod. 3**
x (MPa)	2.38	2.42	2.25	2.83
s_{dev} (MPa)	0.38	0.37	0.43	0.32
v_k (%)	14.5	15.4	15.3	10.9

Considering the obtained results of plywood strength properties, it can be stated that collagen and its modifications are suitable modifiers of UV adhesive compositions. Modifications 1 and 3 at 5% concentration increased the bond strength in comparison with the standard UF adhesive from the value of reference sample 2.38 MPa up to values (2.42 resp. 2.83) MPa.

The addition of modified samples of collagen with reactive amino groups into UF resin was tested on ecologic parameters of wood products. The formaldehyde content in bonded wood material was tested by the perforator method according to EN 120.[6] Results of measurements of the influence of modification and concentration of collagen colloid Nos. 1, 2, and 3 on the formaldehyde content in prepared plywood with UF adhesive mixtures are presented in Table 17.13.

Obtained results of the formaldehyde content in tested plywood were counted from the calibration line presented in Table 17.14.

Results of laboratory and industrial tests confirmed that collagen is suitable modifier for lowering of formaldehyde emission from hardened UF adhesive mixtures. The efficiency of collagen is possible to increase with modification by urea and di-aldehyde. Industrial tests at plywood production confirmed lowering of formaldehyde emission in comparison with standard production down to 50%, stated according to EN 120.[6] Obtained results confirmed the decrease of formaldehyde content in comparison with

TABLE 17.13 Analysis of fd Content from Industrial Test—Five-Layer Plywood.

Sample	E_{xt} (412 nm)	Content fd (mg/1000 ml)	Dry content matter (%)	Weighed moist. (g)	Weighed dry (g)	mg fd/100 g abs. dry board
Standard KRONORES CB 1639F	0.0342	3.8873	94.31	102.9117	97.056	**4.01**
	0.0341					
1	0.0288	3.2680	94.27	102.6993	96.815	**3.37**
	0.0287					
2	0.0251	2.8986	94.96	102.9551	97.766	**2.96**
	0.0259					
3	0.0219	2.4980	94.36	102.804	97.006	**2.58**
	0.0218					

the reference sample from the value of 4.01 mg fd/100 g a.d. board down to the value of 2.58 mg fd/100 g a.d. board at collagen modification No. 3.

TABLE 17.14 Calibration Line of Formaldehyde.

[μg fd] in 10 ml of solution stated with acetyl acetone	E_{xt} (412 nm)
0	0.0000
1.5	0.0128
3	0.0278
6	0.0546
15	0.1346
30	0.2632

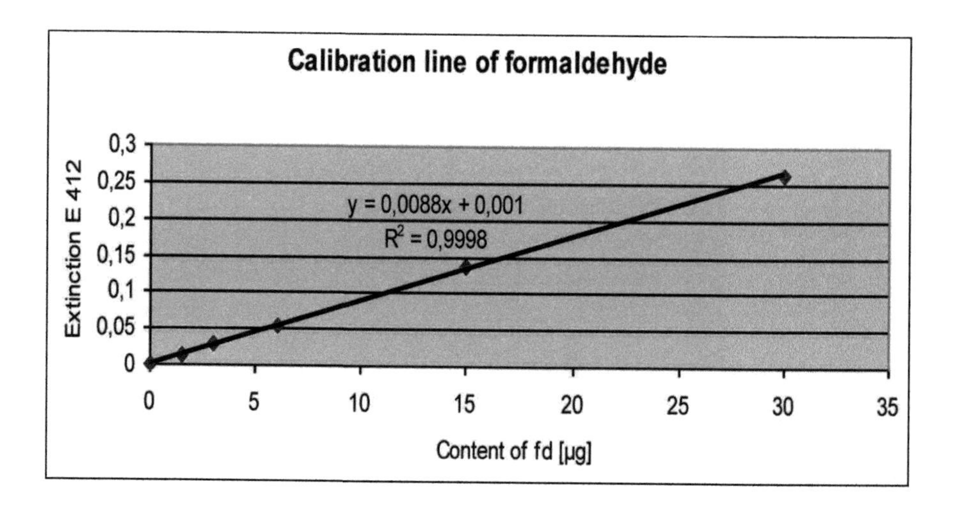

17.4 CONCLUSION

Chemical stability of adhesives is an important property from the point of view of the strength and durability of glued joint and also from the point of view of hygienic requirements regarding the environment where the glued product is used. Modifying adhesives was tried to improve qualitative properties of the adhesives and adhesive joints as well. Three different collagen-based additives of tanned leather waste were added into the UF resin. Physical and chemical properties of the adhesive mixtures were evaluated: shear strength of glued joints, formaldehyde content, and emission. Reference samples were prepared using regular UF adhesives and glued joints. Attained results were evaluated according to the particular technical requirements or standards. Obtained results showed that simultaneous presence of collagen in the UF adhesive mixture improved adhesive properties, increased shear strength of glued joints, and reduced the amount and emission of formaldehyde in laboratory and industrial conditions as well. It was shown that tanned leather waste (hydrolyzed skin collagen) can be used as secondary industrial raw material in UF adhesive production. The partial substitute of the regular adhesive by hydrolyzed collagen results in improved adhesive quality and the quality of glued joints as well. Waste utilization is the positive contribution to the environment.

ACKNOWLEDGMENT

This work was supported by the Slovak Research and Development Agency under the contracts No. APVV-14-0506 "ENPROMO" and APVV-15-0235 "MODSURF."

KEYWORDS

- **viscosity**
- **curing time**
- **shear strength**
- **plywood**
- **resin**

REFERENCES

1. Belbachir, K.; Noreen, R.; Gouspillou, G. Collagen Types Analysis and Differentiation by FTIR Spectroscopy. *Anal. Bioanal. Chem.* **2009,** *395*, 829–837.
2. Bekhta, P.; Proszyk, S.; Krystofiak, T.; Lis, B. Surface Wettability of Short-Term Thermo-mechanically Densified Wood Veneers. *Eur. J. Wood Wood Prod.* **2015,** *73* (3), 415–417.
3. Bryan, M. A.; Brauner, J. W.; Anderle, G.; Flach, C. R.; Brodsky, B.; Mendelsohn, R. FTIR Studies of Collagen Model Peptides: Complementary Experimental and Simulation Approaches to Conformation and Unfolding. *J. Am. Chem. Soc.* **2007,** *129* (25), 7877–7884.
4. Buljan, J.; Reich, G.; Ludvik, J. Mass Balance in Leather Processing. In *Proceedings of the Centenary Congress of the IULCS*, London, 1997; pp 138–156.

5. Dziurka, D.; Mirski, R. Properties of Liquid and Polycondensed UF Resin Modified with pMDI. *Drvna Ind.* **2014,** *65* (2), 115–119.
6. EN 120:1995: Wood Based Panels. Determination of Formaldehyde Content. Extraction Method Called the Perforator Method, 1995.
7. EN 322:1995: Wood-Based Panels. Determination of Moisture Content, 1995.
8. Essawy, H. A.; Moustafa, A. A. B.; Elsayed, N. H. Improving the Performance of Urea–Formaldehyde Wood Adhesive System Using Dendritic Poly(amidoamine)s and their Corresponding Half Generations. *J. Appl. Polym. Sci.* **2009,** *114*, 1348–1355.
9. Kúdela, J.; Ihracký, P. Influence of Diverse Conditions during Accelerated Ageing of Beech Wood on its Surface Roughness. *Acta Facult. Xylolog.* **2014,** *56* (2), 37–46.
10. Kúdela, J.; Wesserle, F.; Bakša, J. Influence of Moisture Content of Beech Wood on Wetting and Surface Free Energy. *Acta Facult. Xylolog.* **2015,** *57* (1), 25–35.
11. Langmaier, F.; Šivarová, J.; Kolomazník, K.; Mládek, M. Curing of Urea–Formaldehyde Adhesives with Collagen Type Hydrolysates under Acid Condition. *J. Therm. Anal. Calorimetry* **2004,** *76*, 1015–1023.
12. Langmaier, F.; Kolomazník, K.; Mládek, M.; Šivarová, J. Curing Urea–Formaldehyde Adhesives with Hydrolysates of Chrome-Tanned Leather Waste from Leather Production. *Int. J. Adhes. Adhes.* **2005,** *25*, 101–108.
13. Mai, C.; Kües, U.; Militz, H. Biotechnology in the Wood Industry. *Appl. Microbiol. Biotechnol.* **2004,** *63*, 477–494.
14. Mamiński, M. Ł.; Król, M. E.; Grabowska, M.; Głuszyński, P. Simple Urea–Glutaraldehyde Mix Used as a Formaldehyde-Free Adhesive: Effect of Blending with Nano-Al_2O_3. *Eur. J. Wood Wood Prod.* **2011,** *69* (3), 505–506.
15. Maminski, M.; Pawlicki, J.; Parzuchowski, P. Improved Water Resistance and Adhesive Performance of a Commercial UF Resin Blended with Glutaraldehyde. *J. Adhes.* **2006,** *82*, 629–641.
16. Matyašovský, J.; Sedliačik, J.; Matyašovský, Jr., J.; Jurkovič, P.; Duchovič, P. Collagen and Keratin Colloid Systems with a Multifunctional Effect for Cosmetic and Technical Applications. *J. Am. Leather Chem. Assoc.* **2014,** *109* (9), 284–295.
17. Matyašovský, J.; Sedliačik, J.; Jurkovič, P.; Kopný, J.; Duchovič, P. De-chroming of Chromium Shavings without Oxidation to Hazardous Cr^{6+}. *J. Am. Leather Chem. Assoc.* **2011,** *106*, 8–17.
18. Novák, I.; Popelka, A.; Luyt, A. S.; Chehimi, M. M.; Špírková, M.; Janigová, I.; Kleinová, A.; Stopka, P.; Šlouf, M.; Vanko, V.; Chodák, I. Valentin, M. Adhesive Properties of Polyester Treated by Cold Plasma in Oxygen and Nitrogen Atmospheres. *Surf. Coat. Technol.* **2013,** *235*, 407–416.
19. Peng, Y.; Shi, S. Q.; Ingram, L. Chemical Emissions from Adhesive-Bonded Wood Products at Elevated Temperatures. *Wood Sci. Technol.* **2011,** *45* (4), 627–644.
20. Pizzi, A. Tannery Row—The Story of Some Natural and Synthetic Wood Adhesives. *Wood Sci. Technol.* **2000,** *34*, 277–316.
21. Pünterer, A. The Ecological Challenge of Producing Leather. *J. Am. Leather Chem. Assoc.* **1995,** *90*, 206–215.
22. Roffael, E.; Johnsson, B.; Engrström, B. On the Measurement of Formaldehyde Release from Low-Emission Wood-Based Panels Using the Perforator Method. *Wood Sci. Technol.* **2010,** *44*, 369–377.
23. Sedliačik, J.; Matyašovský, J.; Šmidriaková, M.; Sedliačiková, M.; Jurkovič, P. Application of Collagen Colloid from Chrome Shavings for Innovative Polycondensation Adhesives. *J. Am. Leather Chem. Assoc.* **2011,** *106* (11), 332–340.

24. Stefke, B.; Dunky, M. Catalytic Influence of Wood on the Hardening Behaviour of Formaldehyde-Based Resin Adhesives Used for Wood-Based Panels. *J. Adhes. Sci. Technol.* **2006,** *20* (8), 761–785.
25. Šmidriaková, M.; Sedliačik, J.; Matyašovský, J. Natural Polymers Based on Modified Collagen as Partial Substitution of UF Adhesive. In *Adhesives in Woodworking Industry*; Zvolen, 2011; pp 14–20.
26. Taylor, M. M.; Bumanlag, L. P.; Brown, E. M.; Liu, C. K. Biopolymers Produced from Gelatin and Chitosan Using Polyphenols. *J. Am. Leather Chem. Assoc.* **2015,** *110* (12), 392–400.

CHAPTER 18

A RESEARCH NOTE ON POLYMERIZATION OF 2-HYDROXYETHYL METHACRYLATE INITIATED WITH VANADYL IONIC COMPLEX

SVETLANA N. KHOLUISKAYA[1*], VADIM V. MININ[2], and ALEXEI A. GRIDNEV[1]

[1]*N. N. Semenov Institute of Chemical Physics, Russian Academy of Sciences, Moskva, Russia*

[2]*N. S. Kurnakov Institute of General and Inorganic Chemistry, Russian Academy of Sciences, Moskva, Russia*

**Corresponding author. E-mail: soho@chph.ras.ru*

CONTENTS

ABSTRACT

Poly(2-hydroxyethyl) methacrylate itself and copolymers have become the important class of biomaterials due to their biocompatibility, hydrophilicity, and nontoxicity. 2-Hydroxyethyl methacrylate (HEMA) can be used for making unique materials, like hyperbranched, stereogradient, and stimuli-responsive polymers. HEMA polymerizes easily by free-radical polymerization, but other types of polymerization can, also, be applied, for example, anionic polymerization.

18.1 INTRODUCTION

Living radical polymerization of 2-hydroxyethyl methacrylate (HEMA) using atom transfer radical polymerization (ATRP) technique was first reported in 1999.[1–9] CuX-based ATRP can be used for living polymerization of HEMA in polar solvents, like methanol, DMSO, acetonitrile, and in the presence of water.[10–12] More recently, poly(2-hydroxyethyl) methacrylate (PHEMA) was prepared by activator-generated by electron transfer ATRP (AGET ATRP)[13] and activator regenerated by electron transfer ATRP (ARGET ATRP).[14] Single-electron transfer living radical polymerization (SET-LRP) was proved to be applicable for synthesis of PHEMA[15]. Thus, because of PHEMA usefulness, intensive development of new techniques for its synthesis continues. We now report the first vanadium-catalyzed synthesis of the polymer of HEMA induced by $VO(DMSO)_5(ClO_4)_2$ (**I**).

18.2 METHOD

Vanadium compound **I** can react with alcohols in two major ways. First, it can promote reduction of dioxygen in the presence of alcohols with formation of free radicals.[16,17] Second, we found compound **I** to be an affective catalyst of Michael addition of alcohols to epoxides and activated double bonds[18]. Thus, **I** catalyzes addition of methanol to double bond of cyclohex-2-en-1-one with formation of ketone **II** (18.1) in 76% yield.

$$\text{cyclohex-2-en-1-one} + CH_3OH \xrightarrow[\text{Ar, } 50^{\circ}C]{(V{=}O)^{2+}} \underset{\mathbf{II}}{\text{3-}(H_3CO)\text{-cyclohexan-1-one}} \quad (18.1)$$

Ability of **I** to catalyze both of these reactions may lead to conclusion that compound **I** can have unusual behavior in polymerization of vinyl monomers bearing hydroxyl groups. Indeed, we found that addition of **I** to HEMA results in relatively fast polymerization of HEMA at ambient temperatures, while methyl methacrylate (MMA), *N*-vinylpyrrolidone or acrylic acid in alcoholic solutions do not polymerize under the same conditions. Remarkably, polymerization of HEMA was conducted in air. Replacement of air with argon does not have much effect on polymerization of HEMA induced by **I**.

Due to its ionic nature, compound **I** is very well soluble in polar media including HEMA. Because of that compound **I** can be used in bulk polymerization of HEMA as well as in polymerization of water or alcoholic solutions of HEMA. In the last case, **I** induces polymerization of HEMA at similar rates as in the bulk HEMA. Low MW alcohols (methanol, ethanol, isopropanol) are preferred solvents as water at concentration over ~30 vol% cases phase separation during polymerization of HEMA.

In bulk polymerization of HEMA concentration of **I** were used in the 0.2–5.0 mM range, or <0.01 wt%. At 1.0 mM concentration of **I** polymerization of HEMA finishes in 22 h at ambient temperatures and 4 h at 50°C to yield transparent, solid material insoluble in DMF even upon continuous heating at 80°C. Apparently, a cross-linked polymer was obtained. The polymerization in both air and Ar have similar rates of polymerization. Formation of cross-linked polymer in bulk HEMA polymerization is a common feature of this monomer. Impurities of dimethacrylate in HEMA are believed to be responsible for insolubility of high MW PHEMA so that no unambiguous conclusion in regard of mechanism of HEMA polymerization induced by **I** can be drawn from the fact of cross-linked polymer formation.

Comparison of IR spectra of the reaction mixture before and after polymerization showed disappearance of C=C bands at $\nu = 1639$ and 817 cm^{-1} (Fig. 18.1). Polymerization both in air and in Ar gave PHEMA with the same IR spectrum.

Michael addition of OH-group of hydroxyethyl substituents of HEMA to double bond of methacrylate in case of HEMA with formation of a polyester **III** cannot be excluded in the light of our previous experiments with compound **I** (18.1). Polyester **III** was obtained repeatedly by other authors in anionic polymerization of HEMA[8,19,20] as the main polymeric product. If HEMA polymerization induced by **I** proceeds through addition of OH groups to C–C double bonds then absorption of OH in IR spectrum of such PHEMA have to be reduced. However, hydroxyl bands in 3200–3700 region remain strong.

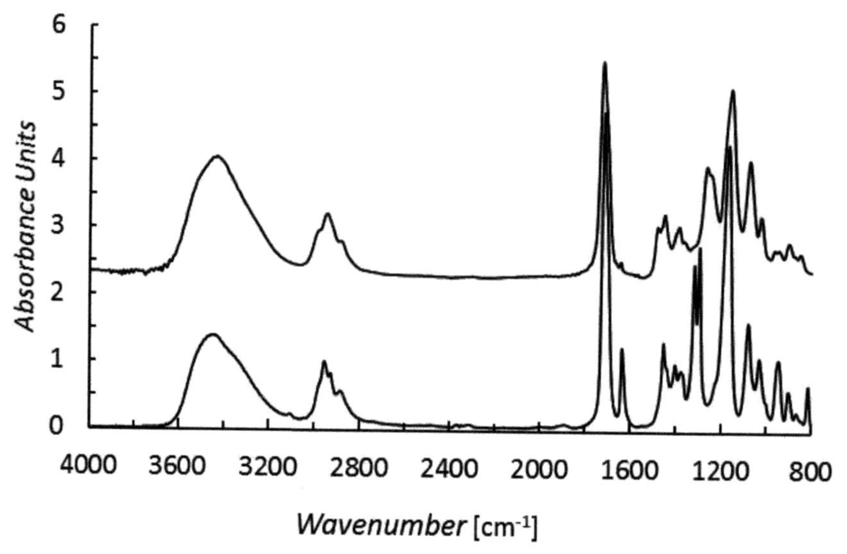

FIGURE 18.1 IR spectra of monomeric HEMA (bottom spectrum) and PHEMA (upper spectrum). Bulk polymerization, [**I**] = 1.0 mM, 22°C, 22 h in air.

III

$$CH_2{=}C(CH_3)-C(=O)-O-CH_2-CH_2-O-\left(CH_2-C(CH_3)-C(=O)-O-CH_2-CH_2-O\right)_n-H$$

Two samples of PHEMA were analyzed by nuclear magnetic resonance (NMR) (Fig. 18.2). First sample was obtained by bulk polymerization in air, whereas the second sample was polymerized in vacuum. ^{13}C NMR spectra of the both samples recorded using the magic angle with cross-polarization of PHEMA are essentially the same. The only found difference were additional minor signals at δ = 32.8 and 78.5 in the ^{13}C NMR spectrum of PHEMA obtained in vacuum.

^{13}C NMR spectra of PHEMA polymerized in air contains broad signals at δ = 16.3; 45.2; 55.2; 60.2, 67.3, and 178.6 ppm, attributed to polymeric methyl group (**a**), quaternary carbon atom (**f**), methylene group (**b**) in the polymer backbone, carbon atoms of ethylene glycol moiety (**c**) and (**d**), and carbonyls (**e**), correspondingly (Fig. 18.2). This data coincides with literature ^{13}C spectra of PHEMA obtained by free-radical polymerization.[20] Polymerization in vacuum or argon provides two additional signals formed at δ = 32.8 (**g**) and 78.5 (**h**) ppm (trace **b** in Fig. 18.2). These signals were

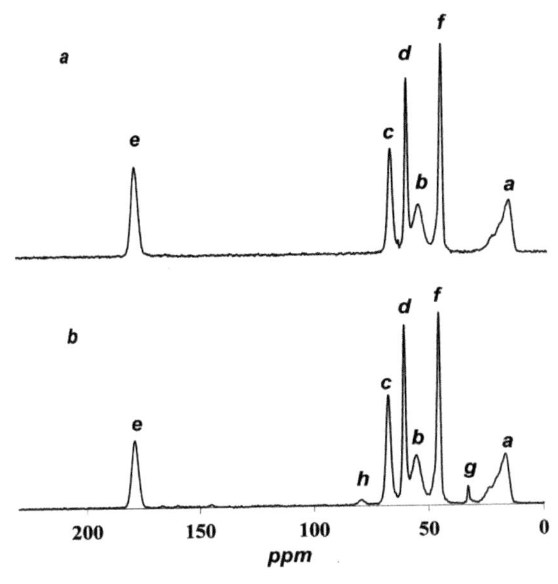

FIGURE 18.2 ^{13}C NMR spectra of PHEMA polymerized in air (a) and in vacuum (b). Bulk PHEMA, [**I**] = 1.0 mM, 50°C, 4 h.

attributed to –$CCH_2\mathbf{C}H(CH_3)$– and –$CO_2CH_2CH_2O\mathbf{C}H_2$– fragments in accord with.[2] Combined integral of (**g**) and (**h**), ^{13}C was found to be about 2% of all ^{13}C. Hence, molecular structure of PHEMA obtained by polymerization in air would be represented by structure **IV**.

IV

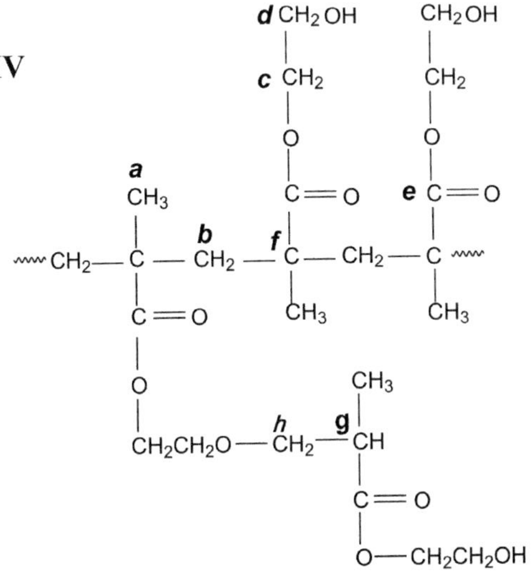

Thus, NMR data confirm IR data that C–C bonds are the major structural element of polymer backbone obtained by HEMA polymerization induced by **I**. Relative amount of polyether fragment is negligible.

In general, methacrylates can be polymerized by free-radical, anionic or coordination mechanism. Cationic polymerization of methacylates is not known. Radical polymerization of HEMA in the presence of **I** is quite possible taking into account our previous findings of formation of free radicals by oxidation of hydrocarbons by **I**. In the literature, free radical copolymerization of MMA and styrene was reported[21] in emulsion initiated by species formed by interaction of $CpVCl_2$ with dioxygen. However, as we mentioned before, neither MMA, nor other vinyl monomers, do not polymerize in the presence of **I**, both in air and argon.

The major method to prove mechanism of polymerization is to add into polymerization mixture species that reacts with propagation center. Most common inhibitors of specific polymerization are used. Unfortunately, inhibitors of radical polymerization, such as nitroxides and hydroquinone, react with **I** so that "inhibitor" approach cannot be applied to elucidate the origin of propagation center of HEMA polymerization induced by compound **I**.

In radical polymerization of methacrylates, cobaloximes are known to catalyze chain transfer to monomer.[22] The catalysis is very effective so that millimolar concentrations of cobaloximes is enough to reduce MW of polymethacrylates to few hundred. Hence, if **I** induces free-radical polymerization of HEMA addition of cobaloximes should lead to substantial reduction of MW of PHEMA. In radical polymerization of HEMA, such concentrations of cobaloximes lead to formation of semiliquid oligomers. Our experiments showed that cobaloximes **V** and **VI** at concentrations 0.3–0.6 mM do not effect polymerization of HEMA induced by **I** (1.0 mM). Rate of polymerization changeless than by 10%, whereas the final product remain the same—solid, glassy, cross-linked polymer.

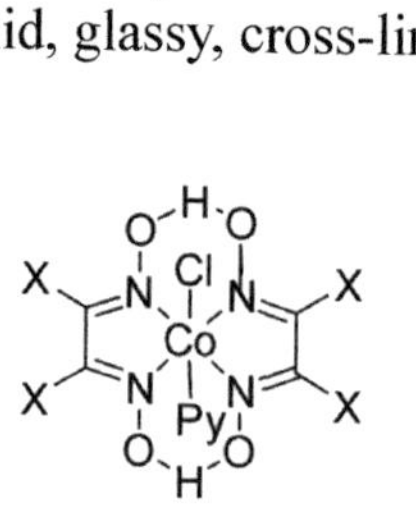

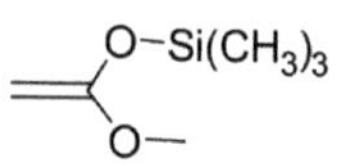

V (X=CH_3) **VII**

VI (X=Ph)

Hence, polymerization of HEMA initiated by **I** proceeds neither by anionic nor by free-radical mechanism. We suggest coordination polymerization as the only mechanism that fits all the experimental data. Since MMA does not polymerize by **I**, we conclude hydroxyl group plays an important role in this coordination mechanism. Coordination polymerization of methacrilates called group-transfer polymerization (GTP) was first conducted by Webster et al.[23] They found that monomeric methacrylates insert into O–Si bond in enolate **VII** along with formation of C–C bond at technologically convenient temperatures (~80°C) so that Si-enolate group keeps at the growing end of polymethacrylate during the polymerization at all the times. Consequent addition of different methacrylates into the GTP reaction mixture leads to the formation of block copolymers.[24]

Although details of GTP mechanism are still under investigation, it is clear that enolates of methacrylates are capable to propogate polymerization of methacrylates. Since ethanol solution of MMA does not polymerize in the presence of compound **I**, one may conclude that OH group of HEMA helps to form some kind of a coordination complex of HEMA with **I** that promotes polymerization. We suggest that formation of complex like **VIII** may the first stage of HEMA polymerization in the presence of **I**. According to our suggestion vanadium–HEMA complex, **VIII** activates HEMA through formation of an enolate. In **VIII**, HEMA completely replaces oxygen atom in the original V=O so that no change in the oxidation state of vanadium is required in formation of the enolate. Indeed, EPR of polymerization system with 5% accuracy showed no change of vanadium oxidation state **(IV)** during the polymerization of HEMA.

If another HEMA molecule approaches **VIII**, a new coordination complex **IX** may form using additional coordination places of vanadium. Newly formed chelate **IX** keeps enolate double bond close to methacrylate double bond of the second HEMA molecule that facilitates addition reaction of the enolate to the second coordinated HEMA and a new cyclic enolate **X** emerges (18.2). Thus, reaction (18.2) proceeds with formation of PHEMA with cyclic alkoxy–vanadium–enolate group at the growing end of the polymeric molecule. Group X in the reaction mechanism (18.2) could be OH. Additional experiments are being undertaken to elucidate chemical origin of the X-group as well as to prove mechanism (18.2) per se.

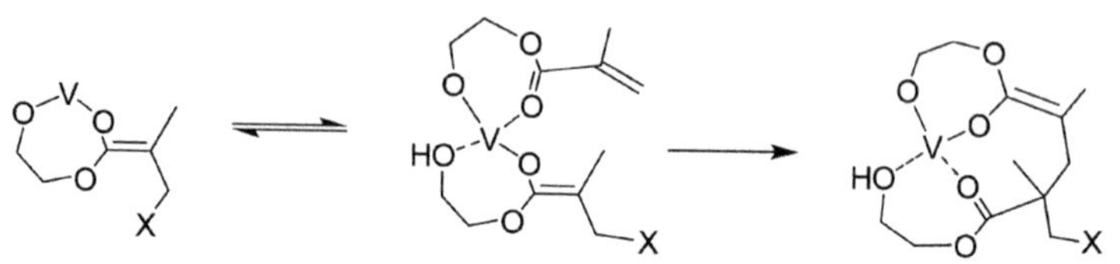

VII IX X

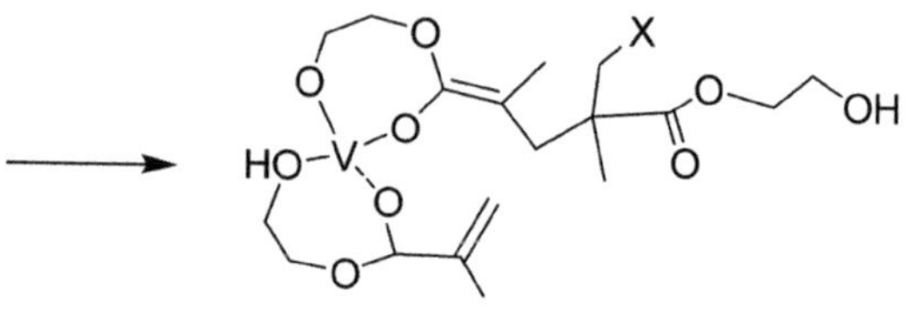

(18.2)

Details of HEMA polymerization induced by **I** and other vanadium complexes will be published elsewhere.

18.3 CONCLUSION

To remove inhibitor, HEMA was passed through column with activated alumina. Ninety-six percent ethanol was redistilled twice. All monomers (Aldrich) were purified by high vacuum distillation immediately prior the experiment. $VO(DMSO)_5(ClO_4)_2$ was prepared according to the reported literature procedure.[25] Reaction mixtures were prepared in air, followed by degassation by three freeze–evacuate–thaw cycles in a vial sealed with rubber septum. After the final thawing, the vial was optionally filled with argon. No difference was found between polymerization in argon or vacuum. In case of polymerization conducted in air, the veil filled with reaction mixture by 10% v/v was sealed by rubber septum without evacuation.

KEYWORDS

- **coordination polymerization**
- **methacrylate**
- **HEMA**
- **vanadium**
- **vinyl monomers**

REFERENCES

1. Hsieh, K.-H.; Young, T.-H. Hydrogel Materials (HEMA-Based). In *Polymeric Materials Encyclopedia*; Salamon, J. C., Ed.; CRC Press: Boca Raton, FL, 1996; vol. 5, pp 3087–3092.
2. Jia, Z.; Yan, D. *J. Polym. Sci., A: Polym. Chem.* **2005,** *43*, 3502–3509.
3. Chen, Y.; Shen, Z.; Barriau, E.; Kautz, Y.; Frey, H. *Biomacromolecules* **2006,** *7*, 919–926.
4. Miura, Y.; Shibata, T.; Satoh, K.; Kamigaito, M.; Okamoto, Y. *J. Am. Chem. Soc.* **2006,** *128*, 16026–16027.
5. Xu, F.-J.; Kang, E.-T.; Neoh, K.-G. *Biomaterials* **2006,** *27*, 2787–2797.
6. Cayre, O. J.; Chagneux, N.; Biggs, S. *Soft Matter* **2011,** *7*, 2211–2234.
7. Weaver, J. V. M.; Bannister, I.; Robinson, K. L.; Bories-Azeau, X.; Armes, S. P.; Smallridge, M.; McKenna, P. *Macromolecules* **2004,** *37*, 2395–2403.
8. Rosenberg, B. A. *Polym. Sci. Ser. C* **2007,** *49*, 355–385.
9. Beers, K. L.; Boo, S.; Gaynor, S. G.; Matyjaszewski, K. *Macromolecules* **1999,** *32*, 5772–5776.
10. Robinson, K. L.; Khan, M. A.; de Paz Banez, M. V.; Wang, X. S.; Armes, S. P. *Macromolecules* **2001,** *34*, 3155–3158.
11. Teoh, R. L.; Guice, K. B.; Loo, Y.-L. *Macromolecules* **2006,** *39*, 8609–8615.
12. Bories-Azeau, X.; Armes, S. P.; van den Haak, H. J. W. *Macromolecules* **2004,** *37*, 2348–2352.
13. Oh, J. K.; Matyjaszewski, K. *J. Polym. Sci., A: Polym. Chem.* **2006,** *44*, 3787–3796.
14. Paterson, S. M.; Brown, D. H.; Chirila, T. V.; Keen, I.; Whittaker, A. K.; Baker, M. V. *J. Polym. Sci.; A: Polym. Chem.* **2010,** *48*, 4084–4092.
15. Nguyen, N. H.; Leng, X.; Percec, V. *Polym. Chem.* **2013,** *4*, 2760–2766.
16. Kholuiskaya, S. N.; Kasparov, V. V.; Rubailo, V. L. *Kinet. Catal.* **1991,** *32*, 1025–1030.
17. Kholuiskaya, S. N.; Rubailo, V. L. *Kinet. Catal.* **1991,** *32*, 1031–1036.
18. Nikitin, A. V.; Kholuiskaya, S. N.; Rubailo, V. L. *J. Chem. Res. (S)*, **1994,** *9*, 358–359.
19. Rozenberg, B. A.; Bogdanova, L. M.; Dzhavadyan, E. A.; Komarov, B. A.; Boiko, G. N.; Gur'eva, L. L.; Estrina, G. A. *J. Polym. Sci., A: Polym. Chem.* **2003,** *45*, 1–10.
20. Rozenberg, B. A.; Boiko, G. N.; Gur'eva, L. L.; Dzhavadyan, E. A.; Komarov, B. A.; Éstrina, G. A. *Polymer Sci. Ser. A* **2004,** *46*, 226–231.
21. Patra, B. N.; Bhattacharjee, M. *J. Polym. Sci., A: Polym. Chem.* **2006,** *44*, 2749–2753.
22. Gridnev, A. A.; Ittel, S. D. *Chem. Rev.* **2001,** *101*, 3611–3659.
23. Webster, O. W.; Hertler, W. R.; Sogah, D. Y.; Farnham, W. B.; Rajan Babu, T. V. *J. Am. Chem. Soc.* **1983,** *105*, 5706–5708.
24. Webster, O. W. *Adv. Polym. Sci.* **2004,** *167*, 1–34.
25. Selbin, J.; Holmes, L. H. *J. Inorg. Nucl. Chem.* **1962,** *24*, 1111–1119.

INDEX

E

G

H

I

L

M

N